Springer Handbook of Auditory Research

For other titles published in this series, go to
www.springer.com/series/2506

Sandra Gordon-Salant • Robert D. Frisina
Arthur N. Popper • Richard R. Fay

The Aging Auditory System

Springer

Editors
Sandra Gordon-Salant
Department of Hearing and Speech Sciences
University of Maryland
College Park, MD 20742,
USA
sgordon@hesp.umd.edu

Robert D. Frisina
Department of Otolaryngology
University of Rochester Medical Center
601 Elmwood Avenue
Rochester, NY14642-8629,
USA
Robert_Frisina@urmc.rochester.edu

Arthur N. Popper
Department of Biology
University of Maryland
College Park, MD 20742,
USA
apopper@umd.edu

Richard R. Fay
Loyola University of Chicago
Chicago, IL 60626,
USA
rfay@luc.edu

ISBN 978-1-4614-2494-9 e-ISBN 978-1-4419-0993-0
DOI 10.1007/978-1-4419-0993-0
Springer New York Dordrecht Heidelberg London

Printed on acid-free paper

Springer is part of Springer Science+Business Media (www.springer.com)

This volume is dedicated to our friend and colleague, Dr. Judith A. Finkelstein, for her sustained interest in the field and her exceptionally strong support for research during her years at the National Institute on Aging (1995-2006) of the National Institutes of Health. Her efforts were instrumental in guiding new and seasoned investigators alike toward securing funding for research on age-related hearing loss and fostering communication between researchers to promote innovative scientific developments in this area. This volume showcases the work of many of the investigators with whom Dr. Finkelstein worked, either as chapter authors or as authors of hundreds of citations that are referenced throughout.

The senior volume editors also wish to express their appreciation to their spouses, Steven and Susan, for their love and support through all of the late-night experiments, the grant and paper deadlines, and travel to sometimes unusual meeting sites.

Series Preface

Springer Handbook of Auditory Research

The Springer Handbook of Auditory Research presents a series of comprehensive and synthetic reviews of the fundamental topics in modern auditory research. The volumes are aimed at all individuals with interests in hearing research including advanced graduate students, postdoctoral researchers, and clinical investigators. The volumes are intended to introduce new investigators to important aspects of hearing science and to help established investigators to better understand the fundamental theories and data in fields of hearing that they may not normally follow closely.

Each volume presents a particular topic comprehensively, and each serves as a synthetic overview and guide to the literature. As such, the chapters present neither exhaustive data reviews nor original research that has not yet appeared in peer-reviewed journals. The volumes focus on topics that have developed a solid data and conceptual foundation rather than on those for which a literature is only beginning to develop. New research areas will be covered on a timely basis in the series as they begin to mature.

Each volume in the series consists of a few substantial chapters on a particular topic. In some cases, the topics will be ones of traditional interest for which there is a substantial body of data and theory, such as auditory neuroanatomy (Vol. 1) and neurophysiology (Vol. 2). Other volumes in the series deal with topics that have begun to mature more recently, such as development, plasticity, and computational models of neural processing. In many cases, the series editors are joined by a coeditor having special expertise in the topic of the volume.

RICHARD R. FAY, Chicago, IL
ARTHUR N. POPPER, College Park, MD

Volume Preface

Age-related hearing loss (ARHL) is one of the top three most common chronic health conditions affecting individuals aged 65 years and older. The high prevalence of age-related hearing loss compels audiologists, otolaryngologists, and auditory neuroscientists alike to understand the neural, genetic, and molecular mechanisms underlying this disorder. A comprehensive understanding of these factors is needed so that effective prevention, intervention, and rehabilitative strategies can be developed to ameliorate the myriad of behavioral manifestations. This volume presents an overview of contemporary research trends on ARHL from interrelated disciplines whose studies aim to meet this compelling need.

The overall objective of this volume is to bring together noted scientists who study presbycusis from the perspective of complementary disciplines for a review of the current state of knowledge on the aging auditory system. In Chapter 1, Gordon-Salant and Frisina provide an overview to the volume and put the material in the perspective of the field in general. In Chapter 2, Schmiedt presents the morphology and physiology of age-related changes in the auditory periphery, with a description of animal models that control for the effects of known acquired disorders (e.g., noise exposure and ototoxicity) on peripheral auditory system function. In Chapter 3, Canlon, Illing, and Walton describe the direct effects of biological aging at each major level of the ascending central auditory nervous system, with a focus on anatomical, physiological, and neurochemical alterations. This is followed by Chapter 4 by Ison, Tremblay, and Allen that reviews definitive evidence of ARHL in animals and compares findings with those obtained from humans, for whom control of diet, environment, genetics, and other factors is not possible. Chapter 5, by Fitzgibbons and Gordon-Salant, begins a series of chapters on behavioral manifestations of presbycusis in human listeners. Fitzgibbons and Gordon-Salant review changes in hearing sensitivity over time as well as age-related alterations in the perception of the spectral, intensive, and temporal attributes of simple and complex nonspeech acoustic signals. Results of masking and suppression studies are also presented. This is followed in Chapter 6 by Eddins and Hall who discuss binaural processing and temporal asymmetries in aging for both speech and nonspeech signals. In Chapter 7, Schneider, Pichora-Fuller, and Daneman present an integrated systems approach to explain the levels of processing required for spoken language comprehension in communication situations encountered in daily life. In Chapter 8, Humes and Dubno review the effects of aging on speech perception, with an effort to distinguish the effects of peripheral hearing loss from those attributed to higher-level processing problems on speech perception performance of older people. The epidemiology of ARHL is presented in Chapter 9 by Cruickshanks, Zhan, and Zhong. Finally, in Chapter 10, Willott and Schacht consider chemical and environmental strategies for delaying the onset and progression of ARHL.

Although this volume focuses on hearing in aging adult humans, there are chapters in other volumes of the Springer Handbook of Auditory Research that provide additional related material. Many chapters in *Auditory Trauma, Protection, and Repair* (Volume 31, edited by Schact, Popper, and Fay) and *Hair Cell Regeneration, Repair, and Protection* (Volume 33, edited by Salvi, Popper, and Fay) consider issues of damage to hearing and to the ear and methods by which some of these problems might arise. One technical intervention for treating changes in hearing with age that is gaining momentum is fitting presbycusic listeners with cochlear implants, and these are considered in depth in *Cochlear Implants: Auditory Prostheses and Electric Hearing* (Volume 20, edited by Zeng, Popper, and Fay). Issues related to general perception sounds by humans are also considered at length in chapters in *Auditory Perception of Sound Sources* (Volume 29, edited by Yost, Popper, and Fay) and in an early volume in the series on *Human Psychophysics* (Volume 3, edited by Yost, Popper, and Fay). Other volumes with considerable bearing on this include *Clinical Aspects of Hearing* (Volume 7, edited by Van De Water, Popper, and Fay), *Speech Processing in the Auditory System* (Volume 18, edited by Greenberg, Ainsworth, Fay, and Popper), and *Plasticity of the Auditory System* (Volume 23, edited by Parks, Rubel, Popper, and Fay).

Sandra Gordon-Salant, College Park, MD
Robert D. Frisina, Rochester, NY
Arthur N. Popper, College Park, MD
Richard R. Fay, Chicago, IL

Contents

Contributors

Paul D. Allen
University of Rochester, Department of Neurobiology and Anatomy,
601 Elmwood Avenue, Box 603, Rochester, NY 14642, USA
pallen@cvs.rochester.edu

Barbara Canlon
Department of Physiology and Pharmacology, Karolinska Institute,
171 77 Stockholm, Sweden
Barbara.Canlon@ki.se

Karen J. Cruickshanks
Department of Ophthalmology and Visual Sciences and Department of Population Health Sciences, University of Wisconsin, School of Medicine and Public Health,
1038 WARF Building, 610 Walnut Street, Madison, WI 53726-2336, USA
cruickshanks@epi.ophth.wisc.edu

Meredyth Daneman
Department of Psychology, University of Toronto Mississauga, Mississauga,
Ontario, Canada L5L 1C6
daneman@psych.utoronto.ca

Judy R. Dubno
Department of Otolaryngology-Head and Neck Surgery, Medical University of South Carolina, 135 Rutledge Avenue, MSA 550, Charleston, SC 29525–5500, USA
dubnojr@musc.edu

David A. Eddins
University of Rochester, 2365 South Clinton Avenue, Suite 200,
Rochester, NY 14618, USA
david_eddins@urmc.rochester.edu

Peter J. Fitzgibbons
Department of Hearing, Speech, and Language Sciences,
Gallaudet University, Washington, DC 20002, USA
peter.fitzgibbons@gallaudet.edu

Robert D. Frisina
Department of Otolaryngology, University of Rochester Medical Center, 601 Elmwood Avenue, Rochester, NY14642-8629, USA
Robert_Frisina@urmc.rochester.edu

Sandra Gordon-Salant
Department of Hearing and Speech Sciences, University of Maryland, College Park, MD 20742, USA
sgordon@hesp.umd.edu

Joseph W. Hall III
Department of Otolaryngology, University of North Carolina, Chapel Hill, 170 Manning Drive, CB: 7070, Chapel Hill, NC 27599, USA
jwh@med.unc.edu

Larry E. Humes
Department of Speech and Hearing Sciences, Indiana University, 200 South Jordan Avenue, Bloomington, IN 47405-7002, USA
humes@indiana.edu

Robert Benjamin Illing
Neurobiological Research Laboratory, Universitäts-HNO-Klinik, D-79106 Freiburg, Germany
E-mail: robert.illing@uniklinik-freiburg.de

James R. Ison
University of Rochester, Department of Brain and Cognitive Sciences, Meliora Hall, Box 270268, Rochester, NY 14627, USA
jison@bcs.rochester.edu

Kathy Pichora-Fuller
Department of Psychology, University of Toronto Mississauga, Mississauga, Ontario, Canada L5L 1C6
k.pichora.fuller@utoronto.ca

Jochen Schacht
University of Michigan, Kresge Hearing Research Institute, 5315 Medical Sciences Bldg I, Ann Arbor, MI 48109-5616, USA
schacht@umich.edu

Richard Schmiedt
Department of Otolaryngology-Head and Neck Surgery, Medical University of South Carolina, 135 Rutledge Avenue, MSC 550, Charleston SC 29425-5500, USA
schmiera@musc.edu

Bruce A. Schneider
Department of Psychology, University of Toronto at Mississauga, Mississauga, Ontario, Canada L5L 1C6
bschneid@utm.utoronto.ca

Kelly L. Tremblay
University of Washington, Department of Speech and Hearing Sciences,
1417 NE 42nd St., Seattle, WA 98105, USA
tremblay@u.washington.edu

Joseph Walton
Otolaryngology and Neurobiology and Anatomy, University of Rochester
Medical Center, Rochester, NY 14642, USA
Joseph_Walton@urmc.rochester.edu

James F. Willott
University of South Florida and The Jackson Laboratory, Department of
Psychology, University of South Florida PCD4118G, Tampa, FL 33620, USA
jimw@niu.edu

Weihai Zhan
Department of Population Health Sciences, University of Wisconsin,
School of Medicine and Public Health, 610 N. Walnut St., 1034 WARF,
Madison, WI 53726-2336, USA
wzhan@wisc.edu

Wenjun Zhong
Department of Population Health Sciences, University of Wisconsin,
School of Medicine and Public Health, 610 N Walnut St 1034 WARF,
Madison, WI 53726-2336, USA
wzhong@wisc.edu

Chapter 1
Introduction and Overview

Sandra Gordon-Salant and Robert D. Frisina

1.1 Introduction

Age-related hearing loss (ARHL) is one of the top three most common chronic health conditions affecting individuals aged 65 years and older (Pleis and Lethbridge-Çejku 2007). Applying a conservative estimate of the prevalence rate of ARHL (50%) among the population 65 years and older to US Census Bureau projections of the population, there are approximately 20 million senior citizens in the United States with significant hearing loss at present, and this number will soar to 36 million by the year 2030 (Agrawal et al. 2008; US Census Bureau 2008; note that in Cruickshanks, Zhan, and Zhong, Chapter 9, the prevalence of ARHL projected for the year 2030 is higher because it includes those aged 45 years and older). The high prevalence of ARHL compels audiologists, otolaryngologists, and auditory neuroscientists alike to understand the neural, genetic, and molecular mechanisms underlying this disorder so that effective prevention, intervention, and rehabilitative strategies can be developed to ameliorate the myriad of behavioral manifestations. This volume presents an overview of contemporary research trends on ARHL from interrelated disciplines whose studies aim to meet this compelling need. The intended audience includes advanced undergraduate and graduate students, basic and applied biomedical and communication sciences researchers, and practicing clinicians who are concerned with understanding auditory mechanisms and improving hearing health care for elderly individuals.

Historically, the term presbycusis has been used to describe hearing loss attributed to the aging process. The impetus for investigations of presbycusis was Harvard Professor Harold Schuknecht's description of four classic types of presbycusis (1955, 1974).

S. Gordon-Salant (✉)
Department of Hearing and Speech Sciences, University of Maryland, College Park, MD 20742,
e-mail: sgordon@hesp.umd.edu

R.D. Frisina
Department of Otolaryngology, University of Rochester Medical Center,
601 Elmwood Avenue, Rochester, NY
e-mail: Robert_Frisina@urmc.rochester.edu

S. Gordon-Salant et al. (eds.), *The Aging Auditory System*,
Springer Handbook of Auditory Research 34,
DOI 10.1007/978-1-4419-0993-0_1, © Springer Science+Business Media, LLC 2010

In his book *Pathology of the Ear*, Schuknecht (1974) attempted to link case histories and hearing loss patterns in elderly individuals with temporal bone analyses demonstrating deterioration of specific structures in the cochlea and auditory nerve. His observations have underscored that presbycusis is a complex phenomenon that is manifested in various forms among individuals. This basic premise continues to be held today; investigators recognize that hearing abilities in advancing age result from a combination of possible factors involving morphological, chemical, physiological, perceptual, and cognitive processes. Application of sophisticated neurobiological and behavioral techniques to the study of presbycusis has led to a broader understanding of the problem. Schuknecht also recognized that the onset of presbycusis may begin during middle adulthood, with progression into advanced age, and men and women tend to exhibit different types of presbycusis. Cross-sectional and longitudinal studies in humans have clarified the nature of the onset and progression of presbycusis throughout the adult life span and underscore gender differences in behavioral manifestations (e.g., Pearson et al. 1995). The observation of gender and heritability differences also points to a genetic component in at least some cases of presbycusis. More recently, animal models have proven essential in explaining the sequence of altered morphology and subsequent physiological events that lead to hearing loss with aging when diet and environment are carefully controlled.

1.2 Overview

The overall objective of this volume is to bring together noted scientists who study presbycusis from the perspective of complementary disciplines for a review of the current state of knowledge on the aging auditory system. Schmiedt (Chapter 2) and Canlon, Illing, and Walton (Chapter 3) have a principal focus on anatomy, neurochemistry, and physiology of the aging auditory system based on animal models. Schmiedt (Chapter 2) presents the morphology and physiology of age-related changes in the auditory periphery with a description of animal models that control for the effects of known acquired disorders (e.g., noise exposure and ototoxicity) on peripheral auditory system function. Compelling evidence is presented to suggest that the principal effects of aging in a leading animal model of ARHL (the Mongolian gerbil) are morphologic changes in the lateral cochlear wall, including the stria vascularis, which degrade the endocochlear potential with age. These changes then can induce additional pathology involving damage or loss of hair cells and reduction of neurons in the spiral ganglion. Schmiedt carefully demonstrates that these changes produce patterns of sensitivity shifts in the gerbil that reflect common audiometric profiles observed with human presbycusis by considering the key roles of the cochlear amplifier, endocochlear potential, and sensorineural transduction process while signals are coded across frequency. Promising methods to halt, or even reverse, the progression of age-related hearing loss at the auditory periphery are also presented in Chapter 2.

In Chapter 3, Canlon, Illing, and Walton describe the direct effects of biological aging at each major level of the ascending central auditory nervous system, with a focus on anatomical, physiological, and neurochemical alterations. Age-dependent changes in the efferent olivocochlear system and secondary effects on the central system of alterations in the auditory periphery are also described. Research on senescent alterations in the central nuclei and pathways derive primarily from two animal models: mice and rats. Contrasts between CBA mice, which exhibit hearing loss near the end of their life span, like most humans, and C57 BL/6J mice, which exhibit early-onset, progressive hearing loss, are extremely useful for distinguishing between direct aging effects in the central pathways and derivative effects of peripheral dysfunction over time, sometimes referred to as peripherally induced central effects (Frisina et al. 2001). The other animal model of central auditory aging, the Fischer 344 rat, has been used by investigators to elucidate the multifaceted neurochemical alterations that accompany the aging process. Data presented in Chapter 3 converge on a theory of reduced inhibitory neurotransmission that is pervasive throughout the aging central auditory pathway. Evidence for age-related changes in auditory neuroplasticity and mitochondrial function and their implications for central auditory system regulation are also presented in Chapter 3. Supportive data from physiological studies of central auditory function in animal models demonstrate the consequences of morphological and neurochemical alterations with age and form a basis for understanding the mechanisms underlying some of the behavioral manifestations of human presbycusis. In particular, coding of intensive and temporal attributes of sound deteriorates in the auditory midbrain (inferior colliculus) of older animals, whereas localization appears to be less affected by age.

Ison, Tremblay, and Allen (Chapter 4) review definitive evidence of ARHL in animals and compare the findings to those obtained from humans for whom control of diet, environment, genetics, and other factors is not possible. Studies are presented in which investigators experimentally manipulated structures in the auditory periphery or the central auditory pathway of healthy animals to produce comparable anatomical changes to those observed with human aging and then to catalog the functional consequences. This strategy attempts to establish cause-and-effect associations underlying ARHL. Additionally, both behavioral data and electrophysiological data are examined, where comparable paradigms have been employed with both animals and humans. Threshold sensitivity data for pure-tone stimuli are available across the life span for monkeys, rats, gerbils, mice, and humans and show remarkably similar patterns of mean changes in audiometric thresholds over time. Similar threshold data are also available from auditory evoked potential studies (i.e., auditory brainstem responses [ABRs]), although some differences across species are noted. Comparable symptoms of ARHL are evident on ABR latency-intensity functions and otoacoustic emissions as measured in humans and animals of varying ages. Despite the correspondence in average data across species, Ison, Tremblay, and Allen are careful to highlight that individual differences in performance are a prominent characteristic of auditory aging and should be exploited for a comprehensive understanding of the aging process. Intriguing comparisons between humans and animals in processing abilities for spectral, temporal, and binaural cues as a function of age are also presented. Overall, the findings from animal and human studies appear to converge on similar manifestations of ARHL, including variation between individuals within a group. This translational chapter therefore provides the critical link between the morphological and neurochemical findings in animal models of auditory aging reported in Chapters 2 and 3 and the psychophysical and evoked potential studies conducted in humans across the life span in the laboratory or clinical setting.

Fitzgibbons and Gordon-Salant (Chapter 5) begin a series of chapters on the behavioral manifestations of presbycusis in human listeners. In Chapter 5, changes in hearing sensitivity over time as well as age-related alterations in perception of the spectral, intensive, and temporal attributes of nonspeech acoustic signals are reviewed. Results of masking and suppression studies are also presented. Performance on basic psychoacoustic measures is thought to subsume speech processing because perception of speech requires the listener to process rapid changes in spectral and intensity cues occurring in a sequence, sometimes in the presence of an interfering background noise (i.e., masking). Thus performance deficits of older listeners on psychoacoustic measures may be useful in explaining the underlying factors related to their difficulty in understanding speech, particularly in degraded conditions such as noise or reverberation. One important issue in the investigations of aging effects on auditory behavior is the possible confounding of acquired hearing loss on performance. That is, differences in performance between young listeners (who usually have normal hearing) and older listeners (who usually have some hearing loss) could be attributed as much to the loss of sensitivity as to other factors associated with age. Chapter 5 considers alternative experimental paradigms that have been used to unravel the independent and interactive contributions of hearing loss and age to auditory performance. In general, the review of senescent changes in signal detection and discrimination presented in Chapter 5 indicates that perceptual judgments of some types of acoustic signals appear to be affected primarily by aging, with minimal impact of peripheral hearing loss. For example, age-related differences, independent of hearing loss effects, have been reported for detection of silent gaps of varying duration (i.e., gap detection) and discrimination of stimulus duration (Fitzgibbons and Gordon-Salant 1994; Snell 1997). Evidence is also presented for reduced suppression within the aging auditory system (Sommers and Gehr 1998; Dubno and Ahlstrom 2001). Such findings underscore the notion that aging produces effects on auditory processing that extend beyond those attributed exclusively to reduced signal audibility. Alterations in central auditory processes, discussed in Chapters 3 and 4, are the likely locus of this age-specific type of deficit.

Eddins and Hall (Chapter 6) review binaural processing and temporal asymmetries in aging for both speech and nonspeech signals. They present a comprehensive tutorial on temporal, spectral, and intensive cues that are necessary for sound source location in the free field. The few investigations of age-related effects on measures of source location are reviewed and indicate changes in judgments of sound localization and discrimination of minimum audible angle as a function of age (e.g., Chandler and Grantham 1991; Abel and Hay 1996). Similarly, the effects of ARHL on binaural processing under earphones are presented, with an emphasis on studies demonstrating a performance deficit for older listeners on the processing of interaural time differences and on the binaural masking-level difference. These findings are interpreted in light of age-related changes in the morphology and neurophysiology of the central auditory system, as presented in Chapters 3 and 4. The influence of sound reflection in enclosed spaces on speech perception also point to age-related performance differences; similar age differences are suggested on measures of the precedence effect. Higher level binaural processes, such as laterality on dichotic listening tasks, also may be affected by aging, although some of the investigations reviewed are difficult to interpret because of the confounding of hearing loss and aging, as noted above. In contrast, studies of the binaural spatial release from informational masking suggest a minimal impact of age. Taken together, the analyses provided in Chapter 6 indicate that although some binaural processing abilities may be preserved in older listeners, at least some diminish with aging. Counseling older people regarding the benefit of specific hearing aid options in light of these findings is also discussed in Chapter 6.

Cognitive decline may occur in advancing age and affect an older individual's ability to perceive and respond appropriately to acoustic stimuli. However, certain cognitive abilities are well preserved in later adulthood, and plasticity may also occur in the older brain. In Chapter 7, Schneider, Pichora-Fuller, and Daneman present an integrated systems approach to explain the levels of processing required for spoken language comprehension in communication situations encountered in daily life. The contributions of sensory, perceptual, and cognitive abilities within this approach are described, with particular emphasis on the cognitive skills that are required for language processing but that may decline with age. These skills include working memory, executive function, and speed of processing. Chapter 7 reviews the age-related effects observed on a range of speech recognition measures, including source segregation, scene analysis, and release from informational masking. The findings suggest that age-related declines in central processing emerge in concert with age-related sensory limitations. However investigations that attempt to identify the independent contributions of sensory decline, cognitive decline, or interactions between the two indicate that deficits in targeted cognitive abilities can influence speech understanding performance among older listeners, even those with normal to near-normal hearing sensitivity, but the strongest effects emerge as the complexity of the listening task increases. The importance of tailoring the auditory rehabilitation process, including amplification, to accommodate the cognitive limitations of older people is also discussed.

Chapter 8 reviews the effects of aging on speech perception, with an effort to distinguish the effects of peripheral hearing loss from those attributed to higher level processing problems on speech perception performance of older people. To that end, Humes and Dubno provide considerable tutorial information on the principles of articulation index (AI) theory (French and Steinberg 1947; ANSI 1969), which quantifies predicted speech recognition performance based on signal audibility across a range of frequency bands that are important for speech, and use the AI as a framework for interpreting the speech understanding problems of older adults as assessed on a range of experimental tasks. They clearly convey the current thinking that audibility issues primarily limit older listeners' performance for speech recognition in quiet and noise. They discuss various central-auditory and cognitive factors that may contribute to observed age-related deficits for understanding specific types of speech materials and listening conditions (i.e., time-compressed or rapid speech, speech with temporally varying noise, and dichotic speech). Chapter 8 culminates with an explanation of the issues limiting older listeners' speech understanding performance while using current hearing aid technology, again by applying principles of AI theory. Areas where further research is needed are also described.

The epidemiology of age-related hearing loss is presented in Chapter 9 by Cruickshanks, Zhan, and Zhong, commencing with a review of large, population-based studies that converge on the prevalence and incidence rates of hearing loss among men and women of advanced age from industrialized societies. Age-related hearing loss may be associated with a host of risk factors, including endogenous (genetic) factors and acquired exogenous factors. Cruickshanks, Zhan, and Zhong review epidemiological data pointing to hereditability patterns for ARHL, although specific genes have not yet been identified in humans. In contrast, numerous modifiable risk factors have been identified from epidemiological research on ARHL. Prominent among these is noise exposure because most people in industrialized societies are exposed to intense noise, either through work-related exposure (e.g., equipment noise, subway noise) or leisure activities (e.g., loud music, sporting events in public arenas, hunting). Cardiovascular disease is another documented risk factor for ARHL. There is some evidence to suggest that other lifestyle issues, such as cigarette smoking, excessive alcohol consumption, and diet, may elevate the relative risk for ARHL. Although certain medications (e.g., aminoglycosides, chemotherapeutic agents such as cisplatin, loop diuretics) have a well-known ototoxic effect, exposure to some solvents and chemicals in the environment may also cause hearing loss. In contrast, some dietary supplements appear to have a protective effect against the onset and progression of ARHL. In addition, there are comorbid medical conditions of aging that may contribute to apparent ARHL, such as Type II diabetes mellitus and hormonal changes in blood chemistry. This array of conditions that may occur over the course of adulthood, reviewed in Chapter 9 from an epidemiological perspective, underscores the observation that age-related hearing loss is a multifactorial disorder in terms of causation, and this likely contributes to the frequent report of considerable variability in the performance of older human participants in listening experiments. Congruence in causative factors of ARHL identified from animal models and epidemiological studies strengthens our understanding of key modifiable risk factors for this disorder.

In Chapter 10, Willott and Schacht consider chemical and environmental strategies for delaying the onset and progression of age-related hearing loss. Chapter 10 reviews some of the known mechanisms of ARHL in the cochlea, auditory nerve, and central auditory pathways, as discussed in Chapters 2–4, to provide a basis for evaluating the range of possible interventions that hypothetically should halt some of these progressive changes with aging. Experimental data from animal models are presented that demonstrate the benefits of the use of antioxidant therapy, hormonal therapy, dietary restrictions, repair of neural circuits in the central nervous system, and augmented acoustic environments. However, for each of these interventions, there are also detrimental effects on auditory system function depending on the specific animal species and strain, gender, and experimental paradigm. Although many of these therapeutic approaches hold great promise for treating ARHL, at present, none of them has emerged as a strong candidate for reversing the course of presbycusis in humans, in part because of the multiple causes and etiological loci of ARHL in humans. Nevertheless, continued research with animal models is essential for accomplishing the ultimate goal of identifying chemical or environmental biomedical interventions that will relieve the extensive and diverse symptoms that characterize human presbycusis.

1.3 Future Research

Research on ARHL has advanced dramatically over the last 20 years as amply demonstrated in this volume. The impetus for these advances has derived from three sources: the support of basic research on the mechanisms of hearing loss from the National Institute on Deafness and Other Communication Disorders at the NIH that has been applied to understanding mechanisms of ARHL, the support of sensory and cognitive research from the National Institute on Aging at the NIH that has permitted extensive assessment of the consequences of aging using behavioral, neuroscientific, and molecular biological experiments, and classic literature in the 1980s and early 1990s that called for research on presbycusis to accommodate the anticipated graying of America in the early 21st century (e.g., Committee on Hearing, Bioacoustics, and Biomechanics [CHABA] 1988; Willott 1991). An extensive array of experimental paradigms has been developed to clarify the morphological changes in the periphery and CNS observed in animal models of ARHL, the manifestations of age-related changes in the cochlea, auditory nerve, auditory brainstem, and auditory cortex utilizing electrophysiological measures of auditory function, and behavioral effects on auditory sensitivity and suprathreshold auditory processing of simple and dynamic nonspeech and speech signals over time, including binaural processing in complex listening environments. The nature of cognitive abilities as people age and the impact of possible changes on speech understanding tasks have also been studied extensively. Each chapter of the present volume provides an overview of these experimental findings and their potential implications in these related areas of inquiry, and it is hoped that the critical background is provided for investigators from a variety of disciplines to identify new avenues of promising research from their own unique perspectives.

What is needed as we look toward the future of research on ARHL is a more definitive analysis of the links between the extensive anatomical, structural, electrophysiological, and molecular genetic findings in animal models and the broad range of behavioral manifestations of ARHL, in addition to formulating a better understanding of the principal sources of individual variation on auditory performance in humans, including cognitive changes with age. It is only through a comprehensive understanding of these factors that better diagnostic procedures for distinguishing different etiologies of presbycusis will be developed and more effective biomedical therapeutic interventions will be introduced that are tailored to individual needs. Such therapeutic techniques span the range from biochemical interventions, including gene therapy and stem cell therapy, to better electroacoustic devices (hearing aids and cochlear implants designed specifically for aged persons) and behavior modification strategies. We hope that this volume provides a renewed impetus toward these visionary objectives.

References

Abel SM, Hay VH (1996) Sound localization. The interaction of aging, hearing loss and hearing protection. Scand Audiol 25:3–12.

Agrawal Y, Platz EA, Niparko JK (2008). Prevalence of hearing loss and differences by demographic characteristics among US adults. Arch Intern Med 168:1522–1530.

ANSI (1969) ANSI S3.5–1969, American National Standard Methods for the Calculation of the Articulation Index. New York: American National Standards Institute.

Committee on Hearing, Bioacoustics, and Biomechanics (CHABA) (1988) Speech understanding and aging. J Acoust Soc Am 83:859–895.

Chandler DW, Grantham DW (1991) Effects of age and auditory spatial resolution in the horizontal plane. J Acoust Soc Am 89:1994.

Dubno JR, Ahlstrom J (2001) Psychophysical suppression measured with bandlimited noise extended below and/or above the signal: Effects of age and hearing loss. J Acoust Soc Am 110:1058–1066.

Fitzgibbons PJ, Gordon-Salant S (1994) Age effects on measures of auditory temporal sensitivity. J Speech Hear Res 37:662–670.

French NR, Steinberg JC (1947) Factors governing the intelligibility of speech sounds. J Acoust Soc Am 19:90–119.

Frisina DR, Frisina RD, Snell KB, Burkard R, Walton JP, Ison JR (2001) Auditory temporal processing during aging. In: Hof PR, Moobbs CV (eds) Functional Neurobiology of Aging. San Diego: Academic Press, pp. 565–579.

Pearson JD, Morell CH, Gordon-Salant S, Brant LJ, Metter EJ, Klein LL, Fozard JL (1995) Gender differences in a longitudinal study of age-associated hearing loss. J Acoust Soc Am 97:1196–1205.

Pleis JR, Lethbridge-Çejku M (2007) Summary Health Statistics for U.S. Adults: National Health Interview Survey, 2006. Washington, DC: Vital Health Stat 10(235), National Center for Health Statistics, US Government Printing Office.

Schuknecht HF (1955) Presbycusis. Laryngosocope 65:402– 419.

Schuknecht HF (1974) Pathology of the Ear. Cambridge, MA: Harvard University Press.

Snell KB (1997) Age-related changes in temporal gap detection. J Acoust Soc Am 101: 2214–2220.

Sommers M, Gehr S (1998) Auditory suppression and frequency selectivity in older and younger adults. J Acoust Soc Am 103:1067–1074.

US Census Bureau (2008) Table 2. Projections of the population by selected age groups and sex for the United States: 2010 to 2050 (Release Date: August 14, 2008), available at http://www.census.gov/population/www/projections/summarytables.html (accessed 1/18/2009).

Willott JF (1991) Aging and the Auditory System: Anatomy, Physiology, and Psychophysics. San Diego, CA: Singular Publishing Group.

Chapter 2
The Physiology of Cochlear Presbycusis

Richard A. Schmiedt

2.1 Introduction

The effects of pure aging on the physiology and morphology of the human peripheral auditory system are difficult to study given the variability inherent in genetics and the environment with which the system must cope. Environmental exposures accumulated over a lifetime often combine mild, continuous noise exposures occurring daily, with occasional punctate episodes of very high decibel trauma associated with loud music, power equipment, and small arms fire. Moreover, the human experience includes many drugs that often have unintended side effects on the auditory periphery. Some drugs have well-known ototoxic properties; others are more insidious, like the continuous high-level use of some narcotics. Noise and drug injuries tend to preferentially damage the hair cells in the cochlea.

Genetics must then respond to an individual's environment, resulting in the very large variability present in the hearing capabilities of elderly humans. It is clear that animal models of age-related hearing loss are required to tease out the effects of aging alone from the effects of environment and genetics. Yet up until ~25 years ago, much of the research in presbycusis was accomplished by using human temporal bones and clinical data (Bredberg 1968; Schuknecht 1974; Gates et al. 1990; Schuknecht and Gacek 1993). Only in the last 30 years or so have animal models been established where the environment, diet, and genetics are strictly controlled (Keithley and Feldman 1979, 1982; Henry 1982; Keithley et al. 1989; Mills et al. 1990; Hequembourg and Liberman 2001; Ohlemiller and Gagnon 2004; for reviews see Willott 1991; Frisina and Walton 2001, 2006; Gates and Mills 2005; Canlon, Illing, and Walton, Chapter 3). Animals raised under these controlled conditions nonetheless show age-related declines in auditory function, consistent with the notion that presbycusis includes effects unique to aging and is not just the result of the combined effects of noise and other ototoxic factors over a lifetime.

R.A. Schmiedt (✉)
Department of Otolaryngology-Head and Neck Surgery, Medical University of South Carolina, 135 Rutledge Avenue, MSC 550, 29425-5500, Charleston, SC 29425-5500
e-mail: schmiera@musc.edu

S. Gordon-Salant et al. (eds.), *The Aging Auditory System*, Springer Handbook of Auditory Research 34,
DOI 10.1007/978-1-4419-0993-0_2,

The deleterious effects of aging are often seen first in highly metabolic tissues in the body, coincident with a degradation of mitochondrial function. Mitochondrial dysfunction with age has been attributed to the buildup of reactive oxygen species (ROS), although this hypothesis still elicits controversy (Gruber et al. 2008). In the cochlea, it is the lateral wall where aerobic metabolism is extremely high because it is needed for maintenance of the K^+gradient between the endolymph and perilymph and the generation of the endocochlear potential (EP). The high K^+ and EP are both present in the endolymph of the scala media. It is not surprising then that there is now substantial evidence that age-related hearing loss uncomplicated by environmental and genetic variables is largely the result of pathologies in the cochlear lateral wall rather than just a general loss of hair cells. This chapter reviews some of the current literature on peripheral presbycusis and how lateral wall dysfunction, leading to a lowered EP, can result in audiograms in animal models that mimic those obtained from elderly humans.

2.2 Overview of Normal Mammalian Auditory Physiology

A concise yet accurate way of understanding normal cochlear physiology and how it breaks down with age is to segregate its functional aspects into three interlocking systems: the cochlear amplifier, its power supply, and the transduction mechanism. The three systems and their relationships are schematized in Fig. 2.1. The discussion here is necessarily brief and relates only to those ideas important for understanding the pathologies relating to presbycusis. Further details can be found in the cited references.

2.2.1 Cochlear Amplifier

The cochlear amplifier relies on an active process located in the outer hair cells (OHCs) to physically amplify the traveling wave vibrations along the basilar membrane (Davis 1983; Russell 1983; Cooper and Rhode 1997; Robles and Ruggero 2001). The amount of amplification is highly dependent on a potential (voltage) between the scala media and scala tympani, thereby present across the OHCs. This voltage is the EP, which is ~90 mV within the scala media when referenced to a neck muscle ground. Indeed, the amplification dependency is logarithmic-linear such that about a 1-dB decrease in amplification (corresponding to a 1-dB increase in threshold) results from a 1-mV decrease in EP (Sewell 1984; Ruggero and Rich 1991; Schmiedt 1993). The basilar membrane amplification from the active OHCs also shows a strongly compressive nonlinearity: vibrations from low-level sounds are amplified most, whereas those from intense sounds are amplified least. This compression of dynamic range at the level of the basilar membrane results in a relatively constant vibratory stimulus exciting the inner hair cells (IHCs) over a wide range of acoustic

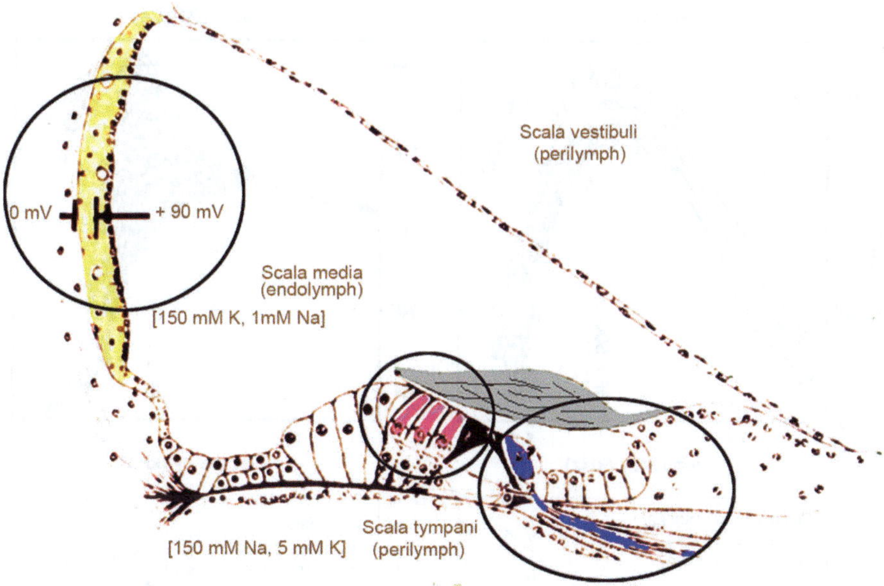

Fig. 2.1 Schematic cross section of a single turn of the cochlea. The three systems underlying basic cochlear function are outlined (circles). The left circle focuses on the lateral wall and stria vascularis and the production of the 90-mV endocochlear potential (EP) present in scala media. The middle circle centers on the outer hair cells (OHCs) and the micromechanics involved in the cochlear amplifier. The right circle is drawn around the inner hair cell (IHC) and the associated primary afferent nerve fibers that make up the transduction process where mechanical vibrations are transduced to neural impulses that are sent to the brain via the auditory nerve. (Adapted with permission from Mills et al. 2006b.)

intensities (Robles and Ruggero 2001) and is also the basis for two-tone suppression. Thus a healthy cochlea is strongly nonlinear in its response to signal intensity and multiple frequencies, resulting in various suppression phenomena and otoacoustic emissions (OAEs). (OAEs are acoustic distortion products that can be measured in the ear canal at frequencies that result when two tones are combined in a nonlinear fashion [Probst 1990]. The strongest in the ear are the cubic difference tones corresponding to frequencies of $2f_1-f_2$.)

A final factor in understanding the normal cochlear amplifier is that its maximum gain varies along the basilar membrane. In the cochlear apex tuned to lower frequencies, the gain is only ~20 dB, yet in the base, the gain can be as high as 50-70 dB (Ruggero and Rich 1991; Mills and Rubel 1994; Cooper and Rhode 1997; Robles and Ruggero 2001; see Fig. 2.2a, b). Thus if cochlear amplification is totally lost, either from OHC loss or from a very low EP, one would expect to see the least effect at low frequencies and the most at high frequencies. This relationship is borne out in gerbil ears treated chronically with furosemide to artificially reduce the EP as well as in quiet-aged ears with a naturally reduced EP (Schmiedt et al. 2002b; Fig. 2.2c, d).

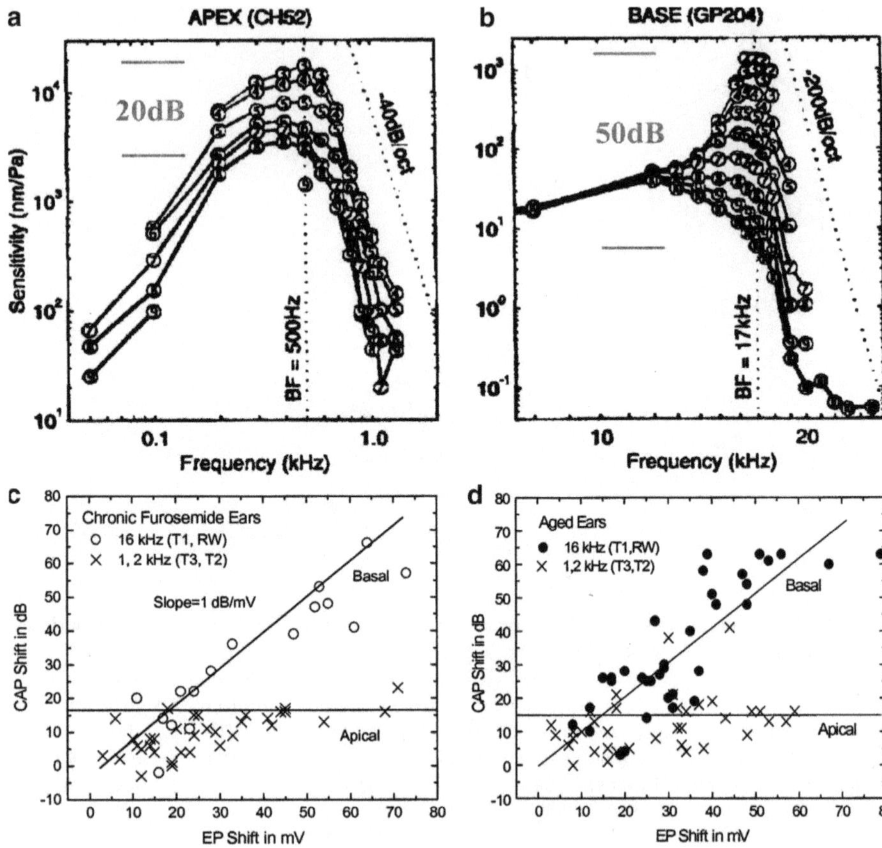

Fig. 2.2 Effects of cochlear location and condition on the vibration amplitudes of the basilar membrane in the apex of a chinchilla cochlea **(a)** and the base of the guinea pig cochlea **(b)**. Two species were used to allow the best recordings from the apex and base. The top lines were obtained first with the OHC amplifier in good condition. The bottom lines were obtained after cochlear death. Note that the apical gain from the OHC amplification is ~20 dB, whereas that from the base is 50 dB. BF, baseline frequency. (Adapted with permission from Cooper and Rhode 1997). Neural shifts in threshold derived from the compound action potential (CAP) response at 1 and 2 kHz (apical) and 16 kHz (basal) are plotted against shifts in the corresponding EP measured in gerbil ears chronically treated with furosemide **(c)** and ears aged 36 months **(d)**. Furosemide is a drug that reversibly decreases the EP and allows studying cochlear function under conditions of lowered EP but with normal hair cells and neurons in a young adult animal (see text; Schmiedt et al. 2002b). **c** and **d** emphasize the points made in a and b; i.e., the OHC amplifier gain at low frequencies is only ~20 dB, whereas at higher frequencies, it is around 50-60 dB. Additionally, **c** and **d** show that the gain in decibels is a linear function of the EP in millivolts at basal locations in the cochlea. The straight lines are not best fits but have slopes of one to show the asymptotic threshold shift for the low-frequency data and unity to show the linear relationship with EP for the high-frequency data. (Adapted with permission from Schmiedt et al. 2002b.)

2.2.2 *Cochlear Power Supply*

The second system is the cochlear power supply comprising the lateral wall tissues, including those of the stria vascularis where the EP is generated. This power supply is intimately related to the K^+-recycling pathway, which actively pulls K^+ back into the endolymph as it is effluxed from the hair cells into the perilymph. The pathway uses a network of supporting cells and fibrocytes (specialized cells that can turnover and often have stem cell precursors) along the basilar membrane and lateral wall, respectively, connected by gap junctions (Spicer and Schulte 1991, 1996; Marcus and Chiba 1999). A final step in K^+ recycling is the actual generation of the EP within the stria vascularis (Salt et al. 1987; Wangemann et al. 1995; Marcus et al. 2002; Wangemann 2002; Schulte 2007).

Note that K^+ recycling works against both concentration and electrical gradients: the K^+ concentration in the endolymph is ~150-170 mM compared with ~1 mM in the perilymph, and the potential present in the endolymph is ~90 mV (the EP) as compared with the 4-mV potential in the perilymph (Salt et al. 1987; Schmiedt 1996). Thus pushing K^+ along this route takes energy that is largely generated by Na^+-K^+-ATPase pumps in concert with the Na^+-K^+-$2Cl^-$ (NKCC) transporter (Wangemann 2002). The NKCC transporter is an important tool in our studies of the effects of EP changes on auditory function in that furosemide, a fairly specific, reversible antagonist against NKCC, provides a means to experimentally turn off and on the recycling pathway and subsequently the EP (Evans and Klinke 1982; Sewell 1984; Schmiedt et al. 2002b; Mills and Schmiedt 2004). Furosemide delivered either intravenously or via a round window application can reduce the EP to near 0 mV, with recovery from a single dose taking between tens of minutes to over a month if osmotic pumps are used for delivery (Sewell, 1984; Mills and Rubel 1994; Schmiedt et al. 2002b).

The EP serves as the cochlear battery. It is generated within the stria across the intrastrial space and is present in the endolymph along the entire cochlear duct (Wangemann 2002). (Note that EP generation in the stria is dependent on the ion flux provided by the fibrocytes in the lateral wall. In this context, strial and lateral wall pathologies can both result in a lowered EP.) The EP is produced largely by the stria in the basal turn where it is the highest and drops by ~10 mV in the more apical turns of the cochlea. Destruction of the stria or lateral wall in the basal turn results in a significantly lowered EP throughout the cochlear spiral, whereas destruction of the stria in the higher turns with an intact basal stria yields relatively minor reductions in the overall EP (Salt et al. 1987; Wu and Hoshino 1999). Thus apical strial pathology, as often seen with presbycusis, does not necessarily correlate with significantly lowered EP values, whereas basal atrophy is highly correlated with a reduced EP.

2.2.3 *Cochlear Transduction*

The third system in the transduction of cochlear vibration to neural impulses comprises the IHCs and the afferent fibers of the auditory nerve (see Fig. 2.1).

The IHCs function as passive detectors of basilar membrane vibration and excite afferent fibers via ribbon synapses around the base of the cell (Robles and Ruggero 2001). IHCs are more resistant to noise and chemical trauma than the OHCs and tend to survive with comparatively less pathology in aged ears. Even so, in animals raised their entire lives in quiet (quiet-aged ears), there is a significant loss and shrinkage of the afferent nerve fibers and their cell bodies, the spiral ganglion cells (SGCs) in Rosenthal's canal. The loss and shrinkage with age occur even with the IHCs present and seemingly normal both in animal models and in humans (Schuknecht 1974; Keithley and Feldman, 1979, 1982; Mills et al. 2006a).

In young healthy ears that have been raised in quiet, afferent fibers can be segregated into two or three groups corresponding to spontaneous rates (spont) and sensitivity (Liberman 1978; Schmiedt 1989). Typically, the most sensitive fibers have high rates of spontaneous activity (high-spont, 18 spikes/s and higher), with somewhat less sensitive fibers forming a middle group with spontaneous rates from 0.5 to 18 spikes/s (medium spont). The third group comprises the low-spont fibers with sensitivities that can be up to 50-60 dB lower than those of the high-spont group and have spontaneous rates below 0.5 spikes/s. Thus the sensitivity range of the three groups of afferents in young ears largely covers an intensity range between 0 and 90 dB SPL.

2.3 Schuknecht's Four Types of Presbycusis

Schuknecht (1974) has described four types of human presbycusis: (1) sensory, mainly affecting the cochlear hair cells and supporting cells; (2) neural, typified by the loss of afferent neurons in the cochlea; (3) metabolic, where the lateral wall and stria vascularis of the cochlea atrophy; and (4) mechanical, where there seemed to be a so-called "stiffening" of the basilar membrane and organ of Corti. To date, no real evidence has been found that the mechanical structure of the organ of Corti stiffens with age. The diagnoses of a mechanical presbycusis was derived from a flat loss of 30-40 dB in hearing threshold and was often coupled with degeneration in the spiral "ligament" along the cochlear lateral wall. The spiral ligament originally was thought to offer structural support to the basilar membrane (thus the descriptive term ligament); however, the spiral ligament is now known to consist largely of ion-transport fibrocytes involved in the recycling of K^+ efflux from the hair cells back to the endolymph. Thus it is very likely the mechanical presbycusis described by Schuknecht is simply a severe case of metabolic presbycusis. Indeed, animals with very low EP often show a flat audiometric loss of 40 dB and greater at low frequencies, similar to that ascribed to mechanical presbycusis.

In a later report, Schuknecht and Gacek (1993) described atrophy of the stria to be the predominant lesion in the temporal bones of elderly humans and sensory cell loss as being the least important cause of hearing loss in older humans, especially if the confounding factors of noise, drug exposure, and genetic defects are eliminated. The recent results of Gates et al. (2002) using distortion product otoacoustic emission (DPOAE) and audiogram data support the conclusion that sensory loss is not as prevalent in the aging population as once thought. Indeed, Gates et al. (2002) and Gates and Mills (2005) conclude that metabolic presbycusis is the predominant cause of human hearing loss with age. Many animal models that exclude noise history or genetic mutations lend support to that conclusion. These models include chinchilla (Bhattacharyya and Dayal 1985), rabbit (Bhattacharyya and Dayal 1989), and CBA mice (Spongr et al. 1997). Even C57 and other mutant mice, if actually aged, develop strial pathologies (Ichimiya et al. 2000; Hequembourg and Liberman 2001; Ohlemiller and Gagnon 2004; Ohlemiller et al. 2008). CBA/J mice, however, seem to show only a hair cell loss with a fairly intact lateral wall with age as discussed in Section 2.4 below (Sha et al. 2008).

2.4 Sensory Presbycusis

Loss of sensory hair cells in the human aging ear is well documented (Bredberg 1968; Schuknecht 1974; Gates and Mills 2005). Indeed, morphologically, hair cell loss is one of the most apparent changes in temporal bones both in humans and in animals of advanced age (Dayal and Bhattacharyya 1989). Species studied include rabbit, guinea pig, cat, rats of various genetic backgrounds, chinchilla, mice of various genetic backgrounds, gerbil, and primate (see Willott 1991 for review). The other universally noticeable pathological change in aged temporal bones is the shrinkage and loss of SGCs in Rosenthal's canal, so it is understandable that presbycusis is commonly thought to be of "sensorineural" origin by many in the field of hearing.

When human audiograms were matched to temporal bone pathologies, it seemed clear that the high-frequency loss so often seen in presbycusis matched the OHC loss in the basal coil of the cochlea (Bredberg 1968; Johnson and Hawkins 1972; Schuknecht 1974). A caveat here is that excess noise exposure is commonplace in western society, and many people, especially men, have been exposed throughout life. Thus the underlying cause of the hair cell loss is problematic. Animals aged in quiet also lose hair cells but more at the apex than at the base of the cochlea. Thus the cochleograms often take on the shape of an inverted "U" (Dayal and Bhattacharyya 1989; Tarnowski et al. 1991). The OHC loss is typically scattered with most, if not all, the IHCs surviving. When neural or behavioral audiograms from the animal models are compared with the OHC loss, there is often a poor correlation.

Fig. 2.3 is an illustration of this last point. Shown in the four panels are the cochleograms and neural threshold shifts of four quiet-aged gerbils. The neural thresholds were obtained with the compound action potential (CAP) response. In all cases, there is a significant scattered OHC loss at the apex, with little or no IHC loss. All the threshold shifts have little relationship to the OHC loss.

Some mouse and rat models do show profound sensory losses with age. They are typically mice with a C57BL/6J background, which has a genetic mutation where hair cell loss begins a few months after birth (Spongr et al. 1997;

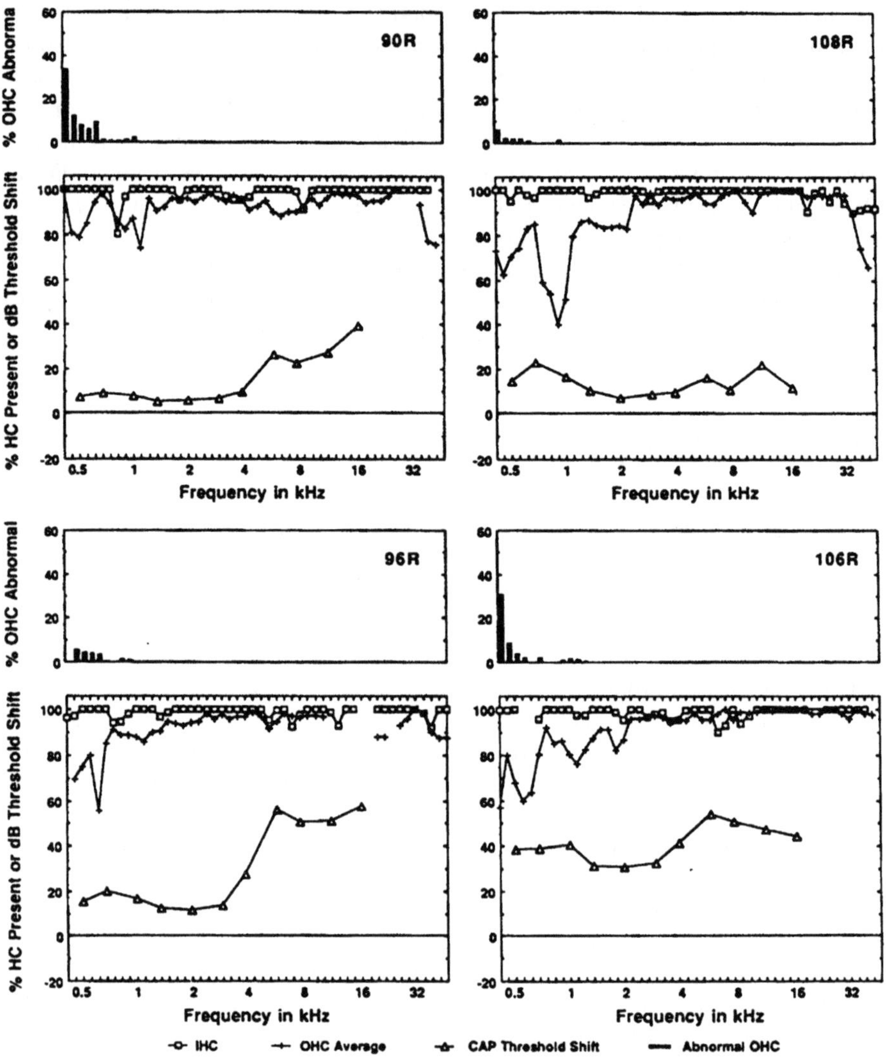

Fig. 2.3 Plots of hair cell condition, hair cell loss, and neural threshold shifts in 4 quiet-aged gerbils at 36 months of age. Percent of IHCs (open circles) and OHCs present (+) and the dB shift of the CAP (open triangles) are plotted against frequency using a gerbil frequency-distance map (Schmiedt and Zwislocki 1977). Note that OHC abnormalities and losses can be significant but are scattered and located mostly in the cochlear apex. IHCs are well preserved throughout the cochlear spiral and CAP shifts have little relationship to hair cell loss, especially with regard to frequencies above 3 kHz. (Adapted with permission from Tarnowski et al. 1991.)

Hequembourg and Liberman 2001). The OHC loss is nearly 100% and progresses from base to apex, with the IHC survival somewhat more robust than that of the OHC. The C57 mouse has been used extensively as a model for sensory presbycusis and the genetic mutation responsible for the hair cell loss is often termed the "age-related hearing loss" mutation (Johnson et al. 1997). Although important as a model of sensory loss, the age-related hearing loss mutant is problematic with regard to being a true aging model. In the human situation, if a teenager is diagnosed with a progressive high-frequency loss caused by sensory cell degeneration, it is unlikely the condition would ever be called presbycusis.

The CBA/J mouse is a model with true sensory presbycusis (Sha et al. 2008). In this model, the sensory cells are progressively lost from the apex with some loss in the base, with little strial involvement (Lang et al. 2002). In both C57BL/6J and CBA/J mice, the EP remains normal throughout the life span of the animal, although subtle changes in the lateral wall of the C57BL/6 mice have been reported (Ichimiya et al. 2000; Hequemberg and Liberman 2001). Both these models with substantial IHC losses have neural losses with age similar to those found in other mutants without the increased IHC loss, i.e., the neural presbycusis seems not to depend directly on the survival of the IHCs. It is interesting to note that the hearing thresholds of the CBA/J model obtained from auditory brainstem recordings (ABRs) are often not well correlated with the hair cell loss, similar to findings obtained from the gerbil (Fig. 2.3). Finally, not all mice exhibit sensory presbycusis. There are some mutants, such as BALB/cj and NOD/ShiLtJ mice, that do show a decrease in EP with age (Ohlemiller et al. 2006, 2008). Given the mutant data, it is of great interest that wild-caught mice have similar patterns of hair cell loss with age as those of gerbils (Dazart et al. 1996).

2.5 Metabolic Presbycusis

2.5.1 Audiometric Data

For reference, audiograms from human subjects between the ages of 50 to more than 85 years of age are shown in Fig. 2.4. The profile of the hearing loss comprises a flat loss of between 10 and 40 dB at frequencies below ~1.5 kHz, coupled with a sloping loss at higher frequencies. In men, the high-frequency hearing loss is greater than in the women with a correspondingly steeper slope. If subjects are screened for noise history, this gender discrepancy is minimized (Jerger et al. 1993). Thus the audiograms from men are probably a mix of pure aging and cumulative noise exposure with concomitant excessive OHC loss. Note from the discussion on the cochlear amplifier that a complete loss of the OHCs in the base should lead to a flat hearing loss of between 50 and 70 dB above ~4 kHz, which is evident in the male audiograms, but not in those of the females.

The audiogram profile found in humans is also found in many animal models. Fig. 2.5 shows audiograms from an aged chinchilla, aged SJL/J mice, and three groups of gerbils raised in quiet. All show a flat loss at low frequencies coupled with a sloping loss at higher frequencies.

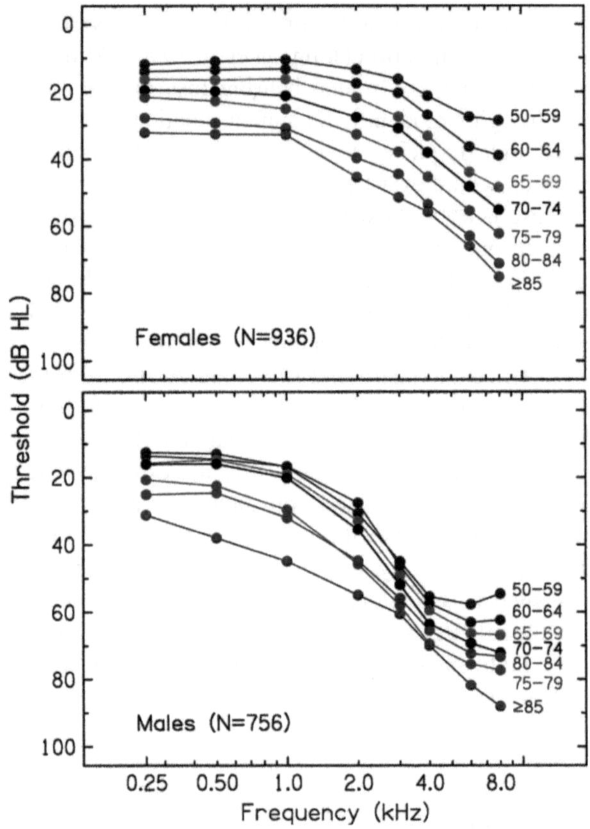

Fig. 2.4 Audiometric mean hearing losses (HL) in female (top) and male (bottom) participants in the ongoing study of age-related hearing loss at the Medical University of South Carolina (Lee et al. 2005; Dubno et al. 2008). The parameter is subject age at the time of enrollment. Note the characteristic profile of human age-related HL: a flat loss at low frequencies coupled with a sloping loss at frequencies above ~1 kHz. These subjects were not screened for noise history, and men typically show more threshold shifts at high frequencies than women, presumably from additional noise exposure (Jerger et al. 1993). Screening for noise history tends to minimize the gender difference. (Adapted with permission from Mills et al. 2006b.)

Individual hearing loss (HL) data are shown for five quiet-aged gerbils in Fig. 2.6 top. (Note that these data have been normalized to young-adult average thresholds that is represented by the 0-dB line and the 90-mV EP.) The EP decreases with age in the gerbil concomitant with a loss of strial volume and Na^{+}-K^{+}-ATPase activity along the lateral wall and stria (Schulte and Schmiedt 1992; Gratton et al. 1996, 1997; Spicer et al. 1997). Again, we see the standard presbycusic profile, which is also evident in some of the threshold shift curves plotted in Fig. 2.3. There is little or no correlation of these curves to the OHC loss in any of these animals. However, if we plot the curves with regard to the EP values found in the basal turns of the individual cochleas, a clear pattern emerges. The high-frequency loss is highly correlated with the amount of EP reduction. It is rare that the EP falls below

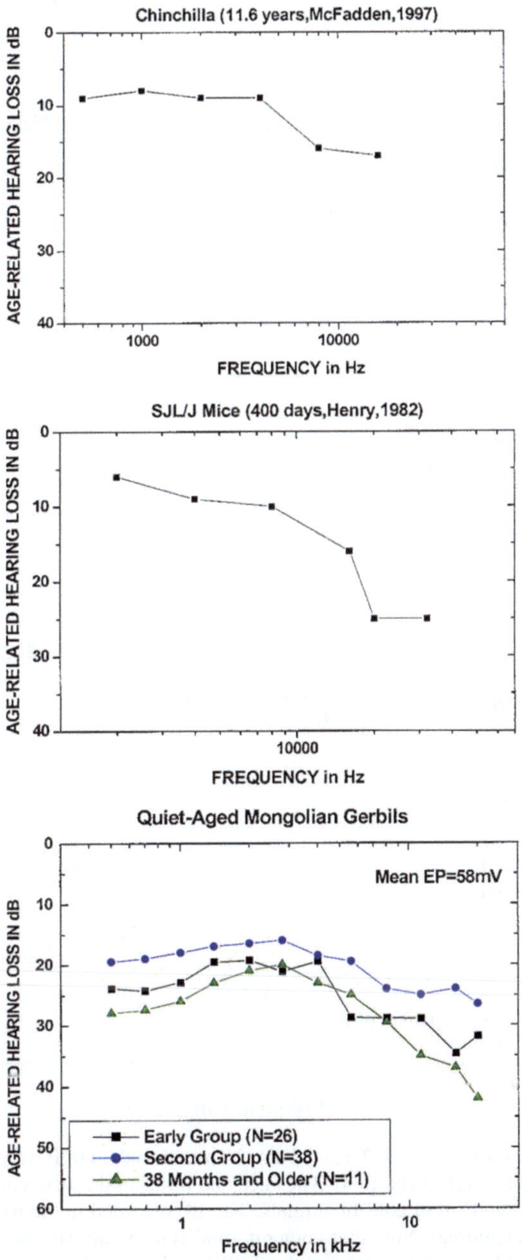

Fig. 2.5 Behavioral and neural hearing losses in three animal models. Top: neural evoked potentials in an 11.6-year-old chinchilla (data redrawn from McFadden 1997a). Middle: mean thresholds from the auditory brainstem response (ABR) in a group of SJL/J mice at 400 days of age (data redrawn from Henry 1982). Bottom: mean CAP thresholds of 3 groups of 36-month-old, quiet-aged gerbils (adapted from Schmiedt et al. 2002b). The early and late groups of gerbils were raised from different genetic stock, whereas the 38-month-old group comprised 11 animals from the late group that were aged an additional 2 months. These models show the classic profile of age-related hearing loss seen in humans: a small flat loss at low frequencies coupled to a sloping loss at higher frequencies. Some of the increased loss at low frequencies in the 38-month group may be the result of excess losses of apical OHCs seen with extreme age (see Fig. 2.3).

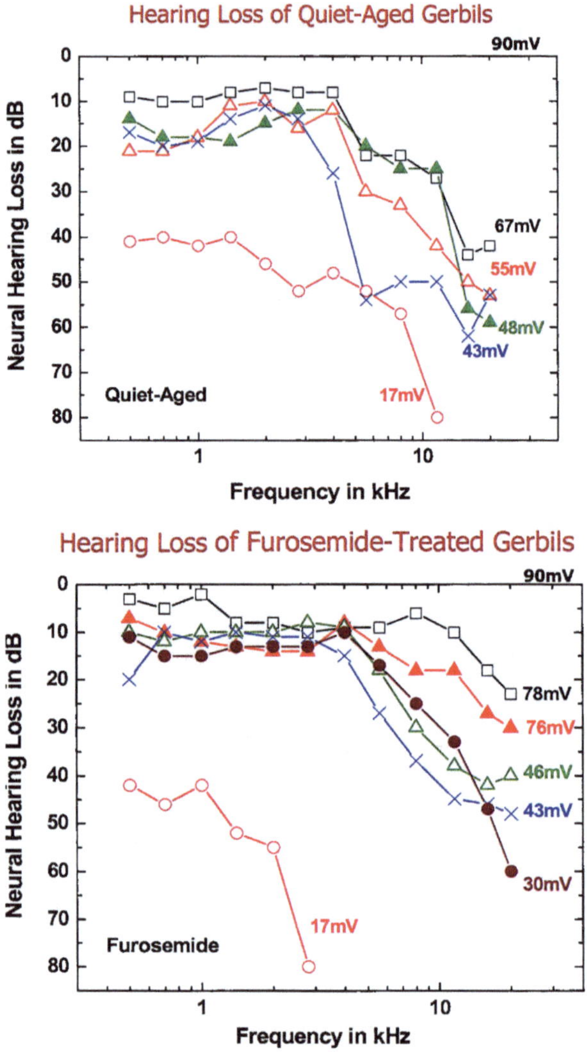

Fig. 2.6 HL profiles for individual ears from 36-month-old gerbils (top) and furosemide-treated ears from young gerbils (bottom). The aged and furosemide curves are normalized to the mean thresholds of young adult control animals. The EP recorded in the base of the cochlea is shown in mV for each animal. Note that young normal control animals have an EP of ~90 mV in the base of the cochlea as represented by the 0-dB HL abscissa. Furosemide was delivered to the round window of one ear via a cannula led from an osmotic pump. The opposite ear served as a control. The HL at high frequencies is well ordered by the amount of EP loss, whereas the loss at low frequencies is largely independent of EP as suggested by Fig. 2.2c, d. When EP drops below 25 mV, the thresholds even at low frequencies correspondingly increase, probably because the very low EP inhibits the IHC transduction process (Schulte and Schmiedt 1992). These 40-dB and greater flat losses may be analogs to the mechanical presbycusis described by Schuknecht (1974).

~40 mV, but when it does, the hearing loss even at low frequencies is greater than the 20-dB loss ascribed to the OHC amplifier. One hypothesis is that this excess loss is probably caused by an IHC transduction process that has been desensitized by the very low EP. The resulting flat 40-dB loss at low frequencies is similar to that seen in some cases of "mechanical" and neural presbycusis. One hypothesis is that these profiles represent an extreme form of metabolic presbycusis.

To test the hypothesis that the EP is the main variable in metabolic presbycusis and in the shaping of the audiogram in age-related hearing loss, furosemide was used to chronically lower the EP in one ear of a young gerbil (Schmiedt et al. 2002b). Furosemide is a potent but reversible inhibitor of the NKCC transporter and is well known to specifically block generation of the EP. Furosemide was applied to the intact round window via a cannula attached to an osmotic pump. The pump was placed between the scapulae, the cannula was led through the bulla, and the bulla was resealed with dental cement. The pumps could be sized to deliver the furosemide for up to one month. Cochleograms obtained from the pump animals showed almost no loss of hair cells and fairly normal strial morphology after several days of chronic exposure. To best mimic the EP loss seen in the 36-month-old gerbils, 5 mg/ml of furosemide were chronically delivered for seven days at a flow rate of 0.25 μl/h. The result is an animal model with essentially one young ear and one old ear. The similarity of the furosemide ear to an aged ear is remarkable in its breadth, from single-fiber responses and otoacoustic emissions to audiometric data (Schmiedt et al. 2002b).

Audiometric data obtained from the furosemide model are shown in the bottom panel of Fig 2.6. The audiometric profiles match those of the quiet-aged data in Fig. 2.6, top. Likewise, the profiles are ordered by the EP parameter. Another interesting point is that the furosemide dose was the same for all the young animals, yet the variation between the treated animals is similar to that seen with quiet-aged gerbils at 36 months of age. Perhaps the resistance or lack thereof to the furosemide threshold shift is somehow predictive of the amount of age-related hearing loss of an individual at a given age?

If the mean CAP threshold shifts obtained from the quiet-aged and furosemide-treated gerbils are compared, there is good quantitative agreement between the two groups (Fig. 2.7, top). The slope of the high-frequency roll off for the furosemide data is −8.4 dB/octave, with a breakpoint of 4.2 kHz. If the hypothesis of the EP-controlled OHC amplifier gain is correct, this slope and breakpoint represent the distribution of the OHC amplifier gain along the cochlear spiral. This gain distribution may be expected to differ among species depending on the frequency-distance map of the particular cochlea.

Fitting the mean furosemide gerbil data to that of the human audiometric profile is shown in the bottom panel of Fig. 2.7. The human data are from two sources and have been screened for noise history to minimize the effects of OHC loss. The breakpoint for the human data appears to be around 0.9 kHz and the slope might be somewhat steeper than that of the gerbil data, suggesting that the OHC amplifier gain distribution along the cochlear spiral for the human is biased toward lower frequencies than in the gerbil. The main point of comparison, however, is that the overall audiometric profiles of the quiet-aged and furosemide-treated gerbils and

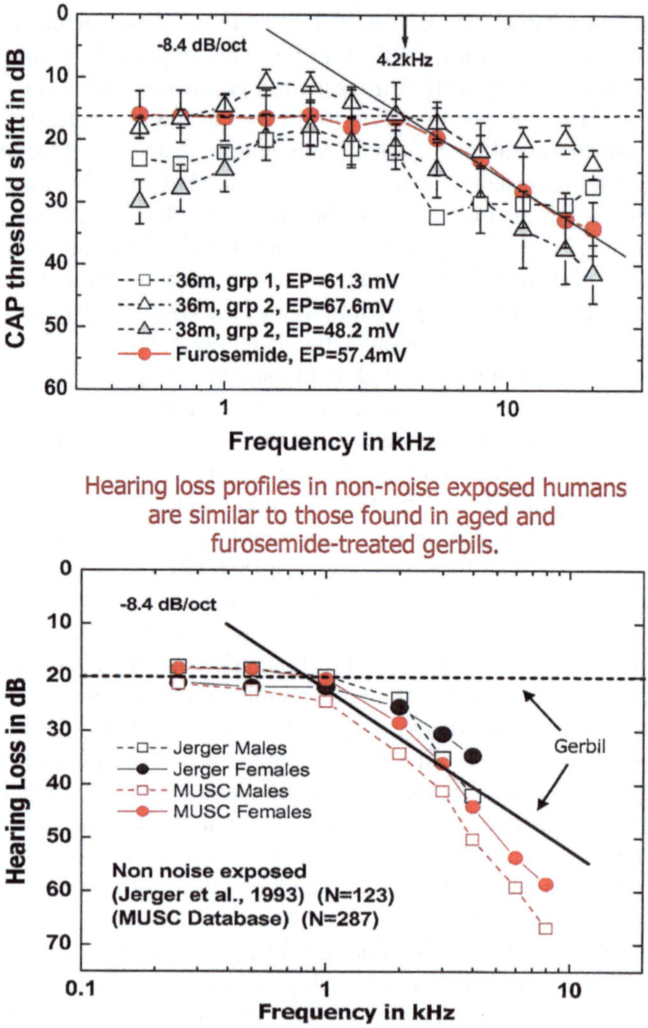

Fig. 2.7 Hearing loss profiles of three groups of quiet-aged and one group of furosemide-treated gerbils (top; adapted with permission from Schmiedt et al. 2002b.) and two groups of non-noise-exposed humans (bottom; adapted with permission from Mills et al. 2006a). The overall profile of the gerbil data arising from EP reduction from chronic furosemide treatment has been fitted to the human data (bottom panel). The flat loss at low frequencies is shifted to 20 dB, and the breakpoint for the shallow high-frequency roll-off has been shifted to ~0.9 kHz. It is clear that the metabolic model comprising EP reduction explains much of the HL profile of human presbycusis screened for noise and genetic histories.

the human data are remarkably similar. These data are strong evidence that the greatest factor underlying human presbycusis is EP loss arising from lateral wall degeneration with age. In other words, true age-related hearing loss in humans is largely of metabolic origin.

2.5.2 Suprathreshold Data

2.5.2.1 Single Tones

Suprathreshold measures obtained from animals are most often in the form of behavioral, neural, acoustic reflex, or otoacoustic data. A very clear suprathreshold result of aging in all animals is seen in the neural CAP response, an evoked electrical waveform recorded from the auditory nerve to either a click or tone pip. The CAP depends on a large number of auditory nerve afferents firing synchronously, resulting in a single negative wave as monitored by a gross electrode near the nerve. Fig. 2.8

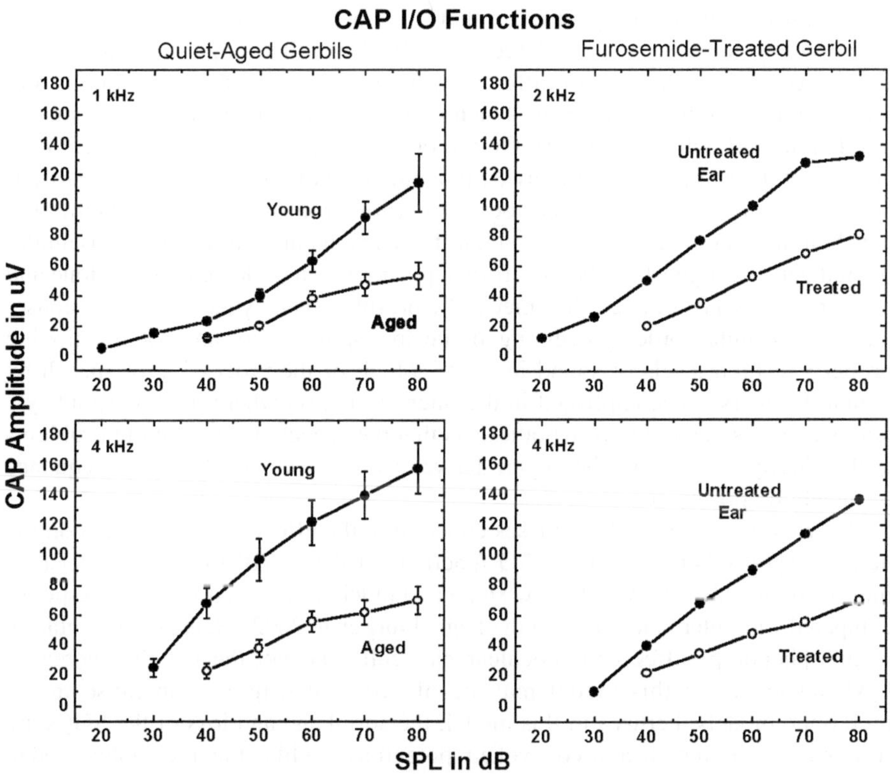

Fig. 2.8 Effects of metabolic presbycusis on suprathreshold measures of neural responses. CAP response amplitudes as a function of tone pip intensity (input/output [I/O] functions) at 1 and 4 kHz are shown for aged gerbils (left; values are means ± SE) and from a gerbil treated with chronic furosemide (right) at 2 and 4 kHz. Characteristics of the I/O function shared by both the aged and furosemide-treated ears are shallow slopes and diminished maximum amplitudes as compared with their control curves. Note that the furosemide-treated ears are in young animals with a full complement of hair cells and primary fibers. Thus the minimized CAP response with lowered EP is not simply from an anatomical reduction in the numbers of fibers available for excitation (see text).

show curves of CAP amplitude as a function of sound intensity in response to tone pips (an input/output function or I/O function) from control, quiet-aged, and furosemide-treated ears. There are the expected threshold shifts along the *x* axis; however, the most apparent changes with age and with EP reduction are the shallower slopes and marked reduction in the maximum amplitudes of the response waveform (Hellstrom and Schmiedt 1990, 1991, 1996; Mills et al. 1990; Schmiedt 1993; McFadden et al. 1997b; Schmiedt et al. 2002b).

Given that the furosemide model is obtained in a young gerbil, the populations of auditory hair cells and nerve fibers are assumed to be intact; however, the reduction in the CAP waveform is similar to that of the aged ear (which almost certainly has an age-related loss of primary fibers). Thus either the number of excited fibers or their synchronicity or both are affected both by age and by a decreased EP. Because the EP is instrumental in driving K^+ through the IHC for the transduction process, it is tempting to hypothesize that it is fiber synchrony that is being compromised by increased age and reduced EP. (Note that single-fiber data from quiet-aged gerbils showed no differences in spike rate intensity functions from young control animals; however, synchrony was not examined in those studies [Hellstrom and Schmiedt 1991]). On the other hand, aging and reduced EP may also affect the responsivity of individual populations of afferent fibers. In particular, the activity of the low-spont fiber population drops out with age and lowered EP (Schmiedt et al. 1996). These single-fiber results are similar for furosemide-treated gerbils, suggesting that it is truly a phenomenon of decreased EP alone and not necessarily just of age. Thus metabolic presbycusis may have dramatic consequences to higher-order processing if the low-spont fibers are not functioning properly at high levels of sound intensity. Much of the processing of speech by human listeners is accomplished in the intensity region where the low-spont system is active, so an obvious hypothesis is that the speech understanding problems in the elderly listener may be in part derived from a lowered EP as a consequence of metabolic presbycusis.

A caveat to the above hypotheses concerning the effects of EP changes on the neural response is that the CAP I/O functions of the C57BL/6J mouse after four months of age also show a shallow slope and much diminished maximal response compared with littermates one month of age (Lang et al. 2002). Because the mutation in this phenotype affects the cochlear hair cells and not the EP, the decreased CAP responses in this model may be of different origins than those in the metabolic models. Perhaps in this model, the shrinkage and loss of the SGCs are responsible for the decreased evoked potentials, unlike the metabolic model. Conversely, the cochlear amplifier may indeed be reduced in gain as a result of the scattered OHC loss, thereby yielding results similar to those of metabolic presbycusis and lowered EP. This hypothesis would assume that decreased OHC amplification implicitly results in CAP I/O functions with flattened slopes. However, in animals with OHC loss from noise or drug exposure, CAP thresholds are shifted, but the slopes of the I/O function can be normal or even steeper than normal (Bobbin 1992).

2.5.2.2 Multiple Tones

OAEs are a direct consequence of a properly functioning cochlear OHC amplifier (Probst 1990). As such, they are an excellent tool for the noninvasive investigation of the OHC system (Schmiedt 1986; Boettcher et al. 1995; Mills 2003, 2006; Mills and Schmiedt 2004). Gerbils have very robust OAEs, and they are only somewhat diminished in furosemide-treated animals with a lowered EP as shown in Fig. 2.9, top. OAEs are similarly diminished in quiet-aged gerbils as shown in Fig. 2.9, middle (Mills et al. 1993; Schmiedt et al. 2002b). Note from Fig. 2.9, bottom, that lowered EP effectively mimics the age-related loss in DPOAEs.

Similarly, emissions are diminished but present in non-noise-exposed elderly humans exhibiting signs of metabolic presbycusis. Indeed, the fine structure of OAEs can survive in these subjects (He and Schmiedt 1996). The point is that the OHC amplifier is still active in metabolic presbycusis with a lowered EP, but it is just not as robust as in the young animal. Conversely, cumulative noise and drug injury to the OHCs as found in many, if not most elderly, humans will drastically decrease OAEs. As a result, there are many reports showing significant loss of OAEs with age (e.g., Dorn et al. 1998).

That the OHC amplifier is still functional under quiet-aged conditions is also borne out in single-fiber studies in the auditory nerve that show tuning curve tips that are reduced somewhat in threshold but are still sharp at high characteristic frequencies (CFs; Schmiedt et al. 1990). Moreover, two-tone suppression is present on the low- and high-frequency sides of the tuning curves in quiet-aged animals. Conversely, animals aged in a continuous, low-level noise field have little OHC function remaining, and single fibers affected by the noise show tuning curves with no sharply tuned tips and no suppression (Schmiedt et al. 1990). Thus metabolic presbycusis is not necessarily as traumatic to the OHC system as sensory presbycusis where the hair cell loss is significantly higher or in noise trauma where the functionality of the stereocilia and OHC motor may be compromised.

2.6 Neural Presbycusis

Neural presbycusis seems to be a universal finding among all aging models, including those exhibiting sensory and metabolic characteristics. As illustrated in Fig. 2.10, top, there is an obvious shrinkage of the SGCs in Rosenthal's canal. Additionally, the number of ganglion cells typically decrease ~15–25% along the entire cochlear duct (Mills et al. 2006a; Fig. 2.10, bottom). Why this neuropathy comes about with age is still unknown, but a model of auditory neuropathy using ouabain is available that may be useful in studying the apoptosis and regeneration of auditory nerve fibers, especially when the hair cells and stria are still intact (Schmiedt et al. 2002a). Ouabain is a cardiac glycoside that irreversibly blocks the activity of Na^+-K^+-ATPase, thereby largely stopping the transport of Na^+ and K^+ across the cell membrane.

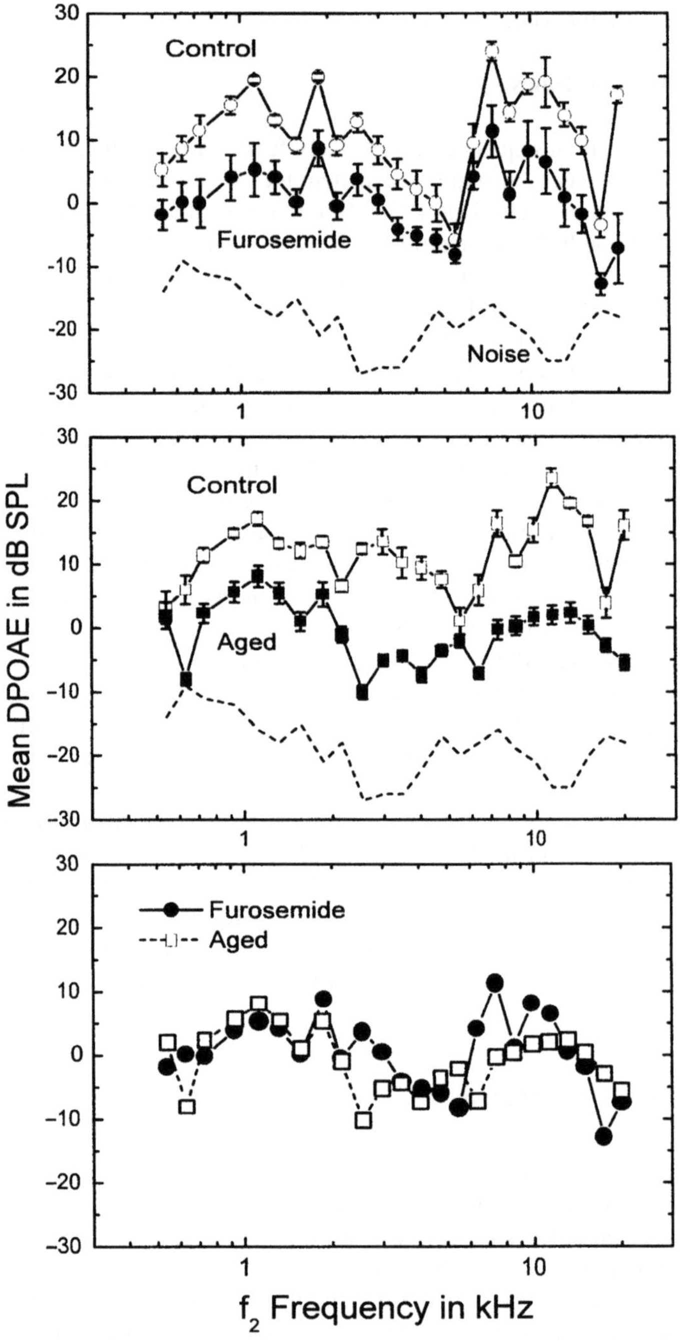

Fig. 2.9 Mean distortion product otoacoustic emission s (DPOAEs) obtained with 50-dB SPL primaries in the gerbil. DPOAEs are reduced but not absent with furosemide treatment (top) or age (middle) as compared with control values. Indeed, the absolute levels of the emissions are almost identical between the quiet-aged and furosemide-treated animals (bottom). Thus whereas the OHC amplifier may be diminished with reduced EP, it is still effective at higher levels with regard to the production of OAEs. (Adapted with permission from Schmiedt et al. 2002b.)

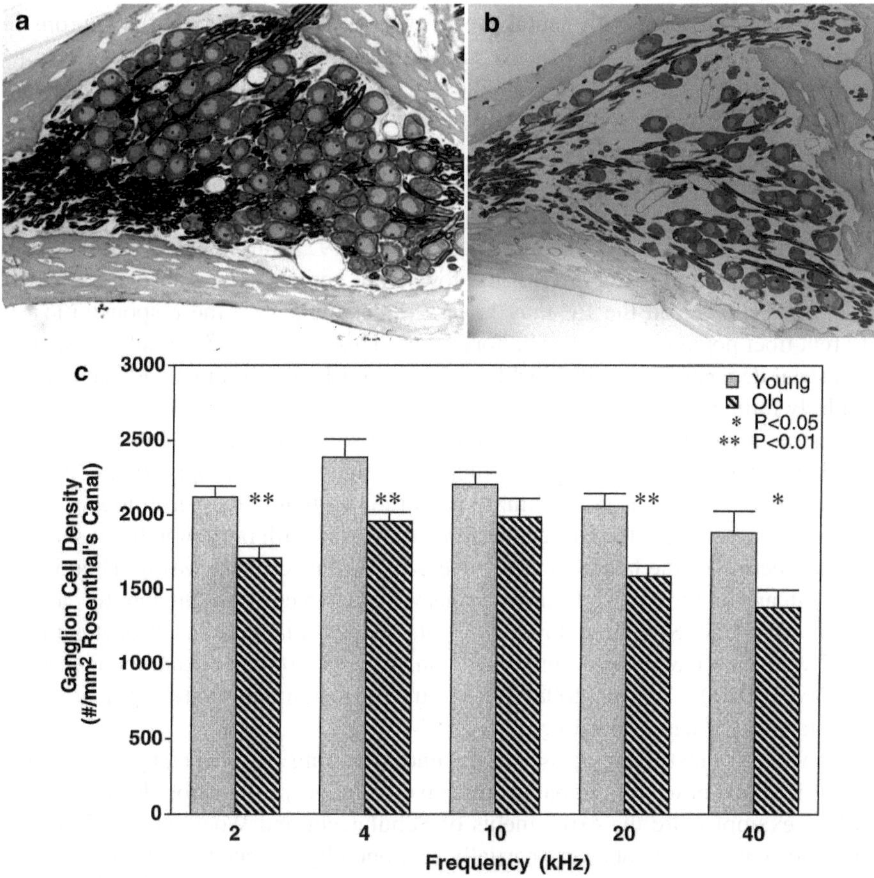

Fig. 2.10 Cross sections of Rosenthal's canal in the first turn of a young adult gerbil (**a**) and a 36-month-old gerbil (**b**). There is a marked reduction in the number and size of the spiral ganglion cells (SGCs) in the aged animal. **c**: SGC counts along the cochlear spiral in young adult and old gerbils. Typically there is about a 20% reduction of SGCs along the spiral at 36 months of age. (Adapted with permission from Mills et al. 2006a.)

In this model, ouabain is applied to the round window niche in gerbils or mice, similar to the furosemide model described previously. Nerve fibers have large amounts of Na^+-K^+-ATPase, and the ouabain selectively targets the type I afferents in the auditory nerve while sparing all hair cells and cochlear lateral wall. All type I afferents undergo apoptosis, whereas the type II fibers going to the OHCs are spared (Lang et al. 2005). This ouabain model also has been used to explore the use of stem cells in the regeneration of auditory ganglion cells (Lang et al. 2006a, 2008). Unfortunately, because all type I afferents degenerate in this model, it is not a good one for neural presbycusis where only 15–25% of the ganglion cells typically degenerate with age.

A very real problem with neural presbycusis is that functionally the neurons in aged animals seem to have grossly similar responses to sound as the neurons in young animals. One of the few studies of single fibers in aged animals revealed the difference between younger and older animals was with regard to the loss of activity of the low-spont system (Schmiedt et al. 1996; Lang et al. 2002). Interestingly, the loss of activity in this fiber group was not related to the shrinkage and disappearance of either the SGCs or the radial fiber population as predicted by another study on cats (Kawase and Liberman 1992; Suryadevara et al. 2001). The only correlation of the loss of the low-spont activity was with decreased EP. Thus aging effects on the EP may have a direct impact on the responsivities of different fiber populations in the auditory nerve. This scenario may possibly associate basal strial degeneration in the cochlea with diminished suprathreshold capabilities in elderly humans.

In another avenue of research, there has been some discussion that the cochlear lateral wall may be trophic (supplies support) to the hair cells, supporting cells, and neurons. This hypothesis arises simply from the fact that in the gerbil model, there was never a case where the hair cells and neurons were still present without survival of some section of the lateral wall. Conversely, whenever there was no lateral wall present in an ear, the hair cells and neurons were completely absent and the ear was functionally dead. Similarly, some new data suggest that certain fibrocytes in the lateral wall may have trophic influences on auditory afferents (Lang et al. 2006b; Adams et al. 2007; Adams 2008). This could be a significant development in our understanding of neural presbycusis.

Among the most difficult aspects in understanding neural presbycusis is that there is still no known functional change given relatively minor fiber losses. One of the best examples are the experiments of Schuknecht and Woellner (1955) where the auditory nerves of cats were partially sectioned before behavioral testing. Most of the treated cats showed little deficit in their thresholds, suggesting that the central nervous system probably needs only a few neurons to detect a sound. The data also speak to the redundancy of the afferent system, especially with regard to threshold loss. On the other hand, older humans show more problems than just their hearing loss, so neural presbycusis could help explain age-related changes in psychophysical performance with suprathreshold stimuli (see Fitzgibbons and Gordon-Salant, Chapter 5). The causes and effects of neural presbycusis are still very much a puzzle at this point.

2.7 Future Directions

2.7.1 Regeneration of Hair Cells

There is now a huge literature of hair cell regeneration (for reviews, see Martinez-Monedero et al. 2007; Stone and Cotanche 2007). These efforts are in part the result of the somewhat erroneous concept that age-related hearing loss in humans is largely sensory in nature. As this review hopefully has shown, human presbycusis is more likely metabolic than sensory. Certainly, it would be of great advantage to regrow hair cells in ears deafened by noise or ototoxic drugs; however, with regard to pure aging, it is the EP that must be regenerated before replacing lost hair cells. Two ways to reestablish the EP come to mind: using external currents to jump start the cochlear battery or regenerating the cells along the lateral wall that are responsible for generation of the EP.

2.7.2 Current Injection for Metabolic Presbycusis

Glass pipettes filled with a K^+ solution can be inserted into the scala media of quiet-aged and furosemide-treated gerbils with the purpose of passing a positive current to bolster the EP (Schmiedt 1993). When this is done, the EP does indeed increase, with concomitant increases in the slopes and maximum amplitudes of the CAP I/O functions as shown in Fig. 2.11, top. A positive current of ~10 μA can increase the EP by ~10-30 mV along the cochlear spiral. Negative currents will decrease the EP and will decrease the CAP response. Fig. 2.11, bottom, demonstrates that thresholds also improve with positive current application in the quiet-aged ear. Unfortunately, the disadvantage to current injection in this manner is that the K^+-recycling pathways function poorly in aged or furosemide-treated animals (which is why the EP is reduced). Passing a large K^+ current through the electrode into the scala media gradually overwhelms the recycling pathways and after some period of time (about an hour at +10 μA), the K^+ overwhelms the homeostatic mechanisms in the cochlea, whereupon the system dies. Thus, although promising at first, the K^+ buildup in the scala media is problematic.

2.7.3 Cell Regeneration

Another way to approach the problem of increasing the EP in metabolic presbycusis lies in maintaining and regenerating the cells responsible for its production. Studies from several labs have shown that the lateral wall cells, including the intermediate cells in the stria vascularis, turn over, i.e., they have a normal cycle of cell renewal (Roberson and Rubel 1994; Lang et al. 2003; Fig. 2.12). If stressed by furosemide, the rate of turnover is greater, whereas aging slows it down. Thus perhaps the most promising way to approach the amelioration of metabolic presbycusis is by understanding lateral wall cell turnover and why it decreases with age. Most of the lateral wall cells are fibrocytes, so maintaining and regenerating these cells should be somewhat easier than finding ways to regenerate terminally differentiated cells like hair cells.

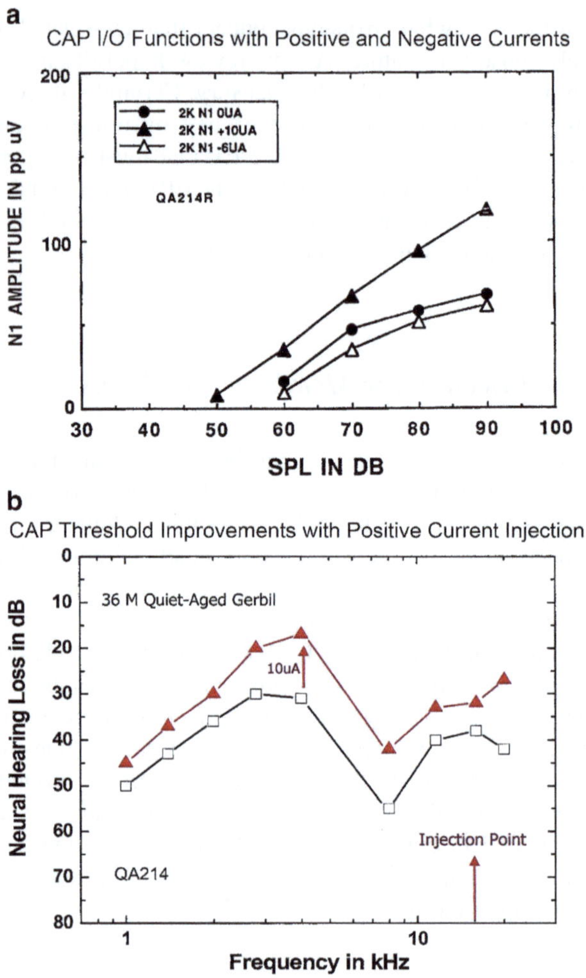

Fig. 2.11 Effects of current injection into the scala media of quiet-aged gerbils. **a**: currents from –6 μA to 10 μA were passed with a constant-current generator through a micropipette filled with 0.1M KCl inserted into the scala media in the basal coil. CAP I/O functions were enhanced with regard to threshold, slope, and maximum amplitudes with positive currents of up to 10 μA, whereas the CAP responses were diminished with negative currents. **b**: effect of current injection on CAP thresholds across frequency in the same animal. Although current was injected into the base, the effect was spread throughout the cochlear spiral (no current, open squares, with current solid triangles). Note that this animal had substantial losses at high and low frequencies before the electrode placement (open squares). Despite these losses, the data suggest that the aged ear is energy starved and boosting the EP would ameliorate much of the age-related hearing loss seen in this model of presbycusis.

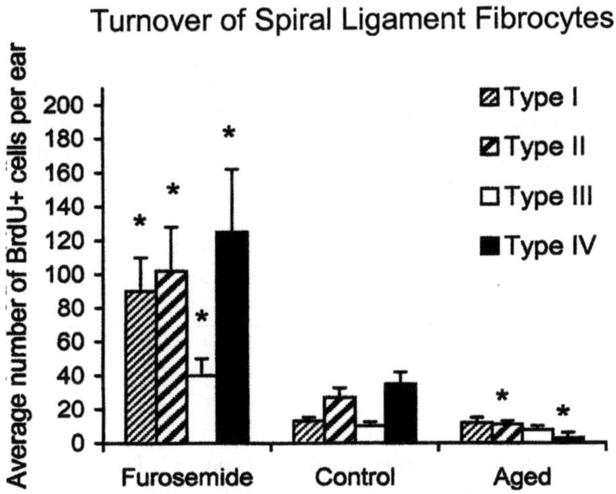

Fig. 2.12 Turnover or proliferation rates of different types of fibrocytes in the cochlear lateral wall. The turnover rates are indicated by the number of counted BrdU$^+$ cells in the four regions of fibrocytes. The rates are increased by cochlear stress (furosemide application) and are decreased with age. There is evidence that these fibrocytes are involved in the recycling of K$^+$ from the hair cells back to the stria vascularis where the EP is generated. Lateral wall fibrocytes are known to degenerate with age; however, they are continually renewed, probably by endogenous progenitor cells or hematopoietic stem cells from the bone marrow (Lang et al. 2006a, 2008). It may be possible to use various trophic factors or encourage their proliferation in the aged ear to restore proper numbers of fibrocytes with age. (Adapted with permission from Lang et al. 2003.)

2.8 Relating Animal Models to the Human Condition

Given what we now know about the physiology of the normal ear and aging ears with different types of presbycusis, is it possible to predict the underlying cause(s) of hearing loss by examination of the human audiogram? In other words, given what is known from animal models of presbycusis, is there enough information in an audiogram, coupled with suprathreshold hearing tests, to discriminate between sensory, metabolic, and possibly neural presbycusis?

Some simple phenotypical rules would include the following: pure OHC loss should result in a maximum of 50- to 60-dB HL at frequencies above ~2 kHz. Punctate OHC lesions should result in audiometric notches. OHC losses may have fairly steep slopes at frequencies above ~2 kHz. A positive noise history would be expected. Moreover, OHC loss in the base should have little or no effect on low-frequency thresholds. IHC loss should be linked to profound deafness as would a total loss of nerve fibers. Finally, OHC loss should profoundly affect cochlear nonlinearities, including compression phenomena and OAEs.

Conversely, metabolic presbycusis should be characterized by flat low-frequency HLs of between 10 and 40 dB, depending on the amount of EP loss. The flat loss should be coupled to a shallow high-frequency roll-off above ~1-2 kHz with a slope

Table 2.1 Phenotypes of age-related hearing loss as indicated by the audiogram.

			Low Frequency (0.25–1.0 kHz)		High Frequency (1.0–8.0 kHz)	
Category	Noise History	Notch (4-8 kHz)	Range (dB HL)	Slope (dB/octave)	Range (dB HL)	Slope (dB/octave)
Older-Normal	No	No	≤10	−5 to 5	0-20	−5 to 5
Premetabolic	No	No	≤10	−5 to 5	≤25	0-10
Metabolic	No	No	10-40	−5 to 5	30-60	10-20
Sensory	Yes	Yes	≤10	>5	>40	≥20
Metabolic+Sensory	Yes	Yes	10-40	−5 to 5	>40	≥20

of ~10-20 dB/octave. No noise history would be expected. There should be no notches. With regard to suprathreshold phenomena, the OHC amplifier should still be functional, albeit somewhat less robust. Thus compression phenomena and OAEs should still be present but reduced.

Table 2.1 uses the above hypotheses and numbers derived from animal results to discriminate five phenotypes describing age-related hearing loss. Obviously, most human presbycusis involves both sensory and metabolic components in the presence of a universal but poorly understood partial neural degeneration seen with neural presbycusis.

As specified in Table 2.1 and illustrated in Fig. 2.13, older subjects who are classified as "normal" or "premetabolic" would have negative noise histories and thresholds ≤ 10-dB HL from 0.25 to 1.0 kHz and ≤ 25-dB HL at higher frequencies. Subjects classified as "metabolic" (mild to severe) would have negative noise histories, flat hearing loss in the lower frequencies ranging from 10 to 40-dB HL, and gradually sloping hearing loss in the higher frequencies, with slopes ranging from 10 to 20 dB/octave. Subjects classified as "sensory" would have positive noise histories, thresholds in the lower frequencies ≤ 10-dB HL, and steeply sloping hearing loss in the higher frequencies with slopes > 20 dB/octave. Notches may be present. Subjects classified as "metabolic+sensory" would have positive noise histories, characteristics of metabolic presbycusis in the lower frequencies (flat loss ranging from 10- to 40-dB HL), and characteristics of sensory loss in the higher frequencies (steeply sloping loss with slopes > 20 dB/octave). The basic profile of these audiometric templates in Fig. 2.13 should be applicable to both human and animal models of hearing loss and presbycusis and is part of ongoing longitudinal studies of human presbycusis (Lee et al. 2005; Dubno et al. 2008).

Further definition of the phenotypes could be done by identifying audiologic or other characteristics beyond the audiograms that differentiate subjects in the five categories, e.g., OAE amplitudes and I/O functions; upward spread of masking; ABR thresholds, latencies, and amplitudes; and, in the future, genetic variations to discriminate the five phenotypes. This work is currently ongoing (Lee et al. 2005; Dubno et al. 2008).

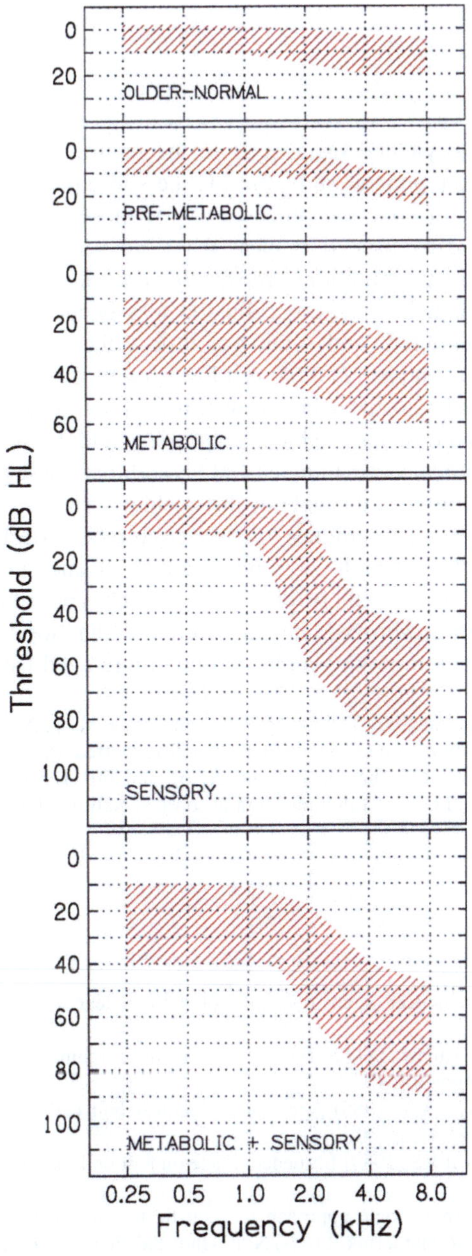

Fig. 2.13 Ranges of audiogram profiles illustrating the five phenotypes of age-related HL in humans. The profiles are schematized from the results of animal models of presbycusis. In general, metabolic presbycusis results in a mild, flat loss at low frequencies coupled with a shallow high-frequency sloping loss above ~1 kHz. The maximum loss is ~60 dB. Sensory loss is typified by steep slopes and notches, with little loss at low frequencies. Maximum loss is ~60 dB if only OHCs are involved; if IHCs are lost, there will be profound deafness. Most humans, particularly men, given their typical noise history, fall under the "metabolic+sensory" profile where the damage is from sensory loss from excess noise exposure coupled with a metabolic loss from aging.

2.9 Summary

Animal models have yielded insights as to the nature of hair cell function, the cochlear amplifier, the generation of the EP, the subsequent recycling of cochlear potassium, and the transduction process. All of those factors have roles in our understanding of presbycusis. It is clear from quiet-aged animal models where noise, drugs, and genetic mutations are strictly controlled, that (1) sensory loss is almost always present, but paradoxically the loss occurs largely at the apex or extreme base, is scattered, and often has little effect on gross auditory nerve thresholds to simple sounds; (2) there is a lowering of the EP as a consequence of the degeneration of the cochlear lateral wall including the stria vascularis; (3) the EP modulates the gain of the cochlear amplifier, yielding an audiometric profile commonly seen in humans and animals; and (4) auditory nerve ganglion cells shrink and are reduced in number throughout the cochlea. Although advances in tissue regeneration and invasive techniques to correct age-related cochlear physiology are of great interest to the research community, a real advance in the amelioration of the effects of presbycusis would be differentiating the types of presbycusis in older adults and relating this improved clinical differential diagnosis to novel biomedical interventions aimed at improving auditory function in aged listeners.

Acknowledgments I thank Judy Dubno, Hainan Lang, Jack Mills, Nancy Smythe, and Diana Vincent for their suggestions and encouragement. Studies reported here were supported by Grants R01 AG 14748 from the National Institute on Aging and P01 DC 00422 from the National Institute on Deafness and Other Communication Disorders, National Institutes of Health.

References

Adams J (2008) Noise induced stress responses of the cochlear lateral wall. Abstr Assoc Res Otolaryngol 31:236.

Adams J, McCaffery S, Kujawa SG (2007) Effects of acoustic trauma in the spiral ligament. Abstr Assoc Res Otolaryngol 30:243.

Bhattacharyya TK, Dayal VS (1985) Age-related cochlear hair cell loss in the chinchilla. Ann Otol Rhinol Laryngol 94:75–80.

Bhattacharyya TK, Dayal VS (1989) Influence of age on hair cell loss in the rabbit cochlea. Anat Rec 230:136–145.

Bobbin RP (1992) Pharmacologic approach to acoustic trauma in the cochlea. In: Dancer A, Henderson D, Salvi R, Hamernik R (eds) Noise-Induced Hearing Loss. St. Louis: Mosby-Year Book, pp. 38–44.

Boettcher FA, Gratton MA, Schmiedt RA (1995) Effects of noise and age on the auditory system. In: Morata T, Dunn D (eds) Occupational Medicine: State of the Art Reviews. Vol. 10, No. 3: Occupational Hearing Loss Philadelphia: Hanley and Belfus, pp. 577–592.

Bredberg G (1968) Cellular pattern and nerve supply of the human organ of Corti. Acta Otolaryngol Suppl (Stockh) 236:1–135.

Cooper N, Rhode W (1997) Mechanical responses to two-tone distortion products in the apical and basal turns of the mammalian cochlea. J Neurophysiol 78:261–270.

Davis H (1983) An active process in cochlear mechanics. Hear Res 9:79–90.

Dayal VS, Bhattacharyya TK (1989) Comparative study of age-related cochlear hair cell loss. Hear Res 15:179–183.

Dazart S, Feldman ML, Keithley EM (1996) Cochlear spiral ganglion degeneration in wild-caught mice as a function of age. Hear Res 100:101–106.

Dorn PA, Piskorski P, Keefe DH, Neely ST, Gorga MP (1998) On the existence of an age/threshold/frequency interaction in distortion product otoacoustic emissions. J Acoust Soc Am 104:964–971.

Dubno JR, Lee FS, Matthews LJ, Ahlstrom JB, Horwitz AR, Mills JH (2008) Longitudinal changes in speech recognition in older persons. J Acoust Soc Am 123:462–475.

Evans EF, Klinke R (1982) The effects of intracochlear and systemic furosemide on the properties of single cochlear nerve fibres in the cat. J Physiol 331:409–427.

Frisina RD, Walton JP (2001) Aging of the mouse central auditory system. In: Willot JP (ed) Handbook of Mouse Auditory Research: From Behavior to Molecular Biology. New York: CRC Press, pp. 339–379.

Frisina RD, Walton JP (2006) Age-related structural and functional changes in the cochlear nucleus. Hear Res 217:216–233.

Gates GA, Mills JH (2005) Presbycusis. Lancet 366:1111–1120.

Gates GA, Cooper JC, Kannel WB, Miller NJ (1990) Hearing in the elderly: the Framingham cohort, 1983–1985. Ear Hear 11:247–256.

Gates GA, Mills DM, Nam B-H, D'Agostino R, Rubel EW (2002) Effects of age on the distortion-product otoacoustic emission growth functions. Hear Res 163:53–60.

Gratton MA, Schmiedt RA, Schulte BA (1996) Age-related decreases in endocochlear potential are associated with vascular abnormalities in the stria vascularis. Hear Res 94:116–124.

Gratton MA, Smythe BJ, Lam CF, Boettcher FA, Schmiedt RA (1997) Decline in the endocochlear potential corresponds to decreased Na,K-ATPase activity in the lateral wall of quiet-aged gerbils. Hear Res 108:9–16.

Gruber J, Schaffer S, Halliwell B (2008) The mitochondrial free radical theory of ageing - where do we stand? Front Biosci 13:6554–6579.

He N, Schmiedt RA (1996) Effects of aging on the fine structure of the 2f1-f2 acoustic distortion product. J Acoust Soc Am 99:1002–1015.

Hellstrom LI, Schmiedt RA (1990) Compound action potential input/output functions in young and quiet-aged gerbils. Hear Res 50:163–174.

Hellstrom LI, Schmiedt RA (1991) Rate-level functions of auditory-nerve fibers have similar slopes in young and old gerbils. Hear Res 53:217–221.

Hellstrom LI, Schmiedt RA (1996) Measures of tuning and suppression in single-fiber and whole-nerve reponses in young and quiet-aged gerbils. J Acoust Soc Am 100:3275–3285.

Henry KR (1982) Age-related auditory loss and genetics: an electrocochleographic comparison of six inbred strains of mice. J Gerontol 37:275–282.

Hequembourg S, Liberman MC (2001) Spiral ligament pathology: a major aspect of age-related cochlear degeneration in C57BL/6 mice. J Assoc Res Otolaryngol 2:118–129.

Ichimiya I, Suzuki M, Mogi G (2000) Age-related changes in the murine cochlear lateral wall. Hear Res 139:116–122.

Jerger J, Chmiel R, Stach B, Spretjnak M (1993) Gender affects audiometric shape in presbyacusis. J Am Acad Audiol 4:42–49.

Johnson KR, Erway LC, Cook SA, Willott JF, Zheng QY (1997) A major gene affecting age-related hearing loss in C57BL/6J mice. Hear Res 114:83–92.

Johnson LG, Hawkins JE Jr (1972) Sensory and neural degeneration with aging as seen in microdissections of the human inner ear. Ann Otol Rhinol Laryngol 81:179–193.

Kawase T, Liberman MC (1992) Spatial organization of the auditory nerve according to spontaneous discharge rate. J Comp Neurol 319:312–318.

Keithley EM, Feldman ML (1979) Spiral ganglion cell counts in an age-graded series of rat cochleas. J Comp Neurol 188:429–442.

Keithley EM, Feldman ML (1982) Hair cell counts in an age-graded series of rat cochleas. Hear Res 3:249–262.

Keithley EM, Ryan AF, Woolf NK (1989) Spiral ganglion cell density in young and old gerbils. Hear Res 38:125–134.

Lang H, Schulte BA, Schmiedt RA (2002) Endocochlear potentials and compound action potential recovery functions in the C57BL/6J mouse model. Hear Res 172:118–126.

Lang H, Schulte BA, Schmiedt RA (2003) Effects of chronic furosemide treatment and age on cell division in the adult gerbil inner ear. J Assoc Res Otolaryngol 4:164–175.

Lang H, Schulte BA, Schmiedt RA (2005) Ouabain induces apoptotic cell death in type I spiral ganglion neurons, but not type II neurons. J Assoc Res Otolaryngol 6:63–74.

Lang H, Schulte BA, Schmiedt RA (2006a) Contribution of bone marrow hematopoietic stem cells to adult mouse inner ear: mesenchymal cells and fibrocytes. J Comp Neurol 496:187–201.

Lang H, Schulte BA, Zhou D, Smythe, N, Spicer SS, Schmiedt RA (2006b) Nuclear factor κB deficiency is associated with auditory nerve degeneration and increased noise-induced hearing loss. J Neurosci 26:3541–3550.

Lang H, Schulte BA, Goddard JC, Hedrick M, Schulte JB, Wei L, Schmiedt RA (2008) Transplantation of mouse embryonic stem cells into the cochlea of an auditory-neuropathy animal model: effects of timing after injury. J Assoc Res Otolaryngol 9:225–240.

Lee FS, Matthews LJ, Dubno JR, Mills JH (2005) Longitudinal study of pure-tone thresholds in older persons. Ear Hear 26:1–11.

Liberman MC (1978) Auditory-nerve response from cats raised in a low-noise chamber. J Acoust Soc Am 63:442–455.

Marcus DC, Chiba T (1999) K^+ and Na^+ absorption by outer sulcus epithelial cells. Hear Res 134:48–56.

Marcus DC, Wu T, Wangemann P, Kofuji P (2002) KCNJ10 (Kir4.1) potassium channel knockout abolishes endocochlear potential. Am J Physiol Cell Physiol 282:C403-C407.

Martinez-Monedero R, Oshima K, Heller S, Edge AS (2007) The potential role of endogenous stem cells in regeneration of the inner ear. Hear Res 227:48–52.

McFadden SL, Campo P, Quaranta N, Henderson D (1997a) Age-related decline of auditory function in the chinchilla (*Chinchilla laniger*). Hear Res 111:114–126.

McFadden SL, Quaranta N, Henderson D (1997b) Suprathreshold measures of auditory function in the aging chinchilla. Hear Res 111:127–135.

Mills DM (2003) Differential responses to acoustic damage and furosemide in auditory brainstem and otoacoustic emission measures. J Acoust Soc Am 113:914–924.

Mills DM (2006) Determining the cause of hearing loss: differential diagnosis using a comparison of audiometric and otoacoustic emission responses. Ear Hear 27:508–525.

Mills DM, Rubel EW (1994) Variation of distortion product otoacoustic emissions with furosemide injection. Hear Res 77:183–199.

Mills DM, Schmiedt RA (2004) Metabolic presbycusis: differential changes in auditory brainstem and otoacoustic emission responses with chronic furosemide application in the gerbil. J Assoc Res Otolaryngol 5:1–10.

Mills DM, Norton SJ, Rubel EW (1993) Vulnerability and adaptation of distortion product otoacoustic emissions to endocochlear potential variation. J Acoust Soc Am 94:2108–2122.

Mills JH, Schmiedt RA, Kulish LF (1990) Age-related changes in auditory potentials of Mongolian gerbil. Hear Res 46:201–210.

Mills J, Schmiedt R, Schulte B, Dubno J (2006a) Age-related hearing loss: a loss of voltage, not hair cells. Semin Hear 27:228–236.

Mills JH, Schmiedt RA, Dubno JR. (2006b) Older and wiser, but losing hearing nonetheless. Hear Health Summer:12–17.

Ohlemiller KK, Gagnon PM (2004) Apical-to-basal gradients in age-related cochlear degeneration and their relationship to "primary" loss of cochlear neurons. J Comp Neurol 479:103–116.

Ohlemiller KK, Lett JM, Gagnon PM (2006) Cellular correlates of age-related endocochlear potential reduction in a mouse model. Hear Res 220:10–26.

Ohlemiller KK, Rybak-Rice ME, Gagnon PM (2008) Strial microvasculature pathology and age-associated endocochlear potential decline in NOD cogenic mice. Hear Res 244:85–97.

Probst R (1990) Otoacoustic emissions: an overview. Adv Otorhinolaryngol 44:1–91.
Roberson DW, Rubel EW (1994) Cell division in the gerbil cochlea after acoustic trauma. Am J Otol 15:28–34.
Robles L, Ruggero MA (2001) Mechanics of the mammalian cochlea. Physiol Rev 81:1305–1352.
Ruggero MA, Rich NC (1991) Furosemide alters organ of Corti mechanics: evidence for feedback of outer hair cells upon the basilar membrane. J Neurosci 11:1057–1067.
Russell IJ (1983) Origin of the receptor potential in inner hair cells of the mammalian cochlea – evidence for Davis's theory. Nature 301:334–336.
Salt AN, Melichar I, Thalmann R (1987) Mechanisms of endocochlear potential generation by the stria vascularis. Laryngoscope 97:984–991.
Schmiedt RA (1986) Acoustic distortion in the ear canal. I. Cubic difference tones: effects of acute noise injury. J Acoust Soc Am 79:1481–1490.
Schmiedt RA (1989) Spontaneous rates, thresholds, and tuning of auditory nerve fibers in the gerbil: comparisons to cat data. Hear Res 42:23–36.
Schmiedt RA (1993) Cochlear potentials in quiet-aged gerbils: does the aging cochlea need a jump start? In: Verrillo R (ed) Sensory Research: Multimodal Perspectives. Hillsdale, NJ: Lawrence Erlbaum and Associates, pp. 91–103.
Schmiedt RA (1996) Effects of aging on potassium homeostasis and the endocochlear potential in the gerbil. Hear Res 102:125–132.
Schmiedt RA, Zwislocki JJ (1977) Comparison of sound-transmission and cochlear-microphonic characteristics in Mongolian gerbil and guinea pig. J Acoust Soc Am 61:133–149.
Schmiedt RA, Mills JH, Adams JC (1990) Tuning and suppression in auditory nerve fibers of aged gerbils raised in quiet or noise. Hear Res 45:221–236.
Schmiedt RA, Mills JH, Boettcher FA (1996) Age-related loss of activity of auditory-nerve fibers. J Neurophysiol 76:2799–2803.
Schmiedt RA, Okamura H-O, Lang H, Schulte BA (2002a) Ouabain application to the round window of the gerbil cochlea: a model of auditory neuropathy and apoptosis. J Assoc Res Otolaryngol 3:223–233.
Schmiedt RA, Lang H, Okamura H-O, Schulte BA (2002b) Effects of furosemide chronically applied to the round window: a model of metabolic presbyacusis. J Neurosci 22:9643–9650.
Schuknecht HF (1974) Presbyacusis. In: Pathology of the Ear. Cambridge, MA: Harvard University Press.
Schuknecht HF, Gacek MR (1993) Cochlear pathology in presbycusis. Ann Otol Rhinol Laryngol 102:1–16.
Schuknecht HF, Woellner RC (1955) An experimental and clinical study of deafness from lesions of the auditory nerve. J Laryngol Otol 69:75–97.
Schulte B (2007) Homeostasis of the inner ear. In: Dallos P (ed) The Senses. Vol. III. Audition. New York: Academic Press, pp. 149–156.
Schulte BA, Schmiedt RA (1992) Lateral wall Na,K-ATPase and endocochlear potentials decline with age in quiet-reared gerbils. Hear Res 61:35–46.
Sewell WF (1984). The effects of furosemide on the endocochlear potential and auditory-nerve fiber tuning curves in cats. Hear Res 14:305–314.
Sha S-H, Kanicki A, Dootz G, Talaska AE, Halsey K, Dolan D, Altschuler R, Schacht J (2008) Age-related auditory pathology in the CBA/J mouse. Hear Res 243:87–94.
Spicer S, Schulte B (1991) Differentiation of inner ear fibrocytes according to their ion transport related activity. Hear Res 56:53–64.
Spicer S, Schulte B (1996) The fine structure of spiral ligament cells relates to ion return to the stria and varies with place-frequency. Hear Res 100:80–100.
Spicer S, Gratton M, Schulte B (1997) Expression patterns of ion transport enzymes in spiral ligament fibrocytes change in relation to strial atrophy in the aged gerbil cochlea. Hear Res 111:93–102.
Spongr VP, Flood DG, Frisina RD, Salvi RJ (1997) Quantitative measures of hair cell loss in CBA and C57BL/6 mice throughout their life spans. J Acoust Soc Am 101:3546–3553.

Stone JS, Cotanche DA (2007) Hair cell regeneration in the avian auditory epithelium. Int J Dev Biol 51:633–647.

Suryadevara A, Schulte B, Schmiedt R, Slepecky N (2001) Auditory nerve fibers in young and aged gerbils: morphometric correlations with endocochlear potential. Hear Res 161:45–53.

Tarnowski B, Schmiedt R, Hellstrom L, Lee F, Adams J (1991) Age-related changes in cochleas of Mongolian gerbils. Hear Res 54:123–134.

Wangemann P (2002). K^+ cycling and the endocochlear potential. Hear Res 165:1–9.

Wangemann P, Liu J, Marcus D (1995) Ion transport mechanisms responsible for K^+ secretion and the transepithelial voltage across marginal cells of stria vascularis in vitro. Hear Res 84:19–29.

Willott JF (1991) Aging and the Auditory System: Anatomy, Physiology, and Psychophysics. San Diego, CA: Singular Publishing Group.

Wu R, Hoshino T (1999) Changes in off-lesion endocochlear potential following localized lesion in the lateral wall. Acta Otolaryngol 119:550–554.

Chapter 3
Cell Biology and Physiology of the Aging Central Auditory Pathway

Barbara Canlon, Robert Benjamin Illing, and Joseph Walton

3.1 Introduction

The most common manifestation of age-induced hearing loss, presbycusis, is the loss of sensitivity for high-frequency sounds, resulting in difficulties in speech perception, hearing in noisy backgrounds, and distorted loudness perception. These abnormalities typically involve progressive damage to the inner ear and spiral ganglion neurons, leading to a diminished input into the central auditory nervous system. The central components of the auditory system can also undergo direct morphological and physiological changes induced by the biological effects of aging. A combination of direct and secondary changes is most likely to contribute to the manifestations of the aging auditory system. This chapter describes the cell biological and physiological changes that occur during aging in experimental subjects.

3.2 Cochlear Nucleus

The cochlear nuclear complex is located on the lateral surface of the brain between the pons and the medulla at the cerebellopontine angle. All auditory afferent fibers from the cochlea terminate in the cochlear nucleus complex. The ascending branch proceeds rostrally to the anteroventral cochlear nucleus (AVCN); the descending

B. Canlon (✉)
Department of Physiology and Pharmacology, Karolinska Institute, 171 77, Stockholm, Sweden
e-mail: Barbara.Canlon@ki.se

R.B. Illing
Neurobiological Research Laboratory, Universitäts-HNO-Klinik, D-79106, Freiburg, Germany
e-mail: robert.illing@uniklinik-freiburg.de

J. Walton
Associate Professor of Otolaryngology and Neurobiology and Anatomy, University of Rochester Medical Center, 14642, Rochester, NY
e-mail: joseph_walton@urmc.rochester.edu

S. Gordon-Salant et al. (eds.), *The Aging Auditory System*,
Springer Handbook of Auditory Research 34,
DOI 10.1007/978-1-4419-0993-0_3, © Springer Science+Business Media, LLC 2010

branch innervates the posteroventral cochlear nucleus (PVCN); and the dorsal cochlear nucleus (DCN) described the neuronal types in the cochlear nucleus and demonstrated that the AVCN contains a variety of neuronal types including spherical bushy cells, globular bushy cells, and multipolar cells. In the PVCN, there are octopus cells, multipolar cells, globular cells, and granule cells. The neurons in the DCN include cartwheel cells, fusiform cells, granule cells, stellate cells, and giant cells. The DCN is a laminated structure with three layers receiving direct input from primary auditory fibers. Layer III receives the majority of nerve fibers and contains large and small multipolar cells.

The morphology of the DCN as a function of age has been evaluated in terms of volume, neuronal number, and synaptic alterations for different species and strains by investigators who employed different analytical methods. The CBA/CaJ (CBA) mouse maintains relatively good hearing across most of its life span, whereas the C57BL/6J (C57) mouse demonstrates morphological and physiological auditory deficiencies earlier in life. Willott et al. (1992) found that the volume in layer III of the DCN decreased with age in C57 mice, whereas in the CBA mouse, a decrease was noted only in the oldest mice investigated. With increasing age, the total number of neurons in the DCN has been shown to decrease in the CBA mouse, whereas the PVCN remained stable as shown in Fig. 3.1. The age-related decrease in DCN neurons was correlated to changes in hair cells and spiral ganglion neurons in the peripheral organ (Idrizbegovic et al. 2001). When morphological analysis was made on the cochlear nucleus from C57 mice, an age-related decrease in the total number of neurons was found in both the DCN and the PVCN. These findings suggest that the more obvious changes observed in the DCN of C57 mice are a reflection of the greater peripheral degeneration that is noted in that strain compared with the CBA mouse.

The spherical and globular cells of the AVCN receive major input from the auditory nerve in the form of end bulbs of Held, whereas multipolar cells do not

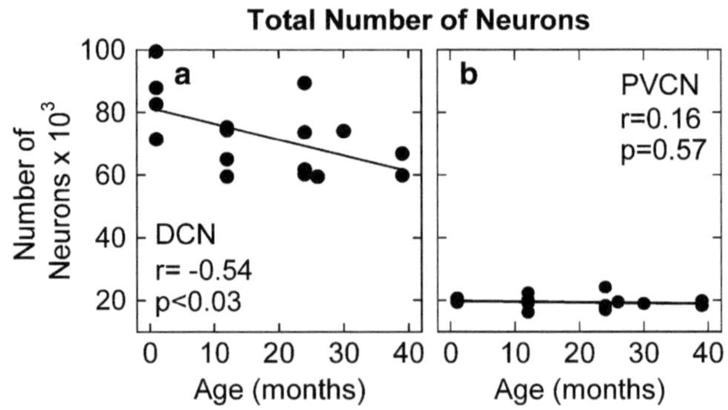

Fig. 3.1 Quantification of the total number of neurons in Nissl-stained sections in the dorsal cochlear nucleus (DCN; **a**) demonstrated a significant decrease in the total number of neurons with increasing age. Quantification of the total cell number in the posteroventral cochlear nucleus (PVCN; **b**) did not show any change with increasing age. (From Idrizbegovic et al. 2001)

(Brawer and Morest 1975; Cant and Morest 1979). The density and number of the AVCN neurons and the size of the neurons were studied in CBA and C57mice (Willott et al. 1987). Neither strain showed any age-related change in AVCN dimensions or volume. The number and packing density of AVCN neurons decreased with age in both strains of mice, yet the C57 mouse showed an earlier degeneration compared with the CBA mouse. The CBA mice showed a tendency toward decreased neuronal size for all cell types in the AVCN, whereas the C57 mice showed a decrease in multipolar cells size and an increase in spherical and globular cells, suggesting that these neurons were swollen. In contrast to the age-related changes found in the mouse cochlear nucleus, only minor changes were found in the rat. Fischer-344 rats showed age-related decreases in the size of the terminals contacting small-caliber dendrites. No changes were noted in this rat strain in the density or number of synaptic terminals or number of neurons in older animals. It remains to be determined why the rat AVCN remains stable in its morphology during aging.

3.2.1 Ion Channels

Neuronal excitability is primarily determined by the interaction between sodium and potassium conductances. Potassium channels are composed of four identical subunits, of which there are at least 23 variants. The specific variation of these subunits determines the dynamic properties of the channel, laying ground for a specific management of membrane currents. In particular, the pattern of expression of potassium channel subunits is thought to contribute to the establishment of the unique discharge characteristics exhibited by cochlear nucleus neurons. Fitzakerley et al. (2000) describe the developmental distribution of mRNA for the three subunits Kv4.1, Kv4.2, and Kv4.3 belonging to one subfamily of voltage-gated potassium channels within the cochlear nucleus of the mouse by in situ hybridization and RT-PCR. They find that Kv4.1 is not present in the cochlear nucleus at any age. Kv4.2 mRNA is detectable as early as postnatal day (P) 2 in all subdivisions of the cochlear nucleus and continued to be constitutively expressed throughout development. Kv4.2 is massively expressed in all of the major projection neuron classes, including octopus cells, bushy cells, stellate cells, fusiform cells, and giant cells. By contrast, Kv4.3 is expressed at lower levels, by fewer cell types, and more abundantly in the ventral than in the dorsal nucleus. Compared with Kv4.2, Kv4.3 expression was significantly delayed in development, emerging only after postnatal day 14. The authors infer from their study of mRNA that the differential distribution of Kv4 transcripts selectively indicates specific profiles of potassium channels among individual neurons throughout development. Looking at potassium channel proteins directly, Jung et al. (2005) found that among all auditory brainstem regions only the PVCN revealed age-related changes. With progressive age, Kv1.1 immunoreactivity is increased in the octopus cell bodies, whereas staining intensity is significantly decreased in the neuropil (unmyelinated neuronal processes within the gray matter).

In contrast, the immunoreactivity for Kv3.1 is decreased in the octopus cells and neuropil of the aged PVCN. The authors conclude that this necessarily affects ion channel activity and signal processing in the central auditory system with progressive age.

3.2.2 Glycine Receptors

The auditory system relies heavily on temporal coding. To achieve precision in the range of a couple of microseconds, extremely reliable and well-balanced neuronal interactions are a prerequisite. This balance rests, among others, on glutamatergic and cholinergic signaling to generate depolarization, and glycinergic and GABAergic interactions to generate hyperpolarization. The distribution of glycine receptors in adult C57 mice showing progressive cochlear pathology compared with aging CBA mice known to retain good hearing. In the cochlear nucleus of 18-month-old C57 mice with a severe hearing loss, the number of glycine-immunoreactive neurons decreased significantly compared with the control animals. Specifically, the number of strychnine-sensitive glycine receptors (GlyR) decreased significantly in the DCN of old C57 mice. Significant effects were not observed in the cochlear nucleus of middle-aged C57 mice that show less severe hearing loss or in very old CBA mice that retain good hearing. The data suggest that the combination of severe hearing loss and old age results in deficits in one or more inhibitory glycinergic circuits in the cochlear nucleus.

A faster deterioration of hearing function and density of glycine receptors in the cochlear nucleus with increasing age is found in Fischer-344 rats compared with Long-Evans rats. This will hamper glycinergic transmission and compromise the balance between excitation and inhibition that is essential for effective processing of auditory stimuli. This observation was confirmed by studying the expression of mRNAs for $\alpha 1$-, $\alpha 2$-, and β-subunits of the GlyR in the AVCN in three age groups (Krenning et al. 1998). Expression of mRNAs for $\alpha 1$- and β-subunits decreased significantly in AVCN in the 18- and 27-month-old age groups compared with younger rats, whereas mRNA expression for the $\alpha 2$-subunit increased. These changes are likely to quantitatively change the protein synthesis of GlyR subunits and may thus alter the function of GlyR, affecting binding to its ligands and changing inhibitory neurotransmission in the aging cochlear nucleus.

3.2.3 GABA

To determine whether there are age-related alterations in the biosynthetic enzyme glutamic acid decarboxylase (GAD), and, the degradative enzyme GABA-transaminase (GABA-T), the uptake system for GABA in the cochlear nucleus of Fischer-344 rats was studied (Raza et al. 1994). In contrast to other regions of the auditory brainstem, no age-related changes were seen for any of these molecules in the cochlear nucleus. Along with their study of GABA metabolism, Raza and coworkers also looked for the acetylcholine-synthesizing enzyme ChAT in the cochlear nucleus but did not find any significant age-dependent changes.

3.2.4 Glutamate Receptors

Glutamatergic neurotransmission was studied at the end bulb of Held in the cochlear nucleus of DBA mice who have an early onset of an age-related hearing loss (Wang and Manis 2005). These authors examined synaptic transmission at the end bulb of Held between auditory nerve fibers and bushy cells in the AVCN. Synaptic transmission in the high-frequency areas of the AVCN, corresponding to the region of peripheral hearing loss, was altered in old DBA mice. The spontaneous miniature excitatory postsynaptic current frequency was substantially reduced, and miniature excitatory postsynaptic currents were significantly slower and smaller in high-frequency regions of old (average age 45 days) DBA mice compared with tonotopically matched regions of young (average age 22 days) DBA mice. Moreover, synaptic release probability was ~30% higher in high-frequency regions in young DBA than in old DBA mice. Auditory nerve-evoked EPSCs showed less rectification in old DBA mice, suggesting recruitment of GluR2 subunits into the AMPA (α-amino-3-hydroxy-5-methyl-4-isoxazolepropionic acid receptor) receptor complex. No similar age-related changes in synaptic release or EPSCs were found in age-matched normal hearing young and old CBA mice. These findings suggest that auditory nerve activity plays a critical role in maintaining normal synaptic function at the end bulb of Held synapse after the onset of hearing. Auditory nerve activity regulates both presynaptic (release probability) and postsynaptic (receptor composition and kinetics) function at the end bulb synapse after the onset of hearing and will be affected during the aging process.

3.2.5 Calcium-Binding Proteins

The number of parvalbumin- (Fig. 3.2) and calbindin- (Fig. 3.3) positive neurons in the DCN have been found to increase with increasing age in CBA mice at a time when they show only moderate sensorineural pathology (Idrizbegovic et al. 2001). There is an age-related decrease in the total number of neurons in the DCN. At the same time, parvalbumin-positive neurons increase against a background of unchanged number of neurons. These changes may, in part, be induced by degenerative changes in the inner ear and are likely to modulate neuronal homeostatsis by increasing calcium-binding proteins in the PVCN and DCN during aging, protecting against age-related calcium toxicity (Idrizbegovic et al. 2003). A statistically significant

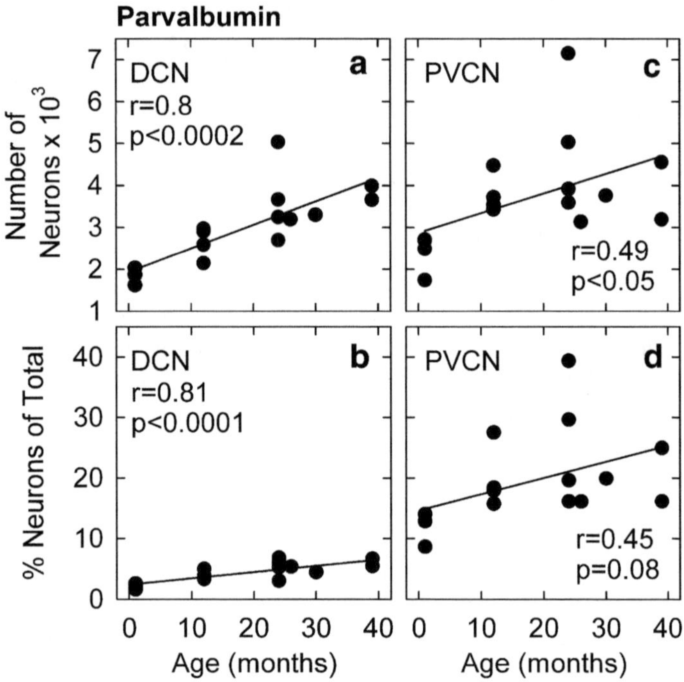

Fig. 3.2 (**a**) There was a significant increase in the total number of parvalbumin-immunopositive neurons in the DCN with increasing age. (**b**) Percentage of parvalbumin-positive of neurons in the DCN, showing a significant positive regression with increasing age. (**c**) Quantitative analysis in the PVCN, showing a significant increase in the total number of parvalbumin-immunopositive neurons. (**d**) The percentage of parvalbumin-positive neurons in the PVCN did not show any significant regression with increasing age. (From Idrizbegovic et al. 2001)

decrease in the density of DCN and PVCN neurons (25%) was found at 24 months of age in CBA/CBA mice compared with 1-month-old animals (Idrizbegovic et al. 2006). The percentage of parvalbumin- and calretinin-positive neurons in the DCN and the PVCN in relation to the density of Nissl-stained neurons showed significant increases in 24-month-old compared with the 1-month-old animals. In the DCN, the percent increase in calretinin and parvalbumin was correlated to the loss of spiral ganglion neurons (SGNs), inner hair cells (IHCs), and outer hair cells (OHCs). In the PVCN, parvalbumin was correlated to SGN, IHC, and OHC loss. The percent increase in calbindin immunoreactivity was not correlated to any peripheral pathology. These data suggest an increase in calcium-binding protein immunoreactivity in the cochlear nucleus in the 24-month-old mice and probably reflect an endogenous protective strategy designed to counteract calcium overload that is prominent during aging and degeneration.

Glial fibrillary acid protein (GFAP) is an intermediate filament protein that helps maintain cell structure. The distribution of GFAP immunoreactivity in the young adult rat cochlear nucleus predominates in the granular cell region and

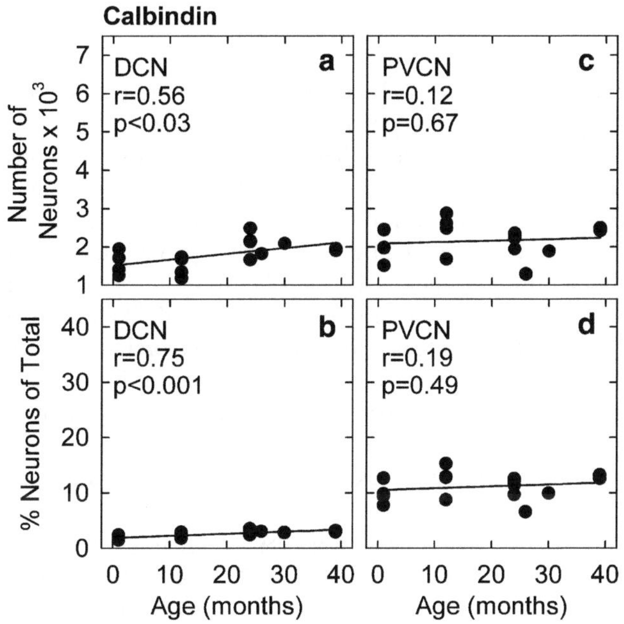

Fig. 3.3 (**a**) There was a significant increase in the total number of calbindin-immunopositive neurons in the DCN with increasing age. (**b**) Percentage of calbindin-positive neurons in the DCN, showing a significant positive regression with increasing age. (**c**) Quantitative analysis in the PVCN, showing a significant increase in the total number of calbindin-immunopositive neurons. (**d**) Percentage of calbindin-positive neurons in the PVCN did not show any significant regression with increasing age. (From Idrizbegovic et al. 2001)

differs from old rats where GFAP immunoreactivity is homogeneously distributed in all parts of the nucleus (Jalenques et al. 1995). There was no change in the total number of GFAP neurons between these two ages in any part of the cochlear nucleus except for its anteroventral subdivision, where a decrease in neuronal number occurs with age. The increase in GFAP immunoreactivity is related to an increase in both GFAP-positive astrocyte number and processes. The increase of GFAP-positive astrocytes is thought to be due to an alteration of auditory nerve fibers, to a change in trophic interactions with postsynaptic cells, or to intrinsic alterations of neurons in the cochlear nucleus and local circuits reflecting age-related impairments. In a follow-up study, a significant increase in the number of GFAP-positive astrocytes was found during the first year of life, as was a significant decrease in GFAP immunoreactivity in the ventral cochlear nucleus of older rats (Jalenques et al. 1997). Aging was found to be associated with a significant increase in GFAP-positive astrocyte sizes, except for immunolabeled astrocytes in the granule cell layer. The different levels of GFAP expression occurring in the ventral cochlear nucleus (VCN) during normal aging are suggested to reflect a progressive decline in cellular activity in the VCN, without severe cell degeneration or synaptic loss.

3.2.6 Growth Factors

The insulin-like growth factor I (IGF-I) is a monomeric peptide with significant homology to proinsulin. It exerts potent effects on cultured neural tissue, including the stimulation of mitosis in sympathetic neuroblasts; the promotion of neurite outgrowth in cortical, sensory, and sympathetic neurons; and the induction of oligodendrocyte differentiation. IGF-I mRNA is localized in regions of synaptic transmission of the developing olfactory, auditory, visual, and somatosensory systems. In the auditory system, IGF-I appears to be involved in maturation and maintenance of spiral ganglion cells (Camarero et al. 2002), but its mRNA is abundant in the cochlear nucleus and other auditory brainstem regions. In these systems, IGF-I gene expression is found predominantly in long-axon projection neurons, appearing during a relatively late stage in their development at a time of maturation of dendrites and synapse formation (Bondy 1991). Mutations of the IGF-I gene with age are therefore expected to contribute to deterioration in all of these systems.

An analysis of the macrophage response using oligonucleotide probes for scavenger receptor-B mRNA indicated that differences in the macrophage response in young and old animals was the likely cause of an age related change in IGF-I and transforming growth factor-β1 mRNA expression patterns. On the basis of their data, Hinks and Franklin (2000) suggest a model of remyelination in which platelet-derived nerve growth factor (PDGF) is involved in the initial phase of oligodendrocyte progenitor recruitment, while IGF-I and transforming growth factor-β1 trigger the differentiation of the recruited cells into myelinating oligodendrocytes (Hinks and Franklin 2000).

Luo et al. (1995) found that the mRNA of the acidic fibroblast growth factor (aFGF) is strongly expressed in the principal neurons of the AVCNs and PVCNs but not in the octopus cells. In the DCN, aFGF mRNA was present only in scattered smaller vertical cells. Developmentally, low levels of aFGF expression appeared in the cochlear nuclei between the day of birth and the sixth day after birth. This expression increased rapidly during the onset of hearing, between P10 and P14 and reached adult level by P14 to P17. Labeling in collicular neurons appeared slightly later. The results suggest that the appearance of strong aFGF mRNA expression is related to the onset of function. An age-related loss would compromise the maintenance of precisely firing neuronal networks in the auditory system.

A decline in brain-derived neurotrophic factor (BDNF) mRNA expression has been found within the dendritic projections from the cochlea toward the cochlear nucleus (Ruttiger et al. 2007). This localization was found to be lost in aged hearing-impaired animals and may suggest that BDNF protein targeting is reduced during aging (Fig. 3.4). These findings show that there is a specific targeting of BDNF protein from the SGNs to the cochlear nucleus and that this interaction is altered by aging. Specific targeting of BDNF protein to activated synapses and BDNF mRNA to dendritic processes has already been described for other systems.

3.2.7 Growth-Associated Proteins

High levels of the growth-associated protein (GAP)-43 are present in most regions of the young brain (Benowitz and Routtenberg 1997), including the cochlear nucleus (Illing et al. 1997). With age, the level of this protein is down-regulated in the cochlear nucleus as it is in most brain regions. Interesting exceptions are regions that have been involved in learning (Oestreicher and Gispen 1986; Benowitz et al. 1988; Masliah et al. 1991a, b). Because GAP-43 is involved in neuroplasticity, different levels of it against the developing brain in the adult are telling. There is evidence that the aged brain is characterized by the inability of GAP-43 biosynthesis. Schmoll et al. (2005) have shown that middle-aged rats are still capable of a sustained, though diminished, GAP-43 response to specifically patterned neuronal activity, whereas old rats lose this ability. Apparently, disruption of the temporal and anatomical coordination of expression of GAP-43 contributes to a general decline in brain plasticity with age.

3.2.8 Mitochondrial DNA Mutations

Mitochondrial respiratory chain dysfunction is an important contributor to human pathology. Several genetic syndromes caused by mutations of nuclear or mitochondrial (mt) DNA are known, and the clinical manifestations often include hearing impairment/deafness as well as symptoms from other organs. Mitochondrial dysfunction is also implicated in the pathophysiology of several age-associated degenerative diseases and in the normally occurring aging process (Fischel-Ghodsian 2003; Pickles 2004). The mitochondrial respiratory chain performs oxidative phosphorylation to generate ATP, the main energy currency for a variety of metabolic processes in mammalian cells. The respiratory chain is located in the inner mitochondrial membrane and consists of 5 large protein complexes containing a total of ~100 different proteins. The compact mtDNA encodes 13 respiratory chain subunits, whereas nuclear genes encode all of the remaining subunits. The biogenesis of the respiratory chain thus depends on the coordinated action of both nuclear and mitochondrial genes. In addition to energy production, mitochondria are directly involved in other processes such as regulation of apoptosis induction, calcium homeostasis, reactive oxygen species generation, and iron-sulfur cluster formation. Several lines of evidence implicate respiratory chain dysfunction in inherited and acquired forms of impaired hearing (Niu et al. 2007; Yamasoba et al. 2007).

Investigations of whether elevated levels of somatic mtDNA mutations affect the auditory system have been studied in mtDNA mutator mice (Niu et al., 2007). These mice have increased levels of somatic mtDNA point mutations causing phenotypes consistent with premature aging and development of a progressive impairment of hearing (Niu et al. 2007). A minor decrease in the number of neurons in the DCN was observed in wild-type mice between 2 and 10 months

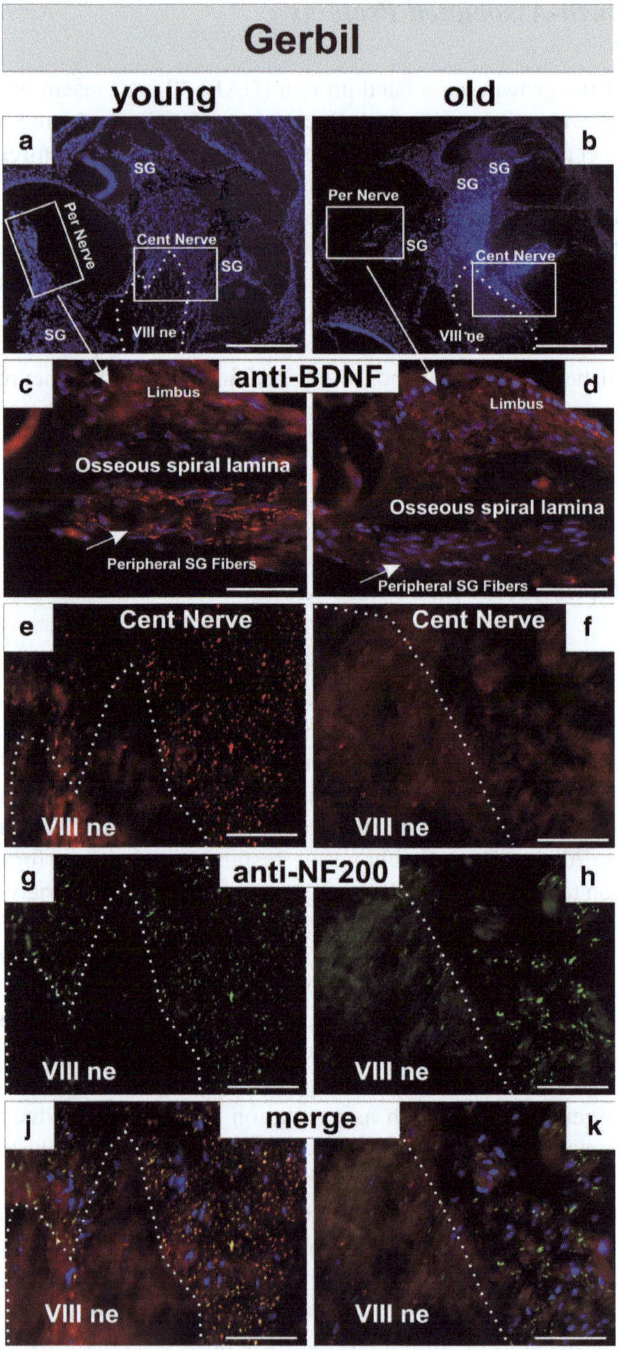
Gerbil
young
old
a
b
SG
SG
SG
Per Nerve
Per Nerve
Cent Nerve
SG
SG
Cent Nerve
SG
VIII ne
VIII ne
anti-BDNF
c
d
Limbus
Limbus
Osseous spiral lamina
Osseous spiral lamina
Peripheral SG Fibers
Peripheral SG Fibers
e
f
Cent Nerve
Cent Nerve
VIII ne
VIII ne
anti-NF200
g
h
VIII ne
VIII ne
merge
j
k
VIII ne
VIII ne

of age. However, a significant reduction in the number of DCN neurons was observed in the mtDNA mutator mice between 2 and 10 months of age (Fig. 3.5). This reduction in DCN neurons in the mtDNA mutator mice was progressive starting at 6 months of age. The total number of neurons from the PVCN of the wild-type mice was constant between 2 and 10 months of age. In contrast, the mtDNA mutator mice showed a progressive loss of neurons in the PVCN with increasing age. Again, there was a significant reduction in the number of PVCN neurons over time. The mtDNA mutator mice developed a progressive hearing impairment with cell loss in both the peripheral and central auditory systems with increasing age (Niu et al. 2007). The observed auditory pathology of mtDNA mutator mice resembles the physiological and anatomical changes seen in human presbycusis. Mitochondrial dysfunction thus has the capacity to cause presbycusis pathology. The hypothesis can thereby be put forward that age-associated somatic mtDNA mutations may contribute to the development of human presbycusis.

3.2.9 Cochlear Nucleus Physiology

The diversity in anatomical and connectional architecture of the cochlear nucleus is mirrored in the diversity of different types of response properties to simple sounds found in each subnucleus. Two coding pathways are postulated to have their origin in the cochlear nucleus. The "what" pathway is composed of the AVCN and PVCN. Neurons in these areas perform simple and complex analyses of sound features such as frequency, intensity, duration, and envelope characteristics (for a review, see Rhode and Greenberg 1994). The "where" pathway originates in the AVCN and DCN. Spherical and globular bushy cells of the AVCN are involved in

Fig. 3.4 Expression of brain-derived neurotrophic factor (BDNF) protein in peripheral and central neurites of gerbil cochlear neurons during aging. Cross section of young (**a**, **c**, **e**, **g**, **j**) and old (**b**, **d**, **f**, **h**, **k**) gerbil cochleae showing spiral ganglia (SG) and peripheral (Per Nerve) and central (Cent Nerve) nerve projections as well as the VIIIth nerve (VIII ne). The dotted line marks the Schwann cell/oligodendrocyte border. (**c** and **d**) Magnification of peripheral nerve area boxed in **a** and **b** showing BDNF antibody staining of the peripheral nervous projections in the osseous spiral lamina of a young (**c**) and old (**d**) gerbil. In young animals, BDNF immunoreactivity (red) is seen in peripheral spiral ganglion nerve fibers (arrow in **c**). In contrast, no immunoreactivity is seen in the corresponding sections in aged animals (arrow in **d**). (**e-k**) Magnification of the central nerve area boxed in A and B showing BDNF antibody staining of the central nervous projections of a young (**e**) and old (**f**) gerbil. BDNF immunoreactivity (red) is present in the central nerve projections of young gerbils (**e**) but absent in old gerbils (**f**). Staining with an antineurofilament antibody (anti-NF200, green) demonstrates that nerve fibers are maintained in aged animals (compare **g** and **h**). (**j** and **k**) Overlay of BDNF and NF200 immunohistochemistry in the central neurites. In young animals, there is a colocalization of BDNF and NF200 immunoreactivity (**j**), whereas the expression of BDNF (but not NF200) in central neurites is lost in aged animals (**k**). Nuclei are stained blue with DAPI. Scale bars: **a** and **b**, 500 μm; C-K, 50 μm. (From Ruttiger et al. 2007.)

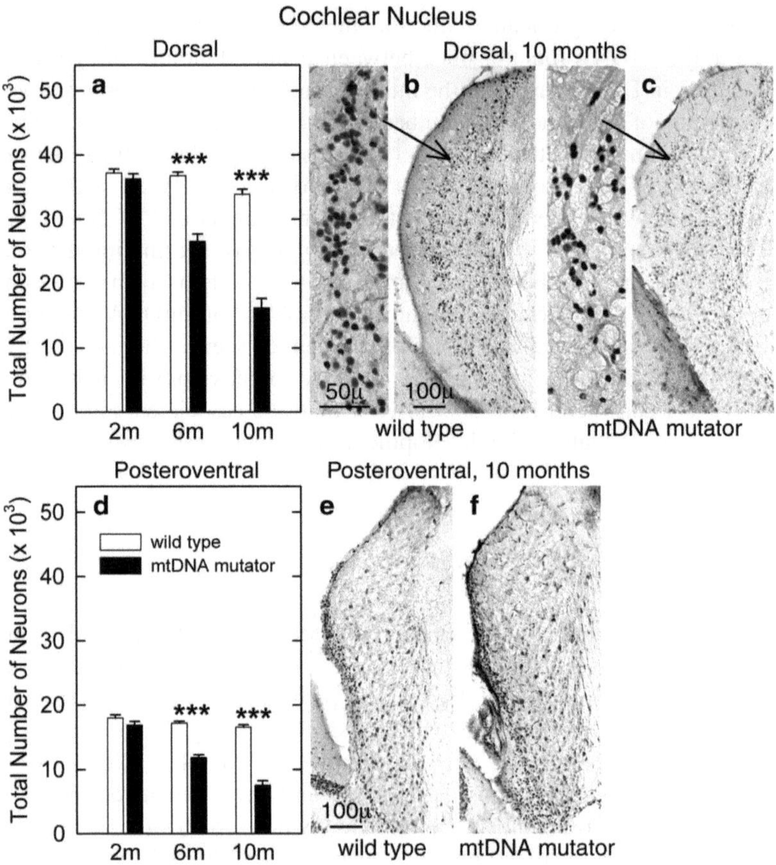

Fig. 3.5 Progressive loss of cochlear nucleus neurons in mitochondrial (mt) DNA mutator mice. (**a**) Quantification of the total number of dorsal cochlear nucleus neurons (DCN) in wild-type and mtDNA mutator mice at 2, 6, and 10 months of age. Representative micrographs obtained from the DCN 10-month-old wild-type (**b**) and mtDNA mutator (**c**) mice are shown. (**d**) Quantification of the total number of PVCNs in wild-type and mtDNA mutator mice at 2, 6, and 10 months of age. Representative micrographs obtained from the PVCNs of 10-month-old wild-type (**e**) and mtDNA mutator (**f**) mice are shown. Values are means ±SE. $^{***}P < 0.001$ by two-sided *t*-test. (From Niu et al. 2007.)

processing localization cues in the horizontal plane, whereas fusiform cells of the DCN have been implicated in the processing of elevation cues important for localization (May 2000). Unfortunately, neurophysiological studies searching for neural correlates of age-related dysfunction in the cochlear nucleus from aged animal models are lacking. There are two major explanations for this: (1) the difficulty and high rate of morbidity due to the surgical exposure necessary for access to the cochlear nucleus in older mice and (2) the oftentimes deleterious effects of general anesthesia in old animals.

After pioneering studies in the inferior colliculus, Willott (1986) were the first to investigate neural encoding of basic sound properties in the cochlear nucleus of old CBA mice. CBA mice are thought to be an appropriate animal model of human aging because they display a very slow, progressive loss of hearing sensitivity beginning in midlife, with thresholds only elevated by 20-30 dB compared with young mice as measured by auditory brainstem responses. Willott et al. (1991) measured simple receptive fields from the VCN and DCN in mice of different ages and reported on alterations in minimum thresholds, best frequencies, and bandwidth of the excitatory frequency response area (FRA). In CBA mice, they reported no significant age-related modifications in minimum thresholds and FRAs in old animals relative to young adults. The distribution of best frequencies, both in the VCN and DCN, was also comparable. They did however find that the width of the FRA measured at 80-dB SPL was decreased by 20-25% in both VCN and DCN units from old CBA mice.

In contrast to the CBA strain, the C57 strain undergoes rapid age-related high-frequency hearing loss, which reaches the profound degree by midlife, due to the presence of the *ahl* gene. This allele (Cdh23 gene) disrupts the normal function of an otocadherin transmembrane adhesion protein located in the hair cell stereocilia, resulting in a rapid peripheral sensorineural hearing loss (Noben-Trauth et al. 2003; Zheng et al. 2005). In comparison to an age-matched CBA group, Willott et al. (1991), found significant increases in the proportion and number of units with thresholds greater than 60 dB in middle-aged (~1 year of age) C57 mice. In middle-aged C57 mice, the sensitivity changes in the C57 VCN were more drastic than in the DCN or inferior colliculus. In contrast to VCN neurons, tuning curves for DCN neurons were statistically indistinguishable from those of the inferior colliculus. Perhaps the stability in the DCN is a reflection of the divergent inputs that sculpt the receptive fields of DCN principal neurons.

A major tenet in studies of age-related alteration in neural coding within the central auditory system is that the interplay between excitation and inhibition alters stimulus evoked activity. Age-related changes in intensity coding in the DCN of old Fischer-344 rats were recently reported by Caspary et al. (2005), and their major findings support the decline in inhibition that acts to shape neurophysiological response properties, such as the shapes of rate-level functions (RLFs) and response areas. Caspary et al. (2006) measured the RLFs from DCN principal cells, presumably fusiform cells, to best-frequency (BF) tones from young and old Fischer-344 rats. They found that the mean discharge rates of units from young rats showed greater nonmonotonicity compared with the mean data from old rats. More specifically, the data show that as intensity increases, the suprathreshold responses from units from older rats increase at a greater rate compared with the young data. The authors conclude that this finding is consistent with the loss of glycinergic inhibitory input from the vertical cells that provide input to fusiform cells. Vertical cells provide on-BF inhibition to fusiform cells via glycinergic circuitry. This important inhibitory output limiter could have consequences for complex sound coding (temporal or spatial processing) in the VCN that also receives glycinergic inputs from the DCN.

3.3 Superior Olivary Complex

The superior olivary complex (SOC) has a variety of functions including localization of sound, encoding temporal features of complex sounds, and modulating descending information to the cochlear nucleus and cochlea (Heffner and Heffner 1989; Masterton 1992; Oliver 2000; Thompson and Schofield 2000). Inputs into the SOC arise in both the ipsilateral and contralateral cochlear nuclei, and the physiological response properties of many of the cells are the results of inputs to both ears (Brugge and Geisler 1978). The SOC is composed of three nuclei with different functional roles. These nuclei are the medial superior olive (MSO), the lateral superior olive (LSO), and the medial nucleus of the trapezoid body (MNTB). These nuclei have clear anatomical boundaries and cell types with distinct morphologies and are discussed separately.

3.3.1 LSO

The LSO receives signals from secondary sensory neurons in the cochlear nucleus on both sides of the brain. Together with neurons of the inferior colliculus, its neurons are the first to process binaural signals, indispensable for sound localization. To achieve the observed precision of auditory localization behavior, it is easily calculated that neuronal signals underlying this function must operate with a temporal precision of a few microseconds.

It has been demonstrated that the ability to localize sound is impaired in aged rats, and morphological changes in the SOC contribute to this deficit (Eddins and Hall, Chapter 6). Neuron counts were performed on the nuclei of the SOC in Fischer-344 rats at ages 3, 12, 24, and 30 months. Neuron number remained stable between 3 and 30 months of age in the LSO and MSO; however, in the MNTB, neuron number was significantly reduced at 24 and 30 months of age (Casey 1990). Neuron loss in the MNTB of 24-month-old Fischer-344 rats is not as prominent as that reported for 24-month-old Sprague-Dawley rats (8% loss vs. 34% loss), indicating a strain difference with regard to aging in the SOC.

During development, survival and maturation of the LSO neurons critically depend on synaptic activity and intracellular calcium signaling. Before hearing onset, glutamatergic synaptic inputs from the cochlear nucleus to the LSO activate metabotropic glutamate receptors, causing calcium release from intracellular stores and massive calcium influx from the extracellular space. During development, the contribution of extracellular calcium influx to metabotropic glutamate receptor-mediated Ca^{2+} responses gradually decreased and was almost abolished by the end of the third postnatal week in mice. Over this period, the contribution of Ca^{2+} release from internal stores remained unchanged. The developmental decrease in transient receptor potential-like channel-mediated calcium influx was significantly less in congenitally deaf waltzer mice. This suggests that early auditory experience is indispensable for the normal age-dependent downregulation of functional transient receptor potential channels (Ene et al. 2007).

In the LSO, the transmitter phenotype of neurons is not established early in development. Instead, its neurons undergo a characteristic transmutation early in development. Kotak et al. (1998) have demonstrated this striking switch during postnatal development in the gerbil. Although GABA and glycine elicit similar postsynaptic ionotropic responses, the results suggest that GABAergic transmission in neonates may play a developmental role distinct from that of glycine.

As data from humans provided evidence for an age-dependent impairment in sound localization, the question arose if this impairment might depend on GABAergic or glycinergic neurons of the LSO. In the gerbil animal model, the size of the LSO as well as the number and density of glycine- and GABA-immunoreactive neurons were not significantly different between young (<15 months) and old (>3 years) gerbils. However, the size of glycine- and GABA-immunoreactive neurons was reported to be significantly reduced in the high-frequency (medial) limb of the LSO in older subjects (Gleich 1994).

3.3.2 MNTB

The formation of the calyx of Held is one of the largest nerve terminals in the mammalian brain (Hoffpauir et al. 2006). In a quantitative ultrastructural study, Casey and Feldman (1985) looked for the effect of aging on axosomatic synaptic terminals in the rat MNTB. According to their measurements, the mean percentage of the surface area of principal cells covered by synaptic terminals is 61.7% in rats 3 months of age. When these animals were 27-33 months old, the coverage went down to 43.7%. Similarly, a highly significant decrease takes place in old rats with respect to the average number of synaptic terminals present along a 100-μm length of principal cell surface, from 28.3 to 18.9. Only terminals derived from calyces of Held are lost in the aged animals, displaying a 37% reduction. Over the same period of aging, the length of apposition by synaptic terminals in the MNTB does not change significantly. These authors concluded that because of a significant loss of calycine synaptic endings, the structure of calyces of Held becomes less complex with advancing age (Casey and Feldman 1988). This is thought to result in an age-related partial deafferentation of principal cells, which entails significant alterations in the processing of auditory information in the MNTB.

Large cavitations emerge in the aging rat MNTB within capillary basal laminae (Casey and Feldman 1985). At the same time, membranous debris accumulated and indicated cellular degeneration within leaflets of the capillary basal lamina. The volume density ratio of capillaries decreases significantly between 6 and 33 months of age.

Much of auditory signal processing depends on the tonotopic organization across almost all regions of the central auditory system in the mammalian brain. Von Hehn et al. (2004) found that the Kv3.1 potassium channel gene is expressed in a tonotopic gradient within the MNTB, with levels of Kv3.1 being highest at its medial end that processes sound stimuli with high frequencies. The promoter for the Kv3.1 potassium channel gene is regulated by a Ca^{2+}-cAMP responsive element, which binds the transcription factor cAMP response element-binding protein (CREB). Levels of Kv3.1, CREB, and phospho-CREB in the mouse strain CBA, which maintains good hearing throughout life, were compared with corresponding measurements on brain tissue of the strain C57BL/6 (BL/6), which suffers from early cochlear hair cell loss. A gradient of Kv3.1 immunoreactivity in the MNTB was detected in young (6 week) and older (8 month) CBA mice and young BL/6 mice. However, by 8 months of age, when hearing is impaired in BL/6 mice, the gradient of Kv3.1 is lost. Moreover, in the older BL/6 mice, there was a decrease in CREB expression along the tonotopic axis, and the pattern of pCREB labeling was random. It appears that ongoing activity in auditory brainstem neurons is necessary for the maintenance of Kv3.1 tonotopicity through the CREB pathway and that the gradient falters when hearing fades.

Calbindin immunoreactivity has been compared in the MNTB of young and old CBA and BL/6 mice (O'Neill et al. 1997; Zettel et al. 1997). Although CBA mice show little change in peripheral sensitivity until very late in life, BL/6 mice exhibit progressively more severe peripheral (sensorineural) hearing loss between 4 and 12 months of age. There was no significant change in the number of calbindin-positive cells or the total number of cells in MNTB of old CBA mice compared with young control mice. However, the mean number of calbindin-positive cells decreased by 11% in middle-aged and 14.8% in old C57 mice. The decline in C57 mice was significant by 6.5-8.5 months of age and could be the consequence of a loss of input from the cochlear nucleus where cell numbers are known to decline by this age in this strain. The total number of neurons in MNTB showed a modest 7.1% decline with age in C57 mice. This suggests that the greater loss of calbindin-immunoreactive cells with age can only partially be attributed to a reduction in the total number of cells.

The soluble gas nitric oxide (NO) also appears to be involved in the aging processes of the SOC. The histochemical detection of NADPH-diaphorase activity (NADPH-d), a marker for neurons containing NO synthase, was used to determine the numbers of NO-producing cells in the SOC of adult and senile Djungarian dwarf hamsters (Reuss et al. 2000). The number of stained neurons almost doubles in the SOC of senile hamsters compared with young animals. The most distinct increase occurs in the MNTB. This increase of NO production in the aging auditory brainstem may be related to hearing impairments with increasing age.

3.3.3 *Olivocochlear Efferent Systems*

Several age-dependent changes in the central auditory system have been found to be a consequence of the deterioration of loss of inner ear hair cells. Conversely, integrity and survival of these hair cells could also depend on their innervation from the auditory brainstem originating in SOC. Both medial olivocochlear (MOC) and lateral olivocochlear (LOC) neurons contain GABA (Fex and Altschuler 1986; Fex et al. 1986; Thompson et al. 1986; Schwarz et al. 1988; Eybalin and Altschuler 1990), and cholinergic and GABAergic markers colocalize in most of their cochlear terminals (Maison et al. 2003). $GABA_A$ receptors are present on cochlear nerve somata and OHCs and in the region of synaptic contact between LOC terminals and cochlear nerve afferents (Plinkert et al. 1989, 1993; Yamamoto et al. 2002). Maison et al. (2006) have suggested that the GABAergic component of the olivocochlear system contributes to the long-term maintenance of hair cells and neurons in the inner ear. $GABA_A$ signaling has important long-term influences on the health of sensory cells and their innervation. The progressive degenerative changes seen after perturbation of GABAergic signaling suggests that GABA plays an important role in sensorineural hearing loss in general and age-related hearing loss in particular.

An age-related decline in Kv3.1b expression that correlates with functional deficits of the MOC system was observed (Zettel et al. 2007) in the mouse AVCN and MNTB (Fig. 3.6). Kv3.1b channel protein is widely distributed in the mammalian auditory brainstem. As temporal processing declines with age, this study was undertaken to determine whether the expression of Kv3.1b also declines and if changes are specific to AVCN and MNTB. Immunocytochemistry using an anti-Kv3.1b antibody was performed, and the relative optical density of cells and neuropils was determined from CBA mice of four age groups. Middle age declines in expression in the neuropils of the AVCN (by 35%), MNTB (26%), and LSO (23%) were found. In contrast, cellular optical density declines were found in the superior paraolivary nucleus (by 24%), VNTB (29%), and LNTB (26%). These trapezoid body nuclei contain the origin of the MOC system. There were no age-related changes in the remaining regions of cochlear nucleus or in the inferior colliculus.

Contralateral suppression of distortion-product otoacoustic emission amplitudes of age-matched littermates declines by middle age as well and suggests a correlation between Kv3.1 expression and MOC function. Aiming for more direct evidence, Kv3.1b knockout mice were examined. Knockouts show poor MOC function compared with wild-type and heterozygous mice. Zettel et al. (2007) suggest that the decline in Kv3.1b expression in MOC neurons by middle age are causally related with functional declines in efferent activity.

GABA is an inhibitory neurotransmitter that is synthesized by two isoforms of GAD, GAD65 and GAD67. Using in situ hybridization and immunocytochemical techniques in hamsters, GAD isoforms within the LSO have been found during postnatal development (Jenkins and Simmons 2006). In the neonatal hamster, GAD67 immunoreactivity, GAD67 transcript labeling, and GABA immunostaining are at low levels. However, robust GAD65 mRNA expression is found throughout the LSO during the early postnatal period. The neonatal GABAergic expression patterns are in strong contrast to those in the adult where the LSO has robust GAD67 mRNA expression and weak GAD65 mRNA expression. By retrograde axonal transport, GAD67-positive cells in the LSO were proven to be part of the lateral olivocochlear system. The late onset of GAD67 expression and intense GABA immunoreactivity in LSO neurons are consistent with a comparatively late maturation of the LOC system. The GABAergic portion of the LOC system is apparently distinguished by preferential GAD67 expression, intense GABA immunoreactivity, and relatively late postnatal onset.

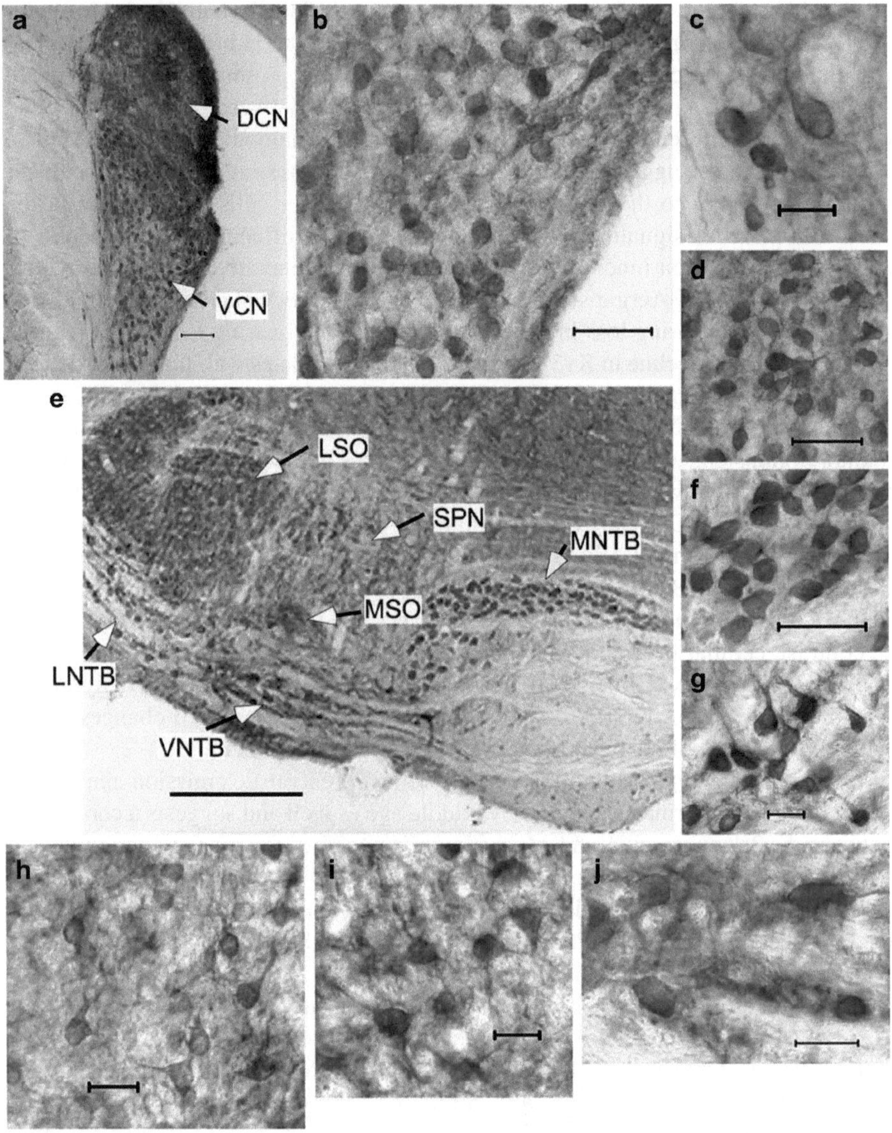

Fig. 3.6 Photomicrographs of Kv3.1β⁺ cells and neuropils in the CBA mouse. (**a**) Cross section of brainstem at the level of the DCN and VCN. DCN has heavily stained neuropil but no immunoreactive cells. VCN has numerous Kv^+ cells and immunopositive neuropil. Scale bar: 75 μm. (**b**) Immunopositive neurons in the VCN. Scale bar: 50 μm. (**c**) Very lightly immunoreactive octopus cells of the VCN. Scale bar: 25 μm. (**d**) Heavily reactive globular bushy cells and neuropil of the anterior division of the anteroventral cochlear nucleus (AVCNa). Scale bar: 50 μm. (**e**) Cross section of the brainstem at the level of the superior olivary complex (SOC). The lateral superior olive (LSO) is large and easy to distinguish by its intensely immunoreactive neuropil but lack of any immunopositive cells. In contrast, the medial superior olive (MSO) is small and indistinct. Scale bar: 200 μm. (**f**) Strongly Kv^+ principal cells and neuropil of the medial nucleus of the trapezoid

3.3.4 SOC Physiology

The SOC is the first major convergence of binaural information in the central auditory nervous system and forms the first processing of differences in time and intensity in binaural hearing. Three major nuclei are involved in the ascending pathway; the LSO, MSO, and MNTB. Neurons in the LSO and MSO receive and compare inputs from both ears after processing by the cochlear nuclei and appear to be especially suited for comparing intensity and timing differences between the two ears. For example, when a sound arises from the ipsilateral ear, the neural input to the LSO is excitatory and when the sound is located contralaterally, the input is inhibitory. Only one study has investigated the effects of aging in single neuron correlates of binaural processing in the LSO. Because neural processing in the LSO of binaural inputs is shaped by the interplay of excitatory and inhibitory circuits arising from each ear, one could hypothesize that a loss in inhibitory drive would alter binaural processing. Recordings from single neurons in the LSO from young and old Fischer-344 rats showed no apparent effects on binaural responses or in coding intensity differences (Finlayson and Caspary 1993). However, although the results did not reach statistical significance, they reported that a greater proportion of neurons from old rats exhibited a lower inhibitory drive when the contralateral ear was stimulated. This trend is supportive of the idea that a decline in inhibition may reduce the contrast in binaural signals, resulting in the deficits observed in elderly human listeners (Fitzgibbons and Gordon-Salant, Chapter 5).

3.4 Inferior Colliculus

The inferior colliculus in the midbrain consists of three major subdivisions: the central nucleus (CIC), the lateral external nucleus (EIC), and the dorsal cortex (DCIC). In a light and electron microscopic study on neurons of the CIC in the CBA mouse across three age groups (young or 3 months, middle aged or 8 months, and old or 25 months), no changes were noted in the size of the principal neurons over the age range examined (Kazee and West 1999). Neither were there changes in synapses on the somata of the principal neurons in number or type nor in the length of synaptic apposition or the size of synaptic terminal area. Conforming to this finding, hearing is quite well preserved across the lifespan in the CBA mouse strain.

Fig. 3.6 (continued) body (MNTB). Scale bar: 50 μm. (**g**) Darkly stained cells of the lateral nucleus of the trapezoid body (LNTB) intercalated between highly myelinated nonreactive fibers of the ventral acoustic stria. Scale bar: 25 μm. (**h**) Kv$^+$ cells of the central division of the inferior colloculus (ICc). Scale bar: 25 μm. (**i**) Large immunoreactive cells of the superior paraolivary nucleus. Neuropil is only lightly stained. Scale bar: 25 μm. (**j**) Kv$^+$ ventral nucleus of the trapezoid body (VNTB) cells intercalated between fibers of the acoustic stria. Scale bar: 25 μm. (From Zettel et al. 2007)

In contrast, a moderately severe synapse loss was observed in the BL/6 mouse strain, a strain that has a genetic deficit producing progressive sensorineural hearing loss starting in young adulthood (Kazee et al. 1995).

A genetic defect of the BL/6 mouse causes a progressive sensorineural hearing loss that starts during young adulthood, or by ~2 months of age, in the high-frequency range of sounds. Over the second year of life, hearing is severely impaired, progressively involving all frequencies. At the same time, a significant decrease in the size of principal neurons in the CIC and a dramatic decrease in the number of synapses of all morphologic types on principal neuronal somas take place. The percentage of somatic membrane covered by synapses decreased by 67%. These synaptic changes are likely to be related to the equally dramatic physiological changes that have been noted in CIC. The synaptic loss noted by Kazee et al. (1995) may represent alterations of the complex synaptic circuitry related to the central deficits of presbycusis. The preservation of synapses on principal neurons in the CBA mouse suggests that synaptic loss is related to the preservation of peripheral auditory function and input to the neurons rather than being a direct effect of genetic malfunction.

3.4.1 Glutamate

As in most brain regions, glutamate functions as the main excitatory neurotransmitter in the central auditory systems. Changes in glutamate and glutamate-related genes with age may be an important factor in the pathogenesis of age-related hearing loss. Glutamate-related mRNA gene expression in the inferior colliculus of CBA mice was investigated using both genechip microarray and real-time PCR molecular techniques for four different age/hearing loss CBA mouse subject groups (Tadros et al. 2007). Two genes showed consistent differences between groups for both the genechip and real-time PCR. Pyrroline-5-carboxylate synthetase enzyme (Pycs) showed downregulation with age and a high-affinity glutamate transporter (Slc1a3) showed upregulation with age and hearing loss. Because Pycs plays a role in converting glutamate to proline, its deficiency in old age may lead to both glutamate increases and proline deficiencies in the auditory midbrain, playing a role in the subsequent inducement of glutamate toxicity and loss of proline neuroprotective effects. The upregulation of Slc1a3 gene expression could reflect a compensatory mechanism of neurons to protect against age-related glutamate or calcium damage.

3.4.2 GABA

There is a substantial, selective, and age-related loss of the putative inhibitory neurotransmitter GABA in the CIC of rats (Caspary et al. 1990). Staining neurons of the CIC against a GABA conjugate in young adult (2- to 7-month-old) and aged

(18- to 29-month-old) Fischer 344 rats showed that the number of GABA-positive neurons decreases significantly by 36% in the ventrolateral portion of the CIC of aged animals (93 neurons/mm^2) compared with matched young adults (145 neurons/mm^2). Concurrently, a significant age-related reduction in both basal (−35%) and K^+-stimulated (−42%) efflux of GABA from tissue slices of the CIC was observed. A corresponding decrease in postrelease tissue content of GABA in the CIC of aged rats was observed (−30%). In contrast, tissue content as well as basal and evoked release of Glu, Asp, Tyr, and ^{3}H-ACh was similar between the two age groups. These data provide evidence that GABA, glutamate, aspartate, and acetylcholine serve neurotransmission in the CIC. In addition, they indicate a pronounced region- and neurotransmitter-selective and age-related reduction of GABA in the CIC. The authors developed the hypothesis that impairment of inhibitory GABAergic neurotransmission in the CIC may contribute to abnormal auditory perception and processing seen in neural presbycusis.

Helfert et al. (1999) determined whether there is an age-related decline in GABAergic inhibition in the CIC and possibly excitant amino acid (EAA)-mediated excitation as well. They applied an electron microscopic method to locate GABA and EAAs known to be essential for signal processing in the inferior colliculus. Their study was done on Fischer-344 rats in three age groups (3, 19, and 28 months) and found an almost equal loss of excitatory and inhibitory synapses in the CIC. Comparing rats aged 28 months with rats aged 3 months, the authors observed a significant reduction in the density of GABA-positive synaptic terminals by ~30% and of GABA-negative terminals by ~24% in CIC. When adjusting these counts for the volume reduction of CIC occurring with age, a change of ~10% still remains. The authors suggest that GABA-positive neurons may have evolved patterns of synaptic and dendritic change during aging in which the distribution of synaptic terminals progressively shifts to dendrites of larger caliber. Because there were no neuronal losses detectable among the three age groups, the decrease in GABA and EAAs identified in the inferior colliculusby previous studies may be attributable to synaptic and dendritic declines rather than to cell loss.

Age-related alterations in the biosynthetic enzyme GAD, the degradative enzyme GABA-transaminase (GABA-T), and the uptake system for GABA in CIC of Fischer-344 rats has been studied (Raza et al. 1994). For comparison, the cholinergic neuronal system identified by choline acetyltransferase (ChAT) was also charted. Considering 3 age groups (young or 3-7 months, mature or 15-17 months, and aged or 24-26 months), the investigators found that GAD activity was high in the young CIC (219 nmol/mg protein/h), whereas ChAT was low. Significant reductions in GAD activity are seen in the CIC of mature (−31%) and aged (−30%) rats. The neurotransmitter selectivity of this deficit becomes obvious by changes in ChAT activity (−22%, aged vs. mature) that occurred after the changes in GAD activity. In contrast, high-affinity uptake for ^{14}C-GABA and ^{3}H-D-aspartate is not significantly affected by ageing. There was also no age-related loss in GABA-T activity. The age-related loss in GABA-mediated inhibition in the CIC of Fischer-344 rats is not attributable to changes in uptake or degradation of GABA but appears to be related to a loss in biosynthetic capacity of GAD.

A selective age-related decrease in both the protein and mRNA levels of the most abundant $GABA_A$ receptor subunits has been revealed in Fischer-344 and Sprague-Dawley rat inferior colliculi (Gutierrez et al. 1994). The number of the native and fully assembled $GABA_A$ receptors assayed by ^{3}H-muscimol binding decreases as well (35-49%). The decrease in GABA receptors was accompanied by a decrease in protein and mRNA of GAD. Among the other regions of the rat brain studied, there was none that showed comparatively large age-related changes in these synaptic molecules. In 24-month-old rats, the combination of ß2 and ß3 peptide subunits was decreased by 55%, whereas the $ß_2$ and $ß_3$ mRNAs were decreased by 31 and 22%, respectively. The γ_{2S} and γ_{2L} subunit proteins decreased by 43 and 21%, respectively, whereas the γ_2 mRNA, including both short and long forms, declined by 61%. The α_1 subunit protein decreased by 26%, whereas the α_1 mRNA decreased by 40%. The GAD protein decreased by 62% and GAD-65 mRNA decreased by 42%. No age-dependent changes occurred in the CIC in the level of expression of glial and/or neuronal markers.

Similarly, an age-dependent decrease takes place for $GABA_B$ receptors (Milbrandt et al. 1994). The 3 major subdivisons of the inferior colliculus were studied in 3 age groups (3, 18-20, and 26 months) of Fischer-344 rats. $GABA_B$ binding sites were localized using [^{3}H]GABA in the presence of a saturating concentration of a selective $GABA_A$ receptor agonist. In all subdivisions, $GABA_B$ receptor binding was significantly reduced in 26-month-old rats when compared with 3-month-old rats (DCC: −44%; EIC: −36%; CIC: −32%).

These studies again reveal those age-related changes in hearing that involve GABA neurotransmitter function. There is a decrease in the number of GABA-immunoreactive neurons, a decreased basal concentration of GABA, a decreased GABA release, decreased GAD activity, decreased $GABA_B$ receptor binding, and subtle changes in $GABA_A$ receptor binding. Knowledge of such deviations from GABA metabolism in a fully functional auditory system is hoped to eventually lead to the development of pharmacotherapy.

3.4.3 Metabolism

Aging, either with or without a severe loss in hearing, is not associated with altered levels of glucose metabolism in the inferior colliculus and AVCN as indicated by the incorporation of 2-deoxyglucose in quiet (Willott et al. 1988).

The age-dependent protein complexity of the inferior colliculus reveals a pattern that is distinctly different from other brain regions (Illing 2004). Molecular complexity in the inferior colliculus is initially high but decreases soon after the onset of hearing to settle on a significantly lower adult level. Looking at specific proteins, Cosgrove et al. (1987) studied the brains of male Fischer-344 rats aged 3-4 months or 28-30 months and found two proteins, 44 kD pI 5.4 and 47 kD pI 5.2, that were present at increased levels in the inferior colliculus of 28- to 30-month animals compared with 3- to 4-month animals. This result supports the concept that the molecular mechanisms causing differential gene expression in different regions of the young adult rat brain are operative and maintained in the brains of senescent rats.

Studies using indicators for oxidative stress have identified biochemical and morphological changes in the inferior colliculus that may contribute to deficits in neurotransmitter functions of Fischer-344 rats in three age groups (Mei et al. 1999). Homogenates obtained from inferior colliculus tissue showed age-dependent reductions in activities of the antioxidant enzymes superoxide dismutase and catalase, with a concomitant increase in lipid peroxidation. Dephosphorylation of inferior colliculus homogenates with alkaline phosphatase reduced the activities of sodium dismutase and catalase in all age groups. This could be restored by protein kinase C (PKC)-dependent phosphorylation. Restoration of enzyme activity was specific to the PKC-α isozyme but not to the β_1, β_2, δ, or γ forms. No age-dependent change in the levels of PKC isoforms (α, β_1, β_2, and γ) was detectable in inferior colliculus homogenates. Morphological analyses made by these authors indicate decreases in mitochondrial density and increase in matricial abnormalities in the somata of both GABA-positive as well as GABA-negative neurons in the colliculi of 19- and 28-month-old rats compared with 3-month-old rats. These data indicate age-related increases in oxidative stress in the inferior colliculus that could be partially counteracted by PKC. Obviously, the progressive increase in oxidative stress may entail morphological and functional impairment of the inferior colliculus with age as described above.

Specific changes in cellular metabolism in the auditory midbrain were observed under the pathology of Alzheimer's disease (AD). Senile plaques and neurofibrillary tangles were distributed throughout the CIC in nine of nine AD patients while adjacent nuclei were consistently spared (Sinha et al. 1993). The degenerative changes pervading the entire inferior colliculus and medial geniculate body (MGB) suggest that the loss of neuronal function affects all frequency ranges in AD. In contrast, typical presbycusis preferentially includes high-frequency loss due to lesions peripherally in the cochlea or auditory centers.

3.4.4 Inferior Colliculus Physiology

The auditory midbrain is composed of the inferior colliculus and represents the first major convergence of both binaural and monaural neural processing in the auditory system. The cytoarchitectonics of the inferior colliculus forms the basis for a neural network devoted to parallel processing and is formed by laminar sheets analogous to the contours of an onion. Each sheet is formed by disc-shaped, principal neurons whose dendrites extend outward in a circular fashion, with the second type of principal neuron, the multipolar cell, acting to integrate processing across the laminar sheets. The incoming afferents form the tonotopic organization that is created by individual laminae, where only a very narrow range of frequencies is represented. Low frequencies are represented dorsally and high frequencies ventrally.

Within a lamina, there is evidence to suggest that both low-level and more sophisticated sound processing occurs. For example, Langner and Schriener (1988) found that amplitude modulation, or the rapid fluctuation in stimulus intensity, was mapped in a systematic fashion onto individual lamina. A similar topographic map is also found for sound intensity, where the most sensitive (lowest thresholds) neurons are located near the center of an isofrequency lamina and neurons with increasingly higher thresholds fall on concentric isointensity contours around the center of the lamina (Stiebler and Ehret 1985). Thus the type of neuronal architecture could provide the basis for parallel processing of complex signals so that multiple sound features, e.g., frequency, intensity, and frequency and amplitude modulation, could be analyzed simultaneously.

A pioneering study by Willott et al. (1988) was the first to examine age-related changes in simple response properties of auditory midbrain neurons using the CBA mouse model of presbycusis. They reported only mild age-related changes in fundamental response properties of inferior colliculus neurons in old CBA mice. Consistent with peripheral measures, a slight reduction in sensitivity across all frequencies was noted while tonotopic organization was maintained. They found an increase in the number of so-called "sluggish" units and neurons that behaved similarly throughout a range of intensities, with <50 action potentials for the entire recording period. Sluggish neurons were found to increase significantly with age from 3% in the young inferior colliculus to 22% in the old inferior colliculus. They also reported a decrease in the proportion of spontaneously active units in the dorsal, low-frequency region of the inferior colliculus and an increase in the number of spontaneously active ventrally located high-frequency units. Several parameters of single-unit response areas were measured, including BF or frequency corresponding to the lowest intensity to which a unit responded, upper and lower frequencies to which a unit was responsive at 80-dB SPL and BF for rate, or the frequency to which the unit exhibited the greatest number of spikes. This implies that the frequency selectivity of these units has been altered without affecting their threshold or rate BF.

Age-related changes in binaural response properties were studied by Palombi and Caspary (1993). In most neurons, no statistically significant differences in coding binaural responses properties were observed when comparing units from young and old animals. The authors did report a trend that there was a decline in the number of units excited by contralateral stimuli and inhibited by ipsilateral stimuli in the aged rat IC compared with units from young mice.

Although alteration in coding simple sound properties appears to be minimally affected by age, several studies have reported altered coding of temporal sound features. Temporal acuity can be defined as the speed at which the auditory system processes incoming acoustic signals and can be measured by one of several temporal processing tasks, one of which is gap detection. Because one of the key factors contributing to poor speech recognition in elderly listeners is a deficit in temporal resolution, neurophysiological correlates of gap detection may provide evidence of a central component (Schneider et al. 1994; Fitzgibbons and Gordon-Salant 1996; Snell et al. 2002). Barsz et al. (2002) and Walton et al. (1997) explored the relationship between behavioral gap detection in mice and the neural correlates of gap detection in inferior colliculus neurons and found that the shortest gaps encoded by inferior colliculus neurons were in the range of behavioral minimal gap thresholds. Subsequently, Walton et al. (1998) reported that unit responses from old CBA mice showed a shift in the proportion of units having the shortest gap thresholds (Fig. 3.7) and were shifted to longer gap duration compared with neurons from young mice. The frequency distributions represent the minimal gap threshold data from over 100 inferior colliculus neurons in young and old CBA mice. The distribution shows that although the smallest gap thresholds of 1-3 ms were observed in units from both young and old animals, the frequency of occurrence of thresholds of the shortest gaps encoded by inferior colliculus neurons was significantly lower in old CBAs. More specifically, roughly 50% fewer neurons had gap thresholds of 2 ms or less in the inferior colliculus in old CBA mice. Whether this magnitude of neuronal decline produces parallel behavioral deficits in old mice is not yet known, although behavioral data (Ison, Tremblay and Allen, Chapter 4) suggest that old mice suffer about the same deficit in detecting silent gaps presented in a noise background.

Walton et al. (1998) also noted a second, more striking age-related difference that is illustrated in the gap functions of Fig. 3.7b. They found that the majority of neurons with onset-type response patterns displayed a deficit in recovery. Neural recovery can be quantified by comparing the spike counts elicited by the control condition with the response elicited by the first noise burst marking the gap. In Figure 3.7, top, gap functions from young mice display rapid recovery, the majority increasing spike counts to near control values (1.0) by 10 ms. In contrast, in many old neurons, the gap response elicited by a second noise burst failed to recover to within 75% of the response to the first noise burst even after 10 ms.

That aging differentially affects the representation of first-spike latencies (presumably) along the tonotopic axis but not the spread of latencies within isofrequency slabs suggests that there are at least two mechanisms influencing latency in the inferior colliculus. One mechanism establishes a latency gradient for determining the shortest latencies along the tonotopic axis, and this mechanism appears to be affected by the aging process. The second mechanism establishes a latency spread within isofrequency slabs, and this representation of response latency appears to be preserved with age. One other metric of response latency, the variance of the first spike latency, was analyzed and remained stable with age. Previous reports indicate that variance in first-spike latency from inferior colliculus units should increase with the mean first-spike latency (Langner and Schreiner 1988). The data indicate that this relationship held true for units from both young and old mice. Finally, alteration in dynamic temporal processing, as measured by encoding of sinusoidal amplitude modulation (SAM), has also been reported (Walton et al. 2002; Simon et al. 2004). In old mice, the upper limit of sinusoidal amplitude modulation encoding is reduced compared with units from young mice.

To summarize the inferior colliculus results, key findings from single-cell recordings across multiple laboratories suggest that both intensity and temporal processing are disrupted in the inferior colliculus of the old rodent. Although

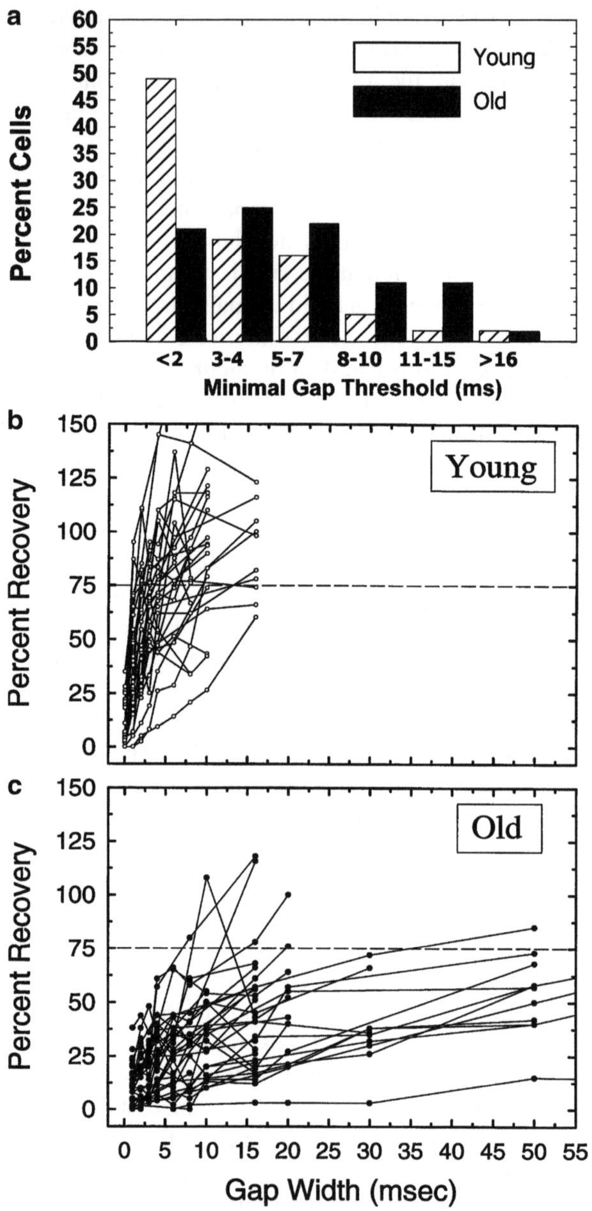

Fig. 3.7 Minimal gap thresholds of IC units recorded from young and old mice to silent gaps embedded in a noise carrier presented at 65 dB SPL (**a**). Thresholds were defined as the shortest gap duration in which a response to the second noise burst was detected by statistical comparison to the control (no gap) response. Gap thresholds tended to shift to longer durations in old mice, where 50% fewer units responded to gaps with durations of 3 ms or less. (**b** and **c**) Recovery functions from units from young and old mice as shown by the magnitude of the response (normalized from spike counts) elicited by the second noise burst. A value of one indicates complete recovery. (From Walton et al. 1997)

Finlayson and Caspary (1993) found no significant age-related changes in binaural response properties, aging does alter monaural temporal processing as seen as a shift to longer gap thresholds, increased recovery from prior stimulation, and alteration in the normal topographic map of first-spike latency. What are the implications of these findings? First, it must be reiterated that the temporal coding deficits noted above are not associated with global changes in all inferior colliculus neurons in aged animals, e.g., some neurons are spared. The mechanisms responsible for the alteration in temporal coding appear to be acting focally and totality. Second, the range of single neuron deficits that appears in gap encoding are on the order of those reported in human listeners (Fitzgibbons and Gordon-Salant, Chapter 5). Schneider et al. (1998), e.g., found that mean gap thresholds were 1-1.5 ms greater in older subjects with near-normal audiometric thresholds than in the younger group. Finally, many of the neurophysiological observations fit within the framework of the loss of inhibition model. Walton et al. (1998) has shown that both inferior colliculus single-unit recovery functions and topographic latency gradient present in the inferior colliculus can be modified by blocking GABAergic and glycinergic circuits. Age-related alteration in recovery and the map of latency in single-neuron recordings could be a reflection of an imbalance in excitation and inhibition needed for normal cellular function.

3.5 Auditory Cortex

In a Golgi study of neocortical pyramidal cells, it was found that in rats more than 30 months old, the density of the dendritic tree decreased significantly within a radius of ~150 μm of the perikaryon compared with that in 3-month-old animals (Vaughan 1977). In contrast, the extent of the dendritic domain did not change appreciably with age. Analysis of the dendritic branching suggests that not only has there been deterioration in the peripheral branches of the dendritic tree but also that entire dendrites have been lost. This loss of primary branches was confirmed through the reconstruction of layer V neuronal perikarya and their proximal dendrites from serial sections of auditory cortex.

An electron microscopic analysis of the size of cell bodies in layers 2 and 5 of the rat auditory cortex revealed that a size increase from 3 to 15 months of age is followed by a decrease falling below the 3-month value by 36 months of age (Vaughan and Vincent 1979). The nuclei of the cells did not change significantly in layer 2. In layer 5, the mean nuclear area decreased significantly in the old animals. The relative volume of dense bodies increased linearly with advancing age, with a slightly more accelerated rate in layer 2 cells. The relative volume of ground substance remained essentially constant through 27 months but decreased later to <90% of the 3-month level. The relative volume of the rough endoplasmic reticulum did not change significantly until after 15 months, at which time it began to occupy increasingly a larger fraction of the perikaryal cytoplasm. The relative volumes of mitochondria, multivesicular bodies, and Golgi apparatus did not show clear trends of change during the 33-month period. Apparently, the major age-related change takes place in dendritic morphology and consequently in synaptic connectivity.

In a study on aged human brains (78 years and older) without any neuropsychic disorders, changes in the cytoplasm, neurocytes, dendrites, spines, axons, and their terminals in the sensory cortex (Brodmann area 40) were found (Iontov and Shefer 1984). The same types of changes were found in the auditory cortex of old cats used for control of the degree of preservation of elements in the human cortex.

Vaughan and Cahill (1984) sectioned the corpus callosum in 3-, 12-, and 24-month-old rats and examined the auditory cortex 3 months later to determine whether there were age-related differences in the morphological response to the partial deafferentation. Analysis focused on those cortical layers known to receive the heaviest callosal projection (layers 2 and 3) and those neurons known to be postsynaptic to callosal afferents (layer V pyramidal neurons). Although no age-related changes in cortical thickness or the relative thickness of the cortical layers were detected in control animals, the apical dendrites of layer V pyramidal neurons did lose dendritic spines and became thinner with age. In all three lesion groups, the cortex became thinner without altering the relative thickness of cortical layers. A decrease in the relative density of apical dendrite spines in layer 3 and an increase in the density of these spines in layer 4 occurred. Both effects varied with age. Spine decreases in layer 3 were greatest in older animals and spine increases in layer IV were greatest in younger animals. The mean diameters of apical dendrites decreased in the youngest group of lesioned animals but increased in the oldest group. These results indicate a complex pattern of age-dependent changes on callosal deafferentation.

During embryonic development, the cells of the brain migrate toward the surface, forming minicolumns in the cortex. These are grouped into larger macrocolumns to form the basis of the mapping of functions across the surface of the brain. Proliferation of radial minicolumnar units of cells was suggested to underlie the expansion of different cortical surface regions during development and across species. The minicolumnar organization of neurons was studied and found to be increasingly disrupted, with cognitive impairment in primates (Chance et al. 2006). These authors measured the minicolumn spacing and organization of cells in Heschl's gyrus (primary auditory cortex A1), the planum temporale (Brodmann area 22), and middle temporal gyrus (Brodmann area 21) of 17 normally aged human adults. An age-associated minicolumn thinning was found in the auditory association cortex but not in the primary auditory cortex. Minicolumn thinning was also associated with greater plaque load, although this effect was present in all areas. The regional variability of age-associated minicolumn thinning reflects the regionally selective progression of tangle pathology in Alzheimer's disease. The authors suggest that plaque load combines with age to increase minicolumn thinning, possibly reflecting an increasing risk for Alzheimer's disease. Because old age is the greatest risk factor for dementia, the transition to dementia may involve an extension of normal aging processes.

3.5.1 Neurochemistry

Neurons containing NADPH-d, a marker of neurons containing NO synthase, are readily detectable in the auditory cortex of rats (Ouda et al. 2003). Ouda et al. examined the auditory cortex of young (3-month-old) and very old (36-month-old) Long-Evans rats. In very old rats, a significant reduction was found in the thickness of the auditory cortex to almost half of that in young animals. They also noticed changes in the shape and configuration of diaphorase-positive nerve cell bodies and dendrites. Quantitative analysis demonstrated an age-related increase in the number of dendritic segments and dendritic branching points. The length of dendrites in NADPH-d-positive neurons and their density increased in very old rats, but the total number of NADPH-d-positive neurons within the Te 1 and Te 3 fields was 13% lower in the old rats than in the young rats.

Neuropeptide Y and NADPH-d in neurons of the cerebral cortex in young (3-month) and aged (24-month) Fischer-344 rats showed that the number of neurons containing both proteins were unchanged in all regions of the cerebral cortex compared with the control group (Huh et al. 1998). However, the number of neurons containing neither of these markers was significantly decreased in frontal association, primary motor, secondary somatosensory, insular, ectorhinal, perirhinal cortex, and auditory cortex in the aged group. These results suggested that neuropeptide Y-containing neurons that do not contain NADPH-d are affected by aging and that aging influences neurons negative for both markers in a region-specific pattern within the cerebral cortex.

NO has been implied in age-related changes of the central auditory pathway (Sanchez-Zuriaga et al. 2007). These authors investigated whether the number of NO-producing cells and their morphometric characteristics in the inferior colliculus and the auditory cortex are changed with the increasing age of Wistar rats. Again, they based their study on the detection of NADPH-d. Their results showed that the cross-sectional area of the somas of stained neurons in the dorsal cortex of the inferior colliculus and a diffuse loss of NADPH-d-positive neurons in the senile IC and primary cortical auditory area (Te1) decreased with age. Because these authors have noticed an increased number of NO-producing cells in different parts of the ageing auditory pathway, age-related changes in NADPH-d-positive cells appear to follow a region-specific route. The observed changes are thought to be related to hearing impairments with increasing age.

Age-related changes within the auditory brainstem typically include alterations in inhibitory neurotransmission and coding mediated by GABA and glycinergic circuits. The impact of aging on neurotransmission in the higher auditory centers was examined. Age-related changes in the GABA synthetic enzyme and different isoforms of GAD, have been found in rat primary auditory cortex (A1), which contains a vast network of intrinsic and extrinsic GABAergic circuits throughout its layers (Ling et al. 2005). Significant age-related decreases in GAD65 and GAD67 mRNA were observed in A1 layers 2-6 of aged rats relative to their young adult cohorts. The largest changes were identified in layer 2 for GAD65 and GAD67. The GAD67 protein expression decreased significantly in parallel with mRNA decreases in all layers of AI. Adjacent regions of parietal cortex showed no significant GAD67 protein changes among the age groups except in layer 6. Age-related GAD reductions likely reflect decreases in both metabolic and presynaptic GABA levels and suggest a downregulation in normal adult inhibitory GABA neurotransmission. Transferring these observations to humans, an age-related loss of normal adult GABA neurotransmission in A1 is likely to alter temporal coding and could contribute to the loss in speech understanding observed in the elderly.

3.5.2 Auditory Thalamus and Cortex Physiology

In a series of studies, Mendelson et al. (2004) investigated the effects of aging on neural coding of frequency modulation (FM) in the medial geniculate body and primary auditory cortex. Certain features of complex signals, including species-specific vocalizations, are composed of various rates of FM. For example, different components of speech contain formant transitions that can be characterized by different types of FM. Two fundamental components of tonal FM are the speed in which frequency is changing and the direction of change, either upward or downward. In comparing neuronal recordings from MGB neurons in young and old rats, Mendelson and Lui (2004) found no effects of age on coding of tonal FM. Neurons from both young and old animals preferred fast FM sweeps and were not found to be directional dependent. This was in agreement with an earlier study showing that there were no significant differences in processing FM in the inferior colliculus of young and old rats (Lee et al. 2002). The results in the inferior colliculus and MGB were in contrast to the finding that significant changes were found in the responses of auditory cortex neurons from old rats. Although neurons from young rats preferred medium and fast rates of FM, neurons from older rats could not encode these rates and were limited to much slower rates of frequency transitions. This temporal processing deficit in FM processing mirrors the deficit found in gap detection in that coding faster FM and shorter gaps is impaired in the aged animal.

More recently, age-related deficits in static frequency selectivity have been reported. Indeed, the frequency receptive fields of cortical layer V neurons in the primary auditory cortex have been shown to undergo major age-related changes in the old rats as shown in Fig. 3.8 (Turner et al. 2005). In addition to a decrease in the number of finely tuned V/U-shaped receptive fields and an increase in the poorly tuned complex receptive fields, the stimulus-driven activity of each unit type was altered with age. Thus aging is associated with functional changes in the AI layer V neurons, but the extent to which the decline in subcortical nuclei contributes to these alterations is unknown.

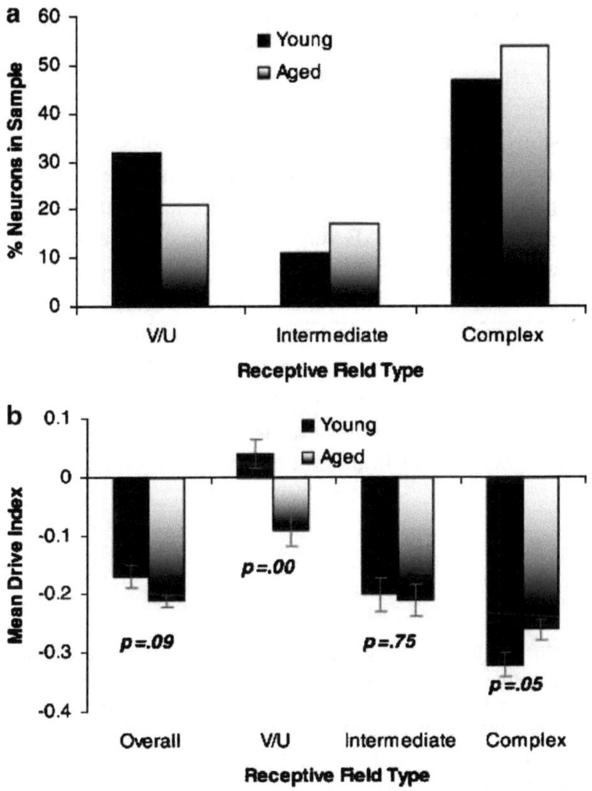

Fig. 3.8 Age-related change in tonal excitatory receptive field properties from layer V auditory cortex neurons in old rats. (**a**) Receptive field shape changes with age. The majority of units (54%) displayed complex-type shapes in the young rat and this increased with aging, whereas the proportion of V/U-type receptive fields decreased with age. (**b**) Shape of the receptive field was more variable in units from old rats and this was independent of shape. Variability was determined by performing correlation analysis between three successive runs of acquiring the receptive field. (Adapted from Turner et al. 2005)

3.6 Summary and Conclusions

Aging is accompanied by peripheral and central hearing loss as well as cognitive decline. Although there is a wealth of information about the cell biology and anatomy within the young central auditory nervous system, there are many questions that remain to be answered for the aging brain. The information that is presently available provides a limited understanding of how the different auditory regions are altered in the aged brain. However, a number of interesting age-related problems still need to be resolved before neuroprotective strategies can be developed. We have seen that a recurrent motive of age-related hearing impairment is the progressive weakening of GABAergic systems, noticeable in both neurochemical and physiological studies on virtually all levels of the central auditory system. It might therefore seem to be an obvious therapeutic approach against age-related deterioration of hearing to support GABAergic transmission pharmacologically. GABA-mimetic drugs have already been used against other neurological ailments (Green et al. 2000). However, an immediate success of such an approach is unlikely to be achieved because a systemic backup of GABAergic synapses will entail inacceptable side effects such as sedation and myorelaxation. Progress is needed for a more local support of selected transmitter systems in specific regions of the brain, and this might come with the maturation of cell and gene therapy.

This review has tried to integrate what is known about age-related central auditory changes and the physiological consequences that follow. In the future, a better understanding of basic molecular mechanism that underlie age-induced central auditory disorders is needed for both experimental and clinical purposes.

References

Barsz K, Ison JR, Snell KB, Walton JP (2002) Behavioral and neural measures of auditory temporal acuity in aging humans and mice. Neurobiol Aging 23:565–578.

Benowitz LI, Routtenberg A (1997) GAP-43: an intrinsic determinant of neuronal development and plasticity. Trends Neurosci 20:84–91.

Benowitz LI, Apostolides PJ, Perrone-Bizzozero N, Finklestein SP, Zwiers H (1988) Anatomical distribution of the growth-associated protein GAP-43/B-50 in the adult rat brain. J Neurosci 8:339–352.

Bondy CA (1991) Transient IGF-I gene expression during the maturation of functionally related central projection neurons. J Neurosci 11:3442–3455.

Brawer JR, Morest DK (1975) Relations between auditory nerve endings and cell types in the cat's anteroventral cochlear nucleus seen with the Golgi method and Nomarski optics. J Comp Neurol 160:491–506.

Brugge JF, Geisler CD (1978) Auditory mechanisms of the lower brainstem. Annu Rev Neurosci 1:363–394.

Camarero G, Villar MA, Contreras J, Fernandez-Moreno C, Pichel JG, Avendano C, Varela-Nieto I (2002) Cochlear abnormalities in insulin-like growth factor-1 mouse mutants. Hear Res 170:2–11.

Cant NB, Morest DK (1979) Organization of the neurons in the anterior division of the anteroventral cochlear nucleus of the cat. Light-microscopic observations. Neuroscience 4:1909–1923.

Casey MA (1990) The effects of aging on neuron number in the rat superior olivary complex. Neurobiol Aging 11:391–394.

Casey MA, Feldman ML (1985) Aging in the rat medial nucleus of the trapezoid body. II. Electron microscopy. J Comp Neurol 232:401–413.

Casey MA, Feldman ML (1988) Age-related loss of synaptic terminals in the rat medial nucleus of the trapezoid body. Neuroscience 24:189–194.

Caspary DM, Raza A, Lawhorn Armour BA, Pippin J, Arneric SP (1990) Immunocytochemical and neurochemical evidence for age-related loss of GABA in the inferior colliculus: implications for neural presbycusis. J Neurosci 10:2363–2372.

Caspary DM, Schatteman TA, Hughes LF (2005) Age-related changes in the inhibitory response properties of dorsal cochlear nucleus output neurons: role of inhibitory inputs. J Neurosci 25:10952–10959.

Caspary DM, Hughes LF, Schatteman TA, Turner JG (2006) Age-related changes in the response properties of cartwheel cells in rat dorsal cochlear nucleus. Hear Res 216–217:207–215.

Chance SA, Casanova MF, Switala AE, Crow TJ, Esiri MM (2006) Minicolumn thinning in temporal lobe association cortex but not primary auditory cortex in normal human ageing. Acta Neuropathol 111:459–464.

Cosgrove JW, Atack JR, Rapoport SI (1987) Regional analysis of rat brain proteins during senescence. Exp Gerontol 22:187–198.

Ene FA, Kalmbach A, Kandler K (2007) Metabotropic glutamate receptors in the lateral superior olive activate TRP-like channels: age- and experience-dependent regulation. J Neurophysiol 97:3365–3375.

Eybalin M, Altschuler RA (1990) Immunoelectron microscopic localization of neurotransmitters in the cochlea. J Electron Microsc Tech 15:209–224.

Fex J, Altschuler RA (1986) Neurotransmitter-related immunocytochemistry of the organ of Corti. Hear Res 22:249–263.

Fex J, Altschuler RA, Kachar B, Wenthold RJ, Zempel JM (1986) GABA visualized by immunocytochemistry in the guinea pig cochlea in axons and endings of efferent neurons. Brain Res 366:106–117.

Finlayson PG, Caspary DM (1993) Response properties in young and old Fischer-344 rat lateral superior olive neurons: a quantitative approach. Neurobiol Aging 14:127–139.

Fischel-Ghodsian N (2003) Mitochondrial deafness. Ear Hear 24:303–313.

Fitzakerley JL, Star KV, Rinn JL, Elmquist BJ (2000) Expression of Shal potassium channel subunits in the adult and developing cochlear nucleus of the mouse. Hear Res 147:31–45.

Fitzgibbons PJ, Gordon-Salant S (1996) Auditory temporal processing in elderly listeners. J Am Acad Audiol 7:183–189.

Gleich O (1994) The distribution of N-acetylgalactosamine in the cochlear nucleus of the gerbil revealed by lectin binding with soybean agglutinin. Hear Res 78:49–57.

Green AR, Hainsworth AH, Jackson DM (2000) GABA potentiation: a logical pharmacological approach for the treatment of acute ischaemic stroke. Neuropharmacology 39:1483–1494.

Gutierrez A, Khan ZU, Morris SJ, De Blas AL (1994) Age-related decrease of GABAA receptor subunits and glutamic acid decarboxylase in the rat inferior colliculus. J Neurosci 14:7469–7477.

Heffner RS, Heffner HE (1989) Sound localization, use of binaural cues and the superior olivary complex in pigs. Brain Behav Evol 33:248–258.

Helfert RH, Sommer TJ, Meeks J, Hofstetter P, Hughes LF (1999) Age-related synaptic changes in the central nucleus of the inferior colliculus of Fischer-344 rats. J Comp Neurol 406:285–298.

Hinks GL, Franklin RJ (2000) Delayed changes in growth factor gene expression during slow remyelination in the CNS of aged rats. Mol Cell Neurosci 16:542–556.

Hoffpauir BK, Grimes JL, Mathers PH, Spirou GA (2006) Synaptogenesis of the calyx of Held: rapid onset of function and one-to-one morphological innervation. J Neurosci 26:5511–5523.

Huh Y, Lee W, Cho J, Ahn H (1998) Regional changes of NADPH-diaphorase and neuropeptide Y neurons in the cerebral cortex of aged Fischer 344 rats. Neurosci Lett 247:79–82.

Idrizbegovic E, Canlon B, Bross LS, Willott JF, Bogdanovic N (2001) The total number of neurons and calcium binding protein positive neurons during aging in the cochlear nucleus of CBA/CaJ mice: a quantitative study. Hear Res 158:102–115.

Idrizbegovic E, Bogdanovic N, Viberg A, Canlon B (2003) Auditory peripheral influences on calcium binding protein immunoreactivity in the cochlear nucleus during aging in the C57BL/6J mouse. Hear Res 179:33–42.

Idrizbegovic E, Salman H, Niu X, Canlon B (2006) Presbyacusis and calcium-binding protein immunoreactivity in the cochlear nucleus of BALB/c mice. Hear Res 216–217:198–206.

Illing RB (2004) Maturation and plasticity of the central auditory system. Acta Otolaryngol Suppl Dec:6–10.

Illing RB, Horvath M, Laszig R (1997) Plasticity of the auditory brainstem: effects of cochlear ablation on GAP-43 immunoreactivity in the rat. J Comp Neurol 382:116–138.

Iontov AS, Shefer VF (1984) The morphological basis of age-induced memory changes. Neurosci Behav Physiol 14:349–353.

Jalenques I, Albuisson E, Despres G, Romand R (1995) Distribution of glial fibrillary acidic protein (GFAP) in the cochlear nucleus of adult and aged rats. Brain Res 686:223–232.

Jalenques I, Burette A, Albuisson E, Romand R (1997) Age-related changes in GFAP-immunoreactive astrocytes in the rat ventral cochlear nucleus. Hear Res 107:113–124.

Jenkins SA, Simmons DD (2006) GABAergic neurons in the lateral superior olive of the hamster are distinguished by differential expression of gad isoforms during development. Brain Res 1111:12–25.

Jung DK, Lee SY, Kim D, Joo KM, Cha CI, Yang HS, Lee WB, Chung YH. (2005) Age-related changes in the distribution of Kvl.1 and Kv3.1 in rat cochlear nuclei. Neurol Res 27(4): 436–440.

Kazee AM, West NR (1999) Preservation of synapses on principal cells of the central nucleus of the inferior colliculus with aging in the CBA mouse. Hear Res 133:98–106.

Kazee AM, Han LY, Spongr VP, Walton JP, Salvi RJ, Flood DG (1995) Synaptic loss in the central nucleus of the inferior colliculus correlates with sensorineural hearing loss in the C57BL/6 mouse model of presbycusis. Hear Res 89:109–120.

Kotak VC, Korada S, Schwartz IR, Sanes DH (1998) A developmental shift from GABAergic to glycinergic transmission in the central auditory system. J Neurosci 18:4646–4655.

Krenning J, Hughes LF, Caspary DM, Helfert RH (1998) Age-related glycine receptor subunit changes in the cochlear nucleus of Fischer-344 rats. Laryngoscope 108:26–31.

Langner G, Schreiner CE (1988) Periodicity coding in the inferior colliculus of the cat. I. Neuronal mechanisms. J Neurophysiol 60:1799–1822.

Lee HJ, Wallani T, Mendelson JR (2002) Temporal processing speed in the inferior colliculus of young and aged rats. Hear Res 174:64–74.

Ling LL, Hughes LF, Caspary DM (2005) Age-related loss of the GABA synthetic enzyme glutamic acid decarboxylase in rat primary auditory cortex. Neuroscience 132:1103–1113.

Luo L, Moore JK, Baird A, Ryan AF (1995) Expression of acidic FGF mRNA in rat auditory brainstem during postnatal maturation. Brain Res Dev Brain Res 86(1–2):24–34.

Maison SF, Adams JC, Liberman MC (2003) Olivocochlear innervation in the mouse: immunocytochemical maps, crossed versus uncrossed contributions, and transmitter colocalization. J Comp Neurol 455:406–416.

Maison SF, Rosahl TW, Homanics GE, Liberman MC (2006) Functional role of GABAergic innervation of the cochlea: phenotypic analysis of mice lacking GABA(A) receptor subunits alpha 1, alpha 2, alpha 5, alpha 6, beta 2, beta 3, or delta. J Neurosci 26(40):10315–10326.

Masliah E, Fagan AM, Terry RD, DeTeresa R, Mallory M, Gage FH (1991a) Reactive synaptogenesis assessed by synaptophysin immunoreactivity is associated with GAP-43 in the dentate gyrus of the adult rat. Exp Neurol 113:131–142.

Masliah E, Mallory M, Hansen L, Alford M, Albright T, DeTeresa R, Terry R, Baudier J, Saitoh T (1991b) Patterns of aberrant sprouting in Alzheimer's disease. Neuron 6:729–739.

Masterton RB (1992) Role of the central auditory system in hearing: the new direction. Trends Neurosci 15:280–285.

May BJ (2000) Role of the dorsal cochlear nucleus in the sound localization behavior of cats. Hear Res 148:74–87.

Mei Y, Gawai KR, Nie Z, Ramkumar V, Helfert RH (1999) Age-related reductions in the activities of antioxidant enzymes in the rat inferior colliculus. Hear Res 135:169–180.

Mendelson JR, Lui B (2004) The effects of aging in the medial geniculate nucleus: a comparison with the inferior colliculus and auditory cortex. Hear Res 191:21–33.

Milbrandt JC, Albin RL, Caspary DM (1994) Age-related decrease in GABAB receptor binding in the Fischer 344 rat inferior colliculus. Neurobiol Aging 15:699–703.

Niu X, Trifunovic A, Larsson NG, Canlon B (2007) Somatic mtDNA mutations cause progressive hearing loss in the mouse. Exp Cell Res 313:3924–3934.

Noben-Trauth K, Zheng QY, Johnson KR (2003) Association of cadherin 23 with polygenic inheritance and genetic modification of sensorineural hearing loss. Nat Genet 35:21–23.

Oestreicher AB, Gispen WH (1986) Comparison of the immunocytochemical distribution of the phosphoprotein B-50 in the cerebellum and hippocampus of immature and adult rat brain. Brain Res 375:267–279.

Oliver DL (2000) Ascending efferent projections of the superior olivary complex. Microsc Res Tech 51:355–363.

O'Neill WE, Zettel ML, Whittemore KR, Frisina RD (1997) Calbindin D-28k immunoreactivity in the medial nucleus of the trapezoid body declines with age in C57BL/6, but not CBA/CaJ, mice. Hear Res 112:158–166.

Ouda L, Nwabueze-Ogbo FC, Druga R, Syka J (2003) NADPH-diaphorase-positive neurons in the auditory cortex of young and old rats. Neuroreport 14:363–366.

Palombi PS, Caspary DM (1996) Responses of young and aged Fischer 344 rat inferior colliculus neurons to binaural tonal stimuli. Hear Res 100:59–67.

Pickles JO (2004) Mutation in mitochondrial DNA as a cause of presbyacusis. Audiol Neurootol 9:23–33.

Plinkert PK, Mohler H, Zenner HP (1989) A subpopulation of outer hair cells possessing GABA receptors with tonotopic organization. Arch Otorhinolaryngol 246:417–422.

Plinkert PK, Gitter AH, Mohler H, Zenner HP (1993) Structure, pharmacology and function of GABA-A receptors in cochlear outer hair cells. Eur Arch Otorhinolaryngol 250:351–357.

Raza A, Milbrandt JC, Arneric SP, Caspary DM (1994) Age-related changes in brainstem auditory neurotransmitters: measures of GABA and acetylcholine function. Hear Res 77:221–230.

Reuss S, Schaeffer DF, Laages MH, Riemann R (2000) Evidence for increased nitric oxide production in the auditory brain stem of the aged dwarf hamster (Phodopus sungorus): an NADPH-diaphorase histochemical study. Mech Ageing Dev 112:125–134.

Rhode WS, Greenberg S (1994) Encoding of amplitude modulation in the cochlear nucleus of the cat. J Neurophysiol 71:1797–1825.

Ruttiger L, Panford-Walsh R, Schimmang T, Tan J, Zimmermann U, Rohbock K, Kopschall I, Limberger A, Muller M, Fraenzer JT, Cimerman J, Knipper M (2007) BDNF mRNA expression and protein localization are changed in age-related hearing loss. Neurobiol Aging 28:586–601.

Sanchez-Zuriaga D, Marti-Gutierrez N, De La Cruz MA, Peris-Sanchis MR (2007) Age-related changes of NADPH-diaphorase-positive neurons in the rat inferior colliculus and auditory cortex. Microsc Res Tech 70:1051–1059.

Schmoll H, Ramboiu S, Platt D, Herndon JG, Kessler C, Popa-Wagner A (2005) Age influences the expression of GAP-43 in the rat hippocampus following seizure. Gerontology 51:215–224.

Schneider B, Speranza F, Pichora-Fuller MK (1998) Age-related changes in temporal resolution: envelope and intensity effects. Can J Exp Psychol 52:184–191.

Schneider BA, Pichora-Fuller MK, Kowalchuk D, Lamb M (1994) Gap detection and the precedence effect in young and old adults. J Acoust Soc Am 95:980–991.

Schwarz DW, Schwarz IE, Hu K, Vincent SR (1988) Retrograde transport of [3H]-GABA by lateral olivocochlear neurons in the rat. Hear Res 32:97–102.

Simon H, Frisina RD, Walton JP (2004) Age reduces response latency of mouse inferior colliculus neurons to AM sounds. J Acoust Soc Am 116:469–477.

Sinha UK, Hollen KM, Rodriguez R, Miller CA (1993) Auditory system degeneration in Alzheimer's disease. Neurology 43:779–785.

Snell KB, Mapes FM, Hickman ED, Frisina DR (2002) Word recognition in competing babble and the effects of age, temporal processing, and absolute sensitivity. J Acoust Soc Am 112:720–727.

Stiebler I, Ehret G (1985) Inferior colliculus of the house mouse. I. A quantitative study of tonotopic organization, frequency representation, and tone-threshold distribution. J Comp Neurol 238:65–76.

Tadros SF, D'Souza M, Zettel ML, Zhu X, Waxmonsky NC, Frisina RD (2007) Glutamate-related gene expression changes with age in the mouse auditory midbrain. Brain Res 1127:1–9.

Thompson AM, Schofield BR (2000) Afferent projections of the superior olivary complex. Microsc Res Tech 51:330–354.

Thompson GC, Cortez AM, Igarashi M (1986) GABA-like immunoreactivity in the squirrel monkey organ of Corti. Brain Res 372:72–79.

Turner JG, Hughes LF, Caspary DM (2005) Affects of aging on receptive fields in rat primary auditory cortex layer V neurons. J Neurophysiol 94:2738–2747.
Vaughan DW (1977) Age-related deterioration of pyramidal cell basal dendrites in rat auditory cortex. J Comp Neurol 171:501–515.
Vaughan DW, Cahill CJ (1984) Long term effects of callosal lesions in the auditory cortex of rats of different ages. Neurobiol Aging 5:175–182.
Vaughan DW, Vincent JM (1979) Ultrastructure of neurons in the auditory cortex of ageing rats: a morphometric study. J Neurocytol 8:215–228.
von Hehn CA, Bhattacharjee A, Kaczmarek LK (2004) Loss of Kv3.1 tonotopicity and alterations in cAMP response element-binding protein signaling in central auditory neurons of hearing impaired mice. J Neurosci 24:1936–1940.
Walton JP, Frisina RD, Ison JR, O'Neill WE (1997) Neural correlates of behavioral gap detection in the inferior colliculus of the young CBA mouse. J Comp Physiol [A] 181:161–176.
Walton JP, Frisina RD, O'Neill WE (1998) Age-related alteration in processing of temporal sound features in the auditory midbrain of the CBA mouse. J Neurosci 18:2764–2776.
Walton JP, Simon H, Frisina RD (2002) Age-related alterations in the neural coding of envelope periodicities. J Neurophysiol 88:565–578.
Wang Y, Manis PB (2005) Synaptic transmission at the cochlear nucleus endbulb synapse during age-related hearing loss in mice. J Neurophysiol 94:1814–1824.
Willott JF (1986) Effects of aging, hearing loss, and anatomical location on thresholds of inferior colliculus neurons in C57BL/6 and CBA mice. J Neurophysiol 56(2):391–408.
Willott JF, Jackson LM, Hunter KP (1987) Morphometric study of the anteroventral cochlear nucleus of two mouse models of presbycusis. J Comp Neurol 260:472–480.
Willott JF, Parham K, Hunter KP (1988) Response properties of inferior colliculus neurons in young and very old CBA/J mice. Hear Res 37(1):1–14.
Willott JF, Parham K, Hunter KP (1991) Comparison of the auditory sensitivity of neurons in the cochlear nucleus and inferior colliculus of young and aging C57BL/6J and CBA/J mice. Hear Res 53(1):78–94.
Willot JF, Bross LS and McFadden SL, (1992) Morphology of the dorsal cochlear nucleus in young and aging C57BL76J and CBA/J mice. J. Comp. Neurol 321:666–678.
Yamamoto Y, Matsubara A, Ishii K, Makinae K, Sasaki A, Shinkawa H (2002) Localization of gamma-aminobutyric acid A receptor subunits in the rat spiral ganglion and organ of Corti. Acta Otolaryngol 122:709–714.
Yamasoba T, Someya S, Yamada C, Weindruch R, Prolla TA, Tanokura M (2007) Role of mitochondrial dysfunction and mitochondrial DNA mutations in age-related hearing loss. Hear Res 226:185–193.
Zettel ML, Frisina RD, Haider SE, O'Neill WE (1997) Age-related changes in calbindin D-28k and calretinin immunoreactivity in the inferior colliculus of CBA/CaJ and C57Bl/6 mice. J Comp Neurol 386:92–110.
Zettel ML, Zhu X, O'Neill WE, Frisina RD (2007) Age-related decline in Kv3.1b expression in the mouse auditory brainstem correlates with functional deficits in the medial olivocochlear efferent system. J Assoc Res Otolaryngol 8:280–293.
Zheng QY, Yan D, Ouyang XM, Du LL, Yu H, Chang B, Johnson KR, Liu XZ (2005) Digenic inheritance of deafness caused by mutations in genes encoding cadherin 23 and protocadherin 15 in mice and humans. Hum Mol Genet 14:103–111.

Chapter 4
Closing the Gap Between Neurobiology and Human Presbycusis: Behavioral and Evoked Potential Studies of Age-Related Hearing Loss in Animal Models and in Humans

James R. Ison, Kelly L. Tremblay, and Paul D. Allen

4.1 Introduction

4.1.1 Contributions of Animal Models to Understanding Human Presbycusis

Any reader who has grown up with a pet dog cannot have failed to notice that the effects of advancing age in dogs are not very different from those apparent in aging grandparents, except that in calendar time they appear more rapidly. Although domesticated animals may present a special case compared with wild animals that hardly survive to the age of sexual maturity, a few wild animals do survive and they also exhibit these common effects of human aging. Very close to human sympathies are the observations of elderly chimpanzees by naturalists who, having followed their stable groups for many years, write that the rare creature that has successfully survived the challenges of the wild exhibits the same thinning hair, slow movements, and sagging and wrinkled facial skin as the elderly human (Hill et al. 2001). And given the laboratory studies of hearing in old monkeys (Bennett et al. 1983) and examinations of cochlear pathology in postmortem studies of aged pet dogs

J.R. Ison (✉)
Department of Brain & Cognitive Sciences, University of Rochester, Meliora Hall, Box 270268, Rochester, NY 14627
e-mail: jison@bcs.rochester.edu

K.L. Tremblay
Department of Speech & Hearing Sciences, University of Washington, 1417 NE 42nd St., Seattle, WA 98105
e-mail: tremblay@u.washington.edu

P.D. Allen
Department of Neurobiology & Anatomy, University of Rochester, 601 Elmwood Avenue, Box 603, Rochester, NY 14642
e-mail: pallen@cvs.rochester.edu

S. Gordon-Salant et al. (eds.), *The Aging Auditory System*, Springer Handbook of Auditory Research 34,
DOI 10.1007/978-1-4419-0993-0_4,

(Shimada et al. 1998), this wrinkled and slowly moving chimpanzee and the graying and arthritic dear old pet must both suffer from poor hearing as do elderly humans.

This chapter is focused on the experimental evidence that will more definitively characterize age-related hearing loss (ARHL) in animals in comparison to human listeners. Both behavioral and noninvasive auditory evoked potential results (AEPs) are included, the latter so that animal data can be assessed relative to human data in very similar experimental paradigms. These laboratory experiments have most often had their rationale founded in the data either from audiology practice and psychophysical laboratory research with human listeners or from neurobiological findings of age-related changes in anatomical structure and connectivity, neurochemical and genetic expression, and physiological evidence observed in laboratory animals. The task of this chapter is to mediate between these separate disciplines. These functional studies of behavior and AEFs in old animals can provide a means of testing hypotheses about presbycusis that are not readily tested in the two flanking disciplines. They cannot be tested in the human psychophysical laboratory because humans cannot be subjected to stringent environmental, surgical, or genetic manipulations, and neither can they be tested in the neurobiological laboratory because neurobiological end points, e.g., evidence of histopathology and loss of neurons or changes in genetic expression with increasing age, have no direct a priori connection to the phenomena of sensory experience. Thus the behavioral and AEP data obtained from noninvasive procedures, often in the awake and behaving animal and in the similarly awake and behaving human research participant, are intended to provide different links back and forth between age-related neurobiological changes in animals using invasive techniques on the one hand and sensory-perceptual measures obtained in human listeners on the other. It is the convergence of their outcomes with each other and with those obtained in clinical practice that will contribute to understanding the common features of presbycusis and their neurobiological bases.

Realistic models of the various features of human presbycusis that are based in animal behavior and neurobiology will become even more valuable in the future with their further evolution and with the continuing development of translational research programs. Besides achieving an integrated multidisciplinary understanding of presbycusis, the ultimate goal of the research on ARHL at each of these levels of analysis is to discover how this progressive deficit may be retarded or prevented or possibly reversed. These translational programs will almost certainly need to assess their validity in animal models of presbycusis before they are ready for clinical testing in humans.

4.1.2 Distinctions Between the Simple and the Complex in Signals and in Listening Environments

In most studies of human presbycusis, the investigators have focused on age-related changes in the audiogram, i.e., the loss of absolute sensitivity as given by the ability to detect pure tones in quiet, measured across the normal frequency range of hearing. In other experiments, subjects are asked to detect, identify, or respond to suprathreshold, clearly audible signals that may be simple tones or noise bursts, may consist of patterned sequences of stimuli that vary in their spectral content and intensity over time, or may contain fragments of speech or sentences, sometimes in complicated listening conditions that include competing messages or reverberation. These latter experiments variously simulate the characteristics of normal speech signals and the nonoptimal listening environments typical of everyday life. The basic finding in the first set of experiments is that absolute thresholds steadily increase with advancing age, and in the second set, that elderly human listeners are often less able than the young to detect or to identify signals when they are presented against a noisy or competing background, even though the signals and the background stimuli would all be audible if they were presented by themselves. Examples of these two types of experiments are shown in Fig. 4.1a (for simple stimuli) from a study by Allen and Eddins (2009) and Figure 4.1b (for complex stimuli) from a study on the effects of reverberation on speech perception published by Nabelek and Robinson (1982). Reduced performance with advancing age is apparent in both sets of data. The appearance of decrements in hearing ability as early as 30 to 40 years of age in these data and in other life-span studies of nonpatient populations such as those of Bergman et al. (1976) and Brant and Fozard (1990) indicate that hearing sensitivity and auditory processing apparently begin a gradual but significant progressive decline in middle-aged adults. Thus even though the word "presbycusis" was initially designed by Roosa (1885) to specifically label the clinically significant

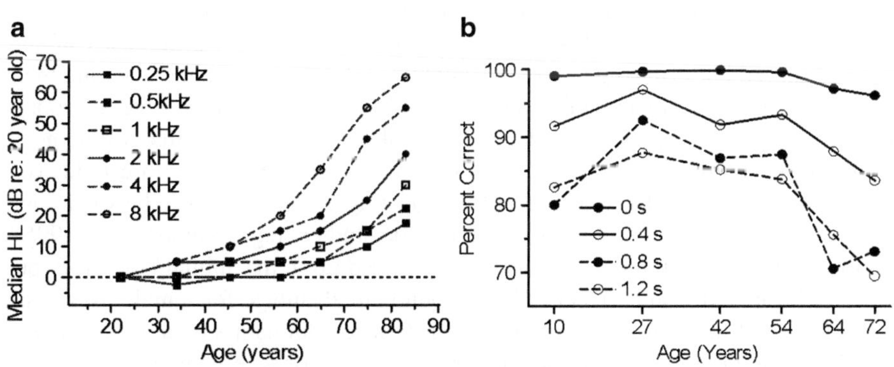

Fig. 4.1 (**a**) Age-related hearing loss (HL) across standard test frequencies from 250 Hz to 8 kHz between the ages of 20 to over 90 years (overall n = 1,209, 689 women, 520 men) presented as median absolute thresholds expressed relative to a group of subjects between the ages of 18 to 20 years (Eddins and Allen 2009). (**b**) Age-related changes in correct word identification presented monaurally in listening environments having different degrees of reverberation (i.e., differences in the duration of a series of diminishing echoes of the stimuli) from 0 to 1.2 s during and after the presentation of a stimulus (groups of 10 for each age group, 46 women, 14 men). (Adapted from Nabelek and Robinson 1982.)

hearing loss of elderly patients, these empirical data suggest the value of including intermediate middle-aged subjects, animals and humans alike, in laboratory studies, to identify and, one hopes, to understand and eventually learn to control the conditions that lead to early onset ARHL.

4.1.3 Chronological versus Biological Aging and Individual Differences in ARHL

A limitation of averaging group data across a time series is that the analysis cannot discriminate between an effect in which subjects are relatively homogeneous in their slow progression of hearing loss, or when each of the subjects displays a rapid increase in hearing loss for some particular frequency but differ in the age at which the deficit first appears. Indeed, all large-scale studies of ARHL have considerable scatter around the mean within an age group, and the distribution of individual differences is most often not a Gaussian distribution but is significantly skewed upward or downward depending on the test frequency and the age group. For example, in the group data depicted in Figure 4.1a, most middle-aged participants have a very small deficit compared with the young, but a few have a serious hearing loss more typical of the elderly. Also, the majority of elderly listeners have a serious hearing loss, especially for the higher test tones, but a few elderly participants have maintained a high level of sensitivity, close to that typical of the young listener. Hence, although increasing chronological (or "calendar") age is obviously associated with greater hearing loss in human listeners, as can be seen most persuasively in the comparison of the extreme age groups, the scatter within the intermediate age groups shows that its onset and progression are highly variable from person to person.

One recurring theme in this chapter is that these individual differences should not be ignored but instead be more thoroughly studied because understanding their biological and environmental bases may provide useful clues to the general properties of presbycusis. One source of the scatter within an age group is that beginning at ~40 to 50 years of age, women tend to have better high frequency thresholds (on the order of 5- to 10- dB) than men of the same age. This difference between men and women has been found in other large-scale studies (e.g., Pearson et al. 1995), and it is one of the findings that has been followed up in the animal laboratory but so far with mixed results. Henry (2004) found an advantage for female over male CBA mice but for male over female C57BL/6 mice; Guimaraes et al. (2004) found an advantage for female CBA mice but only in a group of senescent mice older than 2 years of age, not at the ages at which Henry (2004) had found the CBA female advantage; and Ison and Allen (2007) found no sex difference in large groups of CBA mice at all ages between 2 and 25 months of age when cohorts of males and females were balanced for time of testing.

Other factors that may be responsible for scatter within age groups of humans are variation in the cumulative effect of noise exposure across the life span, the increasing incidence of medical conditions that may affect hearing, variation in the reactions to the treatment for these conditions, dietary variables, and genetic differences that may directly impact hearing (van Eyken et al. 2007). Each of these isolated factors is now being investigated in the animal laboratory, and there is even one very interesting study of an interaction between early noise exposure and later ARHL by Kujawa and Liberman (2006); these investigators showed that an early noise exposure that had little immediate effect accelerated the progress of ARHL as assessed many months later. Other factors that influence ARHL may be similarly subtle and difficult to isolate because they too may interact in as yet unknown ways with other predispositions. For example, stable individual differences in the onset of hearing loss have been reported in inbred C57BL/6 mice (Ison et al. 2007) even though these mice have the same genetic background, were raised in the same controlled environment, and were even tested on the same day. However, it should also be noted that in this experiment, the within-strain variability of C57BL/6 mice was relatively small compared with the distribution of ARHL across populations of mice of different strains and different environmental conditions; i.e. genetic background and obvious environmental differences are responsible for much of the variability in mouse hearing.

Other researchers who work with large groups of human volunteers (e.g., Gates et al. 1999) have suggested that the differences not only in the time of onset of ARHL but also in the pattern of ARHL across spectral frequencies can be obscured in group data. These authors and others have provided data showing that the stereotypic profile of sharply increasing ARHL across frequency in the older age groups shown in Figure 4.1a may be characteristic of many but most certainly not all elderly listeners and, particularly, that many have a more serious low-frequency ARHL than the group data would suggest. To follow up these earlier observations, Allen and Eddins (2009) used an unbiased profile analysis of the data depicted in Figure 4.1a to partition the individuals into seven profiles that provided the best fit. They found that most participants, 80% of the men and 70% of the women, provided profiles that looked like an exaggeration of Figure 4.1a, with minimal ARHL for low-frequency test stimuli and a sharp rise for high frequencies, whereas the remaining participants were distinguished by their having relatively more severe ARHL for low-frequency test stimuli and then a less steep increase in ARHL for higher frequencies. This latter profile was apparent in 30% of the women and just 20% of the men, i.e., ~50% more common in women than in men. Jerger et al. (1993) similarly observed a more even pattern of ARHL across test frequencies in women than in men, in part because the women had a less severe loss for high frequencies but a more severe hearing loss at low frequencies. This latter effect was present in the Allen and Eddins data but was small and not significant. Past observation of this sex link between these two ARHL profiles has spurred attempts to determine their antecedents in the environment or in inherited or familial antecedents (Gates et al. 1999). The full treatment of this endeavor goes beyond the scope of this chapter, but it is clearly relevant to the general topic of animal models of ARHL.

4.2 Are There Different Types of Hearing Loss, Different in Origin and in their Effects?

4.2.1 The Psychophysical Question: Is ARHL Simply a Loss of Absolute Sensitivity?

One presently unresolved issue in interpreting the two manifestations of ARHL shown in Figure 4.1 is whether they have a single source in the loss of absolute sensitivity or, instead, whether the data presented in Figure 4.1b reflect at least a partially independent deficit in complex sensory processing and perception. Tremblay and Burkard (2007) describe experimental approaches that have been used in ARHL research with human listeners to isolate the loss of threshold sensitivity from a possibly additional effect of age on the efficacy of complex auditory processing. Some classic research strategies have been developed to address this issue with human listeners, but their extensive review of the data led them to conclude that as yet there is no single manipulation that cleanly separates the two types of hearing loss. One method is to assemble a single large group of elderly listeners and then subject them to a rigorous testing schedule (e.g. Humes et al. 1994; Humes 2005) in which their scores on the test battery are used to develop a pattern of correlations among them. Humes (2005) tested over 200 elderly listeners with standard clinical audiograms, brainstem AEPs, intelligence scales, several measures of complex auditory processing, including discrimination of tone duration and temporal order, and several measures of speech recognition. The variable that was most important in accounting for individual differences in speech recognition scores was hearing loss as revealed in the audiogram, whereas significant but relatively minor contributions were provided by IQ tests, differences in central processing, and age.

Another commonly used strategy is to compare two groups on some measure of auditory processing, one group of elderly listeners with so-called "golden ears" that have an unusually high sensitivity for at least the range of low- to mid-frequency test stimuli to be used in the experiment and so can be matched one-to-one with a second group of average young listeners (e.g., Snell 1997 for gap detection; Harris et al. 2008 for an evoked potential study of frequency discrimination). And last, two pairs of matched groups can be assembled, including two groups of young and elderly subjects with matched hearing sensitivity within the normal age plus two additional groups with carefully chosen "below average" young listeners with some degree of hearing loss that can be matched to elderly listeners with "average" hearing loss for their age. This classic "2 × 2" factorial design supports a type of statistical analysis that can separate out the independent contributions of hearing loss and of age and also assess the contribution of their interaction. Thus Gordon-Salant and Fitzgibbons (1995), in a study of advancing age and hearing loss on speech perception in reverberant environments with variously time-distorted speech signals, reported that both hearing loss and age have independent effects on performance and that the effect of age increased with the increasing degree of distortion in the stimuli.

This same confounding of age and loss of absolute sensitivity is present in the animal laboratory, and the few experiments that have explicitly recognized the problem have resolved it using the same tools as in the human laboratory, to find similar outcomes. For example, May et al. (2006) found in a longitudinal behavioral study in mice that individual differences in the decline of frequency selectivity with increasing age did not covary with their loss of absolute sensitivity; i.e., age had an independent effect on performance. Barsz et al. (2002) reported that the individual differences in the loss of temporal acuity in a behavioral gap-detection study with aging mice were correlated not only with hearing loss but also with advancing age independent of hearing loss. And Walton and his colleagues found in an electrophysiological study of temporal acuity that the near-senescent CBA mice was impaired compared with young mice (Walton et al. 1998), whereas temporal acuity was not impaired in middle-aged C57BL/6 mice that had more hearing loss than old CBA mice (Walton et al. 2008). All of these researchers cited above have concluded that the effects of age on complex auditory processing are greater than would be expected just on the basis of age-related changes in audibility. Further converging support for this position is provided in neurobiological evidence of histopathology at both peripheral and central sites within the auditory system.

4.2.2 Neurobiology: Might the Type of ARHL Differ According to the Site of the Pathology?

Structural deterioration in the ear has been observed in all animal models of aging and also in humans in temporal bone histopathology in postmortem specimens (Schmiedt, Chapter 2). For example, changes in vascularity and in supporting structures and sensory-neural elements within the cochlea have been reported in humans (Johnsson and Hawkins 1972; Nelson and Hinojosa 2006) and in a great variety of laboratory and domesticated animals (e.g., rats, Keithley and Feldman 1982; rhesus monkeys, Hawkins et al. 1985; squirrel monkeys, Dayal and Bhattacharyya 1986; chinchillas, Bohne et al. 1990; gerbils, Gratton and Schulte 1995; mice, Spongr et al. 1997; Ichimiya et al. 2000; Hequembourg and Liberman 2001; dogs, Shimada et al. 1998; guinea pigs, Ingham et al. 1999). Such peripheral deterioration might be conceived as resulting only in a loss of absolute sensitivity, but it has been suggested that it may also have a direct effect on central processing because it alters the quality of the neural information that is transmitted to the brain through the auditory nerve. Hellstrom and Schmiedt (1991) suggest that peripheral sensory damage could result in a loss of synchrony in spiral ganglion firing and thus indirectly influence central neural processing even in the absence of direct age-related deterioration in the central auditory nervous system (CANS). This is an important and plausible hypothesis, but there is as yet no direct evidence that advancing age reduces auditory nerve synchrony as measured, e.g., in the degree of jitter in its input to the cochlear nucleus.

There is also abundant evidence for the presence of pathological changes in the auditory cortex, midbrain, and brainstem with advancing age (Canlon and Walton, Chapter 3). Changes in the human brain are apparent in postmortem studies; these are reviewed by Mrak et al. (1997). These authors describe the presence of gross changes in the size of the brain, number of cells, and dendritic fields and by more subtle changes in major neurotransmitters in the cholinergic, dopaminergic, serotonergic, and adrenergic systems. The more recent studies using in vivo noninvasive imaging have but very rarely touched on the auditory system; one relevant report (Lutz et al. 2007) found changes in Heschl's gyrus that suggested a disruption of fiber tracts and changes in the inferior colliculus (IC) but no changes in either the lateral lemniscus or the medial geniculate. In the animal laboratory, changes in vascular support structures have been reported in the medial trapezoid body of the auditory brainstem in old rats (Casey and Feldman 1985), and changes in synaptic connectivity have been reported at every level of the auditory system, in the auditory cortex (Vaughn 1977), in the IC (Helfert et al. 1999), in the superior olivary complex (Casey 1990), and in the cochlear nucleus (Helfert et al. 2003). There is also evidence provided by Zettel et al. (2007) that with advancing age in the mouse, there is less expression of potassium ion channels expressed by the *Kcnc1* gene in the medial nucleus of the CBA trapezoid body, an effect that may be related to deficits in downstream control over cochlear sensitivity that these authors found in *Kcnc1* null mutant transgenic mice. Furthermore, in the adult cat, there is evidence that synchrony at the level of the trapezoid body is greater than that provided by the auditory nerve to the cochlear nucleus, this being a beneficial result of central neural processing (Joris et al. 1994). There is as yet no evidence that such enhancement of synchrony by brainstem neural processing deteriorates with age, but the changes in the expression of potassium ion channels in the trapezoid body of the old mouse observed by Zettel et al. (2007) is consistent with this hypothesis.

4.2.3 Peripheral Hearing Loss Can Alter Both the CANS and Central Auditory Processing

There is a significant literature on the degree to which changes in the central auditory system with advancing age may in some cases be directly mediated by peripheral pathology and in other cases by increasing age alone. Most of our understanding of this phenomenon is the product of an extensive program of research conducted over the years by Willott and his colleagues that analyzed the relationships between increasing age, the time course of histopathology, and associated changes in neural activity in CBA and in C57BL/6 mice (reviewed in Willott 1996). For example, cell loss in the high-frequency regions of the anterior ventral cochlear nucleus occurs at ~7 months of age in the C57BL/6 mouse after peripheral hearing loss but not until 2 years of age in the CBA mouse (Willott et al. 1987). In contrast, the loss of octopus cells in the posterior ventral cochlear nucleus becomes apparent in C57BL/6 mice at the same age that it is observed in CBA mice, when the mice are ~2 years old (Willott and Bross 1990). Another very important finding is that the tonotopic maps of the IC (Willott 1986) and the auditory cortex (Willott et al. 1993) are profoundly changed with increasing hearing loss in middle-aged C57BL/6, so that formerly high-frequency areas of these brain regions come to respond with great sensitivity to low-frequency stimuli. This electrophysiological sign of a more extensive representation of low frequencies in the middle-aged C57BL/6 mouse is accompanied by enhanced behavioral responsivity to these stimuli, as demonstrated, e.g., by Willott and Carlson (1995) and Ison et al. (2007).

4.2.4 Experimental Manipulations at Different Sites Can Produce Different Types of Hearing Loss

4.2.4.1 Manipulating the Integrity of the Cochlea

There is considerable evidence for extensive structural degeneration in the ear and the brain in all the animal models of aging that is confirmed in the available human evidence, but it must also be noted that most of these data are scattered across different species, and it is rare that different histopathological end points and functional measures are obtained in the same individual. And even at best, this evidence is correlational. For example, one very interesting set of findings by Kazee et al. (1995) is that regional loss of hair cells along the basilar membrane of the aging C57BL/6 mouse is associated with a frequency-specific increase in auditory thresholds and also with a loss of synaptic endings through the central nucleus of the IC. Although these observations are important, there is no direct evidence that these observations have uncovered a causal chain in which hair cell loss is the primary effective agent that leads directly to the rise in thresholds and hence to the loss of synapses in the IC. The search for causality is the rationale for trying to simulate some effects of age in young animals by directly producing structural or neurochemical changes in the auditory system and then determining the chain of effects of these manipulations.

A study that appears to simulate the effects of aging on regional hair cell loss in the cochlea and thus at least ties hair cell loss causally with absolute threshold changes was provided by Prosen and Moody (1991). These researchers trained young adult chinchillas in which one cochlea had been surgically destroyed to press a lever when test tones were presented, and training continued until the animals were able to generate stable, absolute, and differential thresholds across their range of hearing. Then the researchers destroyed just the apical (low-frequency) hair cells in the remaining cochlea of four of these chinchillas by freezing the area with a cryoprobe. The restricted effect of this manipulation on apical hair cells was later verified in a histopathological study, while the functional effect of the manipulation was assessed by the within-subject changes in behavioral thresholds that resulted after the second surgery, in quiet or in the presence of a high-pass masking noise. As could be expected given the known distribution of best frequencies along the basilar membrane, both the absolute and differential thresholds for low- but not high-frequency test stimuli were increased by apical hair cell loss. However, the animals could still detect these low-frequency stimuli when they were presented at higher levels, and, to a limited extent, the animals were able to discriminate between different low-frequency test tones. An important additional finding was that high-frequency masking stimuli further reduced low-frequency hearing, an indication that the residual low-frequency sensitivity is mediated in part by more basal high-frequency regions of the cochlea. This finding agrees with the spread of excitation and receptivity of high-frequency hair cells in the basal areas of the cochlea to relatively intense low-frequency test stimuli reported by Cody and Russell (1987). It may also explain the common observation that there is not a simple one-to-one relationship between the position of a hair cell along the basilar membrane and its sensitivity to particular tonal stimuli, especially at high stimulus levels. It would also have been interesting to see if this manipulation would have changed the patterns of connectivity with the IC as has been observed in the C57BL/6 mouse.

The most recent method of manipulating the auditory periphery is by genetic engineering focused on specific stages of sensory processing in the ear. McCullough and Tempel (2004) studied the effects of deleting several related alleles of the gene that encodes the protein that is responsible for extruding calcium from stereocilia and spiral ligament cells (plasma membrane calcium ATPase isoform 2 [PMCA2]). They discovered that these different alleles produced phenotypes with varied degrees of hearing loss appearing in ABR thresholds at different onset times. These data certainly demonstrate the causal relationship between hair cell dysfunction and ABR thresholds and further raise the possibility that one cause of ARHL may result from a changing expression of PMCA2 with advancing age.

4.2.4.2 Manipulating the Central Auditory Nervous System

There are other reports that specific experimental administration of drugs or destruction of regions within the CANS can have substantial specific sensory-perceptual effects that are at least qualitatively similar to some aging effects. For example, scopolamine is a cholinergic antagonist best known in behavioral pharmacology for its deleterious effects on memory, but it has also been shown that the systemic administration of scopolamine adversely affects gap-detection measures of temporal acuity in human volunteers (Caine et al. 1981) and in rats (Ison and Bowen 2000). However, it does not affect absolute threshold sensitivity. Intracranial application of another cholinergic antagonist (atropine) directly onto the cochlear nucleus of cats increased the threshold at which these subjects could successfully perform a behavioral task in which tones were presented in noise (Pickles and Comis 1973). It is also important that this manipulation had little effect on responses to tones presented in quiet. Turning to another neurochemical manipulation, vigabatrin is a pharmacological compound that prevents the breakdown of GABA in the synaptic cleft and has been used clinically to alleviate epilepsy. Gleich et al. (2003) found a dose-sensitive enhancement of behavioral gap-detection thresholds by systemic administration of vigabatrin in old gerbils with unusually poor temporal sensitivity before drug treatment, suggesting an important role for GABA in this task. It has also been shown that temporal acuity as measured by gap detection is very much diminished by bilateral lesions of the auditory cortex in human patients after vascular accidents (Buchtel and Stewart 1989) and in rats after surgery (Bowen et al. 2003). These lesions also did not affect absolute thresholds. And furthermore, there are reports (e.g., Kopp-Scheinpflug et al. 2003) that neurons of the medial nucleus of the trapezoid body in *Kcna1* transgenic mice that lack the Kv1.1 ion channel show deficits in onset responding that would certainly degrade the neuronal synchrony of firing in this nucleus. There are no changes in absolute threshold in the null mutant mice. However, these same null mutant mice, like old mice, also have behavioral deficits in sound localization (Allen et al. 2003), which is in part mediated by the trapezoid body.

These experimental data show in both adult humans and laboratory animals that threshold sensitivity and frequency discriminations depend on the integrity of peripheral auditory structures and that the integrity of central neural and neurochemical mechanisms is necessary for complex auditory processing but does not affect absolute threshold measures. They show, at a minimum, that altering the integrity of these structures does affect hearing, but they do not speak to the degree that, e.g., blocking cholinergic receptors by scopolamine simulates an age-related loss of cholinergic neurons or changes in the sensitivity of cholinergic receptors. These experimental data also suggest the possibility that age-related central processing deficits may take a number of different forms depending on where deterioration has occurred in the CANS. For the periphery, there are arguments that hearing loss attendant on stria vascularis dysfunction has a fundamentally different functional signature than does hair cell loss (Boettcher 2002). It is no less plausible to suspect that the anatomical targets of aging could vary among individuals, and if so, it is then conceivable that some elderly listeners could have a deficit only in the functions mediated by specific sites, such as spatial localization mediated by the nuclei of the superior olivary complex, but not in functions mediated by octopus cells in the cochlear nucleus, perhaps gap detection. Although it may seem more likely that the central degenerative effects of aging would appear consistently throughout the CANS rather than being restricted to one nucleus or another, this is an idea that has never been tested. It is possible that physiological/gene-expression examinations of the CANS carefully combined with the results of a battery of auditory tasks tests would provide a useful test of this hypothesis.

4.3 Noninvasive Objective Methods for Studying ARHL in Animals and Humans

4.3.1 Introduction to the Methods

This section provides a closer examination of the nonverbal behavioral and noninvasive AEP methods that have been used in both animal and human research. Following the earlier organization, the first part deals with age-related changes in absolute thresholds related to Figure 4.1A, and the second examines changes in suprathreshold hearing, including both high-intensity simple test stimuli and the more complex hearing paradigms related to Figure 4.1B.

Psychoacoustic and audiological investigations rely on the willingness of research subjects to follow instructions and indicate that they heard a stimulus by making some voluntary response, such as pressing a button or pointing to a location. The experimenter's careful control over the stimulus conditions ensures that the subject is listening for the intended stimuli. Elaborate statistical analyses are used to convert the responses into a single measure of sensitivity, these being complicated because even cooperative human volunteers will fail to respond to stimuli that were previously detected (a miss) or will report hearing a sound when there was none presented (a false alarm) and, at worst, human research participants sometimes appear to lose interest in the task at hand and may even fall asleep.

Each of these problems is magnified in animal research, beginning with the necessity of developing an indicator response that is sensitive to stimulus presentation. Procedures for measuring stimulus-evoked overt responding in the awake animal have all been developed over the course of about the last 100 years, starting with work by Yerkes on hearing in frogs (1905) and mice (1907). Some indicator responses are directly elicited by auditory stimuli as a built-in reflexive reaction to an acoustic stimulus, such as the Preyer reflex (studied in mice by Jero et al. 2001). Others are obtained by the modification of a reflex response by an auditory stimulus that has no built-in relationship to the response, this being an example of the prepulse inhibition phenomenon common to many reflexes and many species; its use in developmental and comparative sensory research has been described by Hoffman and Ison (1992). A third behavioral approach requires training the animal to make some more-or-less arbitrary response when an acoustic stimulus is presented and to refrain from responding in its absence. These training methods have been described, e.g., by Stebbins (1990) and Heffner et al. (2006), and their many contributions to auditory processing in animals have been described by Fay (1988) and Long (1994). The experiments by Pickles and Comis (1973) and Prosen and Moody (1991) cited above used training procedures, the experiments by Willott and Carlson (1995), Bowen et al. (2003), and Ison and Bowen (2000) used a prepulse inhibition paradigm, and the results of Ison et al. (2007) used the method of reflex elicitation.

Noninvasive electrophysiological responses that are evoked by auditory stimuli and now often used in both human and animal studies of hearing have evolved over the course of the last 40 years, beginning with the demonstration by Jewett et al. (1970) that very small bioelectric potentials can be measured with surface electrodes in response to simple acoustic pulses. (This is most often called the ABR). Similar methods have been refined so that AEPs and neuromagnetic auditory evoked fields (AEFs) can be recorded not only to simple tone bursts in experiments intended to measure absolute thresholds in animals and in human infants too young to cooperate with the audiologist but also to examine the physiological reactions to complex stimuli such as speech signals (Tremblay et al. 2002) and interaural timing cues (Ross et al. 2007). Most commonly, AEPs are collected when animals are anesthetized rather than awake, while in convenient contrast, adult human research participants can be awake and infants can be asleep. Depending on the stimulus paradigm, brain activity can be measured while participants are actively engaged in a listening task or, in contrast, passively exposed to the ongoing stimuli. Because of the different variations in AEP methods, it is possible to measure the physiological capacity of the human listener and animals either independent of the cognitive and emotional contributions to a behavioral response or dependent on these internal states, so as to conform to the needs of the experiment.

4.3.2 The Application of These Methods to the Study of ARHL

As described above, the most rapid behavioral technique to study hearing in animals is to take advantage of a built-in reflexive response to an abrupt and relatively intense sound burst, e.g., the rapid twitch of the ears that is called the Preyer reflex or the whole body startle reflex. Jero et al. (2001) measured the startle reflex in a group of young adult albino mice of the FVB strain that have substantial individual differences in hearing ability to determine whether startle elicitation could be used as a fast screening test of hearing. The startle reflex was recorded as "positive" or "negative" depending on a mouse making at least two visible startle responses to three presentations of either hand claps or the sound of two hammers hitting together and validated these scores against ABR click thresholds. They reported that the reflex test was successful in detecting all mice with a severe hearing loss, defined as an ABR threshold over 80 dB. The amplitude of the acoustic startle reflex (ASR) declines with age in humans as shown by Ford et al. (1995) in a study comparing two groups of subjects, one with a mean age of 22 years, the second a mean of 69 years. They also found that two components of AEPs evoked by startle stimuli were reduced in amplitude in the older subjects, N1 with a latency of 50 to 150 ms and P3 with a latency of ~300 to 600 ms. Similar declines in the ASR have been observed in old rats (Krauter et al. 1981) and in three strains of mice, with the time course of the decline in the ASR corresponding to the time course of their developing hearing loss as measured by ABRs (Ison et al. 2008). However, age-related decrements in reflex strength by themselves do not necessarily implicate changes in auditory function without converging evidence from sensory threshold measurements because the motor system loses motor axons and muscle fibers with increasing age (Einsiedel and Luff 1992). And furthermore, at least in the middle-aged C57BL/6 mouse, hearing loss for high-frequency test stimuli as shown in ABR measures is correlated with exaggerated startle reflexes for low-frequency stimuli (Ison et al. 2007); this is thought to be a result of central tonotopic reorganization as described by Willott (1996).

Behavioral measures of absolute thresholds in trained animals often provide measures that rival the sensitivity of human psychophysical procedures. Fig. 4.2 shows absolute thresholds obtained for young adult human listeners using the verbal instructions standard in clinical audiology (data kindly provided by David Eddins) and by adult laboratory animals: monkeys (Bennett et al. 1983),

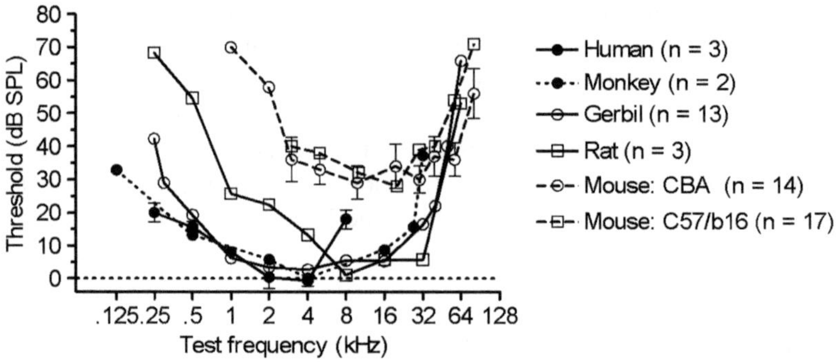

Fig. 4.2 Absolute thresholds obtained by the standard methods of clinical audiometry in young adult human listeners and by operant conditioning techniques in monkeys, gerbils, rats, and two mouse strains. (Animal data replotted from data in Fay 1988; human data kindly provided by David Eddins.)

gerbils (Ryan 1976), rats (Kelly and Masterton 1977), CBA mice (Birch et al. 1968), and from the same laboratory, C57BL/6 mice (Mikaelian et al. 1974). All of these studies used training techniques in which the animals either worked for food or escaped shock, with the acoustic stimuli signaling the availability of food or predicting the presence of shock. The standard mammalian U-shaped audiogram is evident in these data, with greatest sensitivity in a mid-frequency region that differs across species but is correlated with the species-specific spectral frequency of their vocalization, then a progressive increase in thresholds for lower and higher frequencies. It may be noticed that humans, monkeys, rats, and gerbils all hear stimuli presented at ~0-dB SPL at their best frequency, whereas mice appear to have relatively poor sensitivity. This difference could indicate that mice have a species-specific hearing impairment, but the real problem may have been that the training/testing method was still being developed at that time and was not yet well suited for these mice. Birch et al. (1968) described the mice as spending much of their time grooming (a classic displacement activity in response to conflict; Spruijt et al. 1992) and that when they were grooming, they did not respond to the tones.

Two behavioral experiments have followed changes in absolute thresholds across the adult life span (Mikaelian et al. 1974 in C57BL/6 mice; Bennett et al. 1983 in rhesus monkeys). Mikaelian et al. (1974) used the same methods as Birch et al. (1968) but tested their C57BL/6 mice for many months. The results of these two experiments are presented in Fig. 4.3. Both show a profound effect of advancing age on auditory thresholds, and both exhibit the sharply rising hearing loss for high frequencies, with more modest loss at low frequencies. Mikaelian et al. (1974) documented the increasing cochlear damage in his mice by sacrificing small numbers of mice at various time points to show how the behavioral changes corresponded to pathological changes in the ear. The monkeys tested by Bennett et al. (1983) were a subset from a group that were later sacrificed for an anatomical study that showed increasing cochlear damage in older animals (Hawkins et al. 1985). Unfortunately, the potential for studying the association of the individual differences depicted in the functional

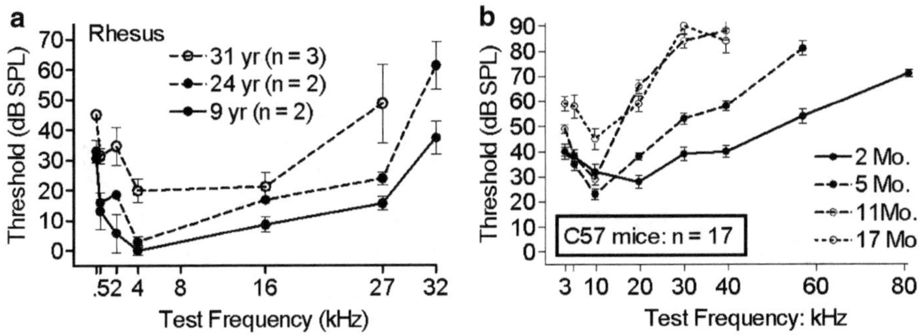

Fig. 4.3 Changes in auditory thresholds as a function of frequency of the test stimulus and age in rhesus monkeys (**a**; replotted from Bennett et al. 1983) and C57BL/6 mice (**b**; replotted from Mikaelian et al. 1974). The mice were also tested at other ages, omitted here for the sake of clarity.

data provided by Bennett et al. (1983) with the individual differences in the cochleograms reported by Hawkins et al. (1985) seems not to have attracted attention. It is important to note that the over all profile of these two sets of behavioral data was similar to that shown for humans in Figure 4.1a, as was the cochlea pathology in old animals similar to those found in temporal bones from elderly humans.

A major obstacle to the greater use of training methods for assessing auditory thresholds in older animals has been the prolonged investment in training the animal to do the task so that it is best employed in life span experiments for which animals can be trained and then tested many times over with increasing age (Mikaelian et al. 1974; Brown 1984; May et al. 2006). In contrast, the ABR measures are much more efficient, so that the entire spectrum of hearing in an anesthetized animal can be collected in ~1 hour. The validity of the ABR measures is shown in their providing an audiogram that has the same shape as the behavioral audiogram, save often for a constant offset that probably reflects the fact that behavioral thresholds benefit from greater temporal integration because behavioral test stimuli have a longer duration than ABR test stimuli. In general, ABR thresholds across species are near identical in their region of best sensitivity; this confirms the hypothesis that the higher behavioral thresholds of mice compared with those in gerbils and rats seen in Figure 4.2 resulted because it was difficult to control competing grooming behavior of the mouse, not because the mice were normally insensitive to sound. Fig. 4.4 shows ARHL for absolute thresholds measured by the ABR across age and across frequency in three laboratory animals commonly used in hearing research, the C57BL/6 mouse (adapted from Ison et al. 2007), the CBA mouse (adapted from Rivoli et al. 2005), and the Mongolian gerbil (adapted from Mills et al. 1990). The C57BL/6 and CBA mice obviously present very different profiles of ARHL. As noted in the behavioral data of Mikaelian et al. (1974) presented in Fig. 4.3, C57BL/6 mice undergo a progressive high-frequency hearing loss, but CBA mice have only a modest 10-dB hearing loss up to 20 months of age and then an additional 5- to 10-dB loss as the mice approach their median life span at ~24 months of age. The small frequency dependence of the hearing loss for the CBA mice is seen in the

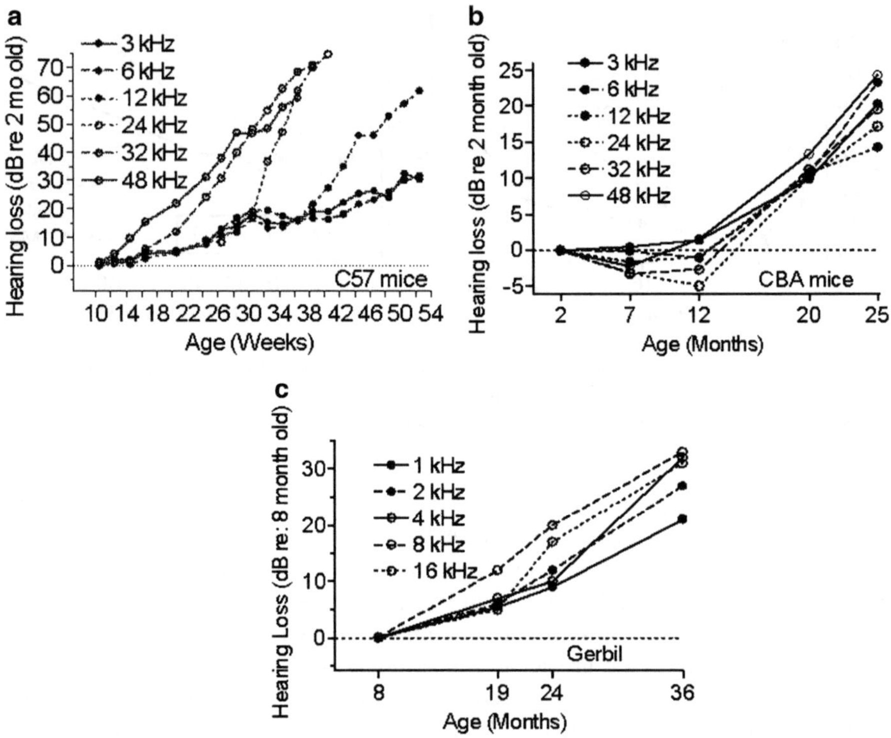

Fig. 4.4 Hearing loss with advancing age provided by changes in auditory brainstem response thresholds across frequency. (**a**) Longitudinal study of the C57 mouse (replotted from Ison et al. 2007). (**b**) Cross-sectional study of the CBA mouse (replotted from Rivoli et al. 2005). (**c**) Cross-sectional study of the gerbil (replotted from Mills et al. 1990).

comparison of the only slightly greater threshold loss for high- and low-frequency hearing compared with mid-frequencies. Erway et al. (1993) examined the ABR for 5 inbred mouse strains and all 10 of the F1 hybrid mice at different ages and found that the CBA mice had none of the three recessive alleles that lead to early ARHL, whereas the C57BL/6 mice had one such allele.

4.4 Peripheral Degeneration: A Major Source of ARHL in These Rodent Models

4.4.1 Cochlea Pathology and Its Relationship to Threshold Measures of ARHL

Spongr et al. (1997) provided quantitative measures of the patterns of hair cell loss in C57BL/6 and CBA mice that correspond with their disparate patterns of hearing loss evident in Figure 4.4. Hair cell loss began at the high-frequency base of the cochlea in the 3-month-old C57BL/6 mice, then progressed towards the low-frequency apex, whereas in the 18-month-old CBA mice, there was little evidence of hair cell loss, and at 25 months, hair cell loss was largely confined to the base and to the apex. Hearing loss in the gerbil was similar to that of the CBA mouse in its modest extent compared with the C57BL/6 mouse but had more definite monotonic frequency dependence than that of the CBA mouse, with 10 dB separating their best hearing at the lowest frequency of 1 kHz from the highest at 16 kHz (Mills et al. 1990). This seemingly "high-frequency" test stimulus of 16 kHz is still in the broad range of best hearing in the gerbil (see Fig. 4.2), and so this experiment may be thought to have missed any more serious impairment for higher-frequency hearing in these animals. However, Henry et al. (1980) tested gerbils up to frequencies of 64 kHz (but only up to 2 years of age) and did not find additional evidence of ARHL at high frequencies.

Tarnowski et al. (1991) tested ARHL in gerbils from the same colony used by Mills et al. (1990) and examined both hair cell loss and compound action potentials in the auditory nerve (roughly equivalent to wave I of the ABR) of young adult and 3-year-old gerbils. The old gerbils had variable hearing loss, but as a group, the hearing loss was relatively flat at ~20 dB for 4 kHz and below, with a further increase in hearing loss of ~10 dB for the higher frequencies. Unlike the C57BL/6 mouse (Spongr et al. 1997), the domestic dog (Shimada et al. 1998), or the monkey (Hawkins et al. 1985) but more like the CBA mouse (Spongr et al. 1997) and the guinea pig (Ingham et al. 1999), the old gerbil had its hair cell loss most pronounced at the apex, then next at the base, and only modest loss in the middle turns. Tarnowski et al. (1991) reported that although there was no precise correspondence between the frequencies that were lost with age and the regional hair cell loss as there was in the C57BL/6 and CBA mice, the overall loss of hearing across individuals was correlated with their overall loss of hair cells. Later studies by this group confirmed the conclusion of Tarnowski et al. (1991) that age-related degenerative changes in the stria vascularis were primarily responsible for hearing loss in the gerbil (Schulte and Schmiedt 1992). Schmiedt (1993) showed that strial degeneration in the old gerbil produced a reduction in the endocochlear potential (EP), which serves as the electrochemical "battery" for outer hair cell (OHC) activity. He concluded that the reduction in the EP was responsible for reduced OHC activity in response to acoustic input, which in turn reduced sensory-neural activity in the inner hair cells and thus activity in the auditory nerve (see Schmiedt, Chapter 2, for further details).

In contrast to the gerbil, the EP of the C57BL/6 mouse does not change over the period of rapid hearing loss (Lang et al. 2002; Ohlemiller et al. 2006), although McCulloch and Tempel (2004) suggest that changing the ionic composition of the EP may also impair its function. And although the C57BL/6 mouse appears to provide a very different animal model than the gerbil in its most obvious faster onset and greater severity of high-frequency hearing loss as shown in Fig. 4.4, these ABR threshold data indicate the middle-aged C57BL/6 mouse appears to present the two developmental profiles simultaneously, so that the slow rate of development of ARHL for relatively low-frequency test stimuli gives way progressively to the high rate of ARHL for the relatively high-frequency stimuli. Consistent with these observations, Hequembourg and Liberman (2001) found that there are two types of cochlear degeneration in the C57BL/6 mouse, one in the spiral ligament that is associated in time with a slowly developing non-frequency-specific hearing loss and a later degeneration of regional OHC loss that is associated with a more severe frequency-specific loss of hearing. Other investigators have also found degenerative changes in the stria vascularis of the C57BL/6 mouse (e.g., Ichimiya et al. 2000; Di Girolamo et al. 2001). Considering all these data, it may be very reasonable to use this mouse model to study simultaneously both common phenotypes of ARHL.

There is one other evoked response called an "otoacoustic emission" that is now the basis of a frequently used test of hearing in human infants and animal subjects. The test is based on the discovery by Kemp (1978) that weak sounds that originate in the inner ear can be recorded in the outer ear canal, with delays of several milliseconds after the presentation of an acoustic stimulus. These emissions are the result of an active increase in basilar membrane vibration as the OHCs contract and relax in phase with acoustic input, and their spectral frequencies as recorded in the outer ear canal include not simply the presented tone pips but also the frequency of the several distortion products produced by the OHCs. One particular large "distortion product otoacoustic emission" (DPOAE) is generated when two tones, f_1 and a higher f_2, are simultaneously presented, and the DPOAE has the frequency $[2f_1 - f_2]$. This is the DPOAE that is the primary focus of hearing tests because it is understood to be a valid index of OHC activity. Age-related declines in DPOAEs have been observed in older laboratory animals, most extensively by Jimenez et al. (1999) who measured DPOAEs in four mouse strains (including CBA and the C57BL/6) and showed the correspondence of these measures with the different degrees of susceptibility to ARHL of these mice. Guimaraes et al. (2004) observed that female CBA mice maintained better DPOAE levels with advancing age than males, even though their samples of male and female mice had no differences in ABR thresholds. Loss of DPOAEs has also been observed in middle-aged rhesus monkeys by Torre and Fowler (2000) and in middle-aged chinchillas by McFadden et al. (1997). The major benefits of DPOAE measures are that they measure OHC function in intact organisms, and in combination with the ABR, they provide an analysis of the relative contributions of OHCs, inner hair cells, and the auditory nerve fibers to ARHL. DPOAEs have also been studied in elderly human listeners, most often to find that the loss of absolute thresholds is primarily correlated with OHC loss (Oeken et al. 2000).

4.4.2 Suprathreshold Measures of ARHL

The ABR evoked by suprathreshold stimuli has been used to test the hypothesis that age and/or hearing loss alters central efficiency as measured by age-related changes in peak amplitude or in the latencies and interpeak intervals that are believed to represent ascending levels of brainstem processing. In their review of these data, Tremblay and Burkard (2007) conclude that these suprathreshold measures of amplitude and latency may be sensitive to age-related differences in peripheral sensitivity and central processing efficiency. One effect of a loss in audibility is a slowing of the first wave generated at the level of the auditory nerve, and the subsequent waves may be affected by both intrinsic central delays and by their being "inherited" from the auditory nerve. Central changes have been hypothesized to be more noticeable when very high rates of stimulation are used, much higher than the usual ABR rate of a constant 10–20 brief tone pips or clicks in each second. A special presentation pattern called the maximum length sequence (MLS) allows for very fast repetition rates in quasi-random sequences of brief stimuli interspersed with short periods of silence, so that linear effects (for isolated stimuli) or nonlinear effects (for dyads or triads of stimuli) can be recorded. Using MLS, Burkard and Sims (2001) were able to test subjects with modest hearing loss using click rates as high as 500/s and found that both wave I and wave V were delayed by ~0.1 ms in elderly subjects, with no age difference in the central delay between waves I and V.

A comparison of the age effects for a regular presentation ABR and a MLS ABR for click rates up to 250/s was provided by Lavoie et al (2008), with results again suggesting that ARHL can be observed in middle-aged humans. These investigators studied three groups of women: young women and girls, a young middle-aged group, and an elderly group with a range of just 44 to 62 years of age that was much younger than most elderly groups of subjects. This last group had only a small (but significant) hearing loss at 8 kHz compared with the youngest group. In the regular ABR, the oldest group showed a significant amplitude decrement on wave V compared with the younger groups (perhaps a sign of their small but real high-frequency hearing loss), whereas in the more rapid presentation of the MLS condition, the middle group had a smaller wave I compared with the youngest group. The authors also reported that the interval between the nonlinear waves I and V was longer for the oldest group compared with the middle group. Overall, these data support three conclusions: age effects on AEPs may be most apparent when the CANS is stressed by high repetition rates, central as well as peripheral processing is disrupted with age, and age effects can be observed in relatively young middle-aged listeners. A research program that repeated these age-sensitive MLS tests in a battery of standard psychophysical measures might illuminate the types of functional deficits that correlate with these apparent deficits in very rapid central processing.

The results obtained with conventional ABR methods in laboratory animals reinforce the conclusions provided by human subjects that when age effects are found on ABR amplitude and latency measures, they can most parsimoniously be attributed to peripheral hearing loss (in the old cat, Harrison and Buchwald 1982; in C57BL/6 and CBA mice, Hunter and Willott 1987; in the old gerbil, Boettcher et al. 1993; in the guinea pig, Ingham et al. 1999). As yet, there seems to be no study on the effects of MLS presentation in laboratory animals, and this represents an important gap in the comparative literature that attempts to link the effects found in humans with those of animals. But it is also important to keep in mind that both ABR methods are selective measures of fast synchronous activity in response to simple acoustic transients and that all of this neural activity is completed within just 10 ms of the onset of the stimulus. Thus although the ABR provides a very useful noninvasive measure of peripheral hearing loss and additional insights concerning brainstem processing, its scope is limited even for brainstem activity and the ABR cannot illuminate function in more rostral regions.

4.5 Central Processing Deficits: Temporal, Spectral, and Spatial Dimensions

Speech signals are complex acoustic tokens that undergo rapid spectral and amplitude modulations over time, and it is reasonable to hypothesize, as many have, that a diminished ability to track these modulations must affect speech perception. The psychophysical literature reveals that temporal acuity and frequency resolution are affected by both hearing loss and advancing age, with the effect of age increasing with stimulus complexity and background conditions. An example is gap detection, especially when combined with a spectral shift, if the task has an added fluctuating background, or when signals and noise are presented from different locations (see Fitzgibbons and Gordon-Salant, Chapter 5).

4.5.1 Neurobiological Measures of Complex Processing

The neurobiological literature concerning age effects on more complex spectrotemporal and spatial auditory processing is limited. As described previously, two single-unit studies that focused on onset neurons in the IC of young vs. old CBA mice (Walton et al. 1998) and young vs. middle-aged C57BL/6 mice (Walton et al. 2008) showed that neurons in the old CBA mouse had poorer gap thresholds and slower recovery functions than in the young mouse, whereas the cells of the middle-aged C57BL/6 mouse, with more hearing loss than the old CBA mouse, were not different in these measures of temporal acuity from those of the young C57BL/6 mouse. Finlayson (2002) measured recovery functions to pairs of tone bursts in single units of the IC of young adult and old rats with minimal hearing loss and reported that the initial suppressive effect of the first stimulus on the second was not affected by age, but the subsequent recovery functions were ~50% delayed in the old rats. A more complex method of assessing temporal resolution is to present a stimulus that does not have a single gap but consists of a series of amplitude-modulated (AM) waves that can be varied over time in their frequency or depth of modulation. Walton et al. (2002) measured single-unit activity in the IC of young and old CBA mice in response to AM stimuli and discovered that the younger cells were able to respond to faster rates. One additional finding in this experiment was that the older CBA mice had more vigorous neural responses than the young mice, this being understood as another instance of central auditory hyperactivity associated with advancing age. In a similar experiment that was focused on spectral rather that amplitude modulation, Mendelson and Ricketts (2001) measured single-unit activity in the rat auditory cortex in response to frequency-modulated (FM) sweeps that changed in speed and extent from trial to trial in young adult and old rats. Differences that could be due to audibility were minimized by eliminating any animal with ABR thresholds more than 10-dB SPL above that of the average young rat and then by presenting all stimuli at 30 dB above threshold for each of the tested single units (this being very similar to the protocol of Nabelek and Robinson 1983 that was described above). These authors reported that the majority of neurons of young rats responded best to faster rates of modulation than did the neurons in old rats, the majority of which responded best to slow rates of modulation.

There are just two published neurobiological studies of the effects of age on spatial location and on binaural unmasking, both from the same laboratory and both showing a spatial deficit in neurons of the IC in the middle-aged C57BL/6 mouse with high-frequency ARHL (McFadden and Willott 1994a, b). Specifically, they discovered a loss of directional sensitivity to best-frequency tone pips in the middle-aged mouse, a greater masking effect overall, and no benefit from providing a greater separation between the locations of the signal and the masking noise. It should be noted that mice depend on very high frequency hearing for distinguishing between sound source locations (Heffner et al. 2001), and these critical frequencies were no longer audible to the middle-aged C57BL/6 mouse.

4.5.2 AEP Studies of Complex Auditory Processing

The AEPs recorded at intervals of ~50 to 300 ms after stimulus onset are generated rostral to the brainstem, in the thalamus, in the thalamocortical pathways, and in the auditory cortex, and they are reliably evoked by stimuli having a complex time-varying spectral structure. One of the passive listening tasks is called "mismatch negativity" (MMN), which is seen in the AEP as a negative response that follows N1 ~200 to 300 ms after the presentation of a deviant stimulus (the "mis-match") that is occasionally presented in a series of otherwise identical sounds. In a study of gap detection, Bertoli et al. (2002) compared elderly humans with near-normal hearing up to 3 kHz (mean age, 72 years) with young subjects (mean age, 26 years) on an active psychophysical task and in a passive listening MMN task in which the AEPs to test tones were recorded while the subjects read a book. Most of these tones (85%) had no gap, whereas other tones (15%) contained brief quiet gaps having durations of 6 to 24 ms. In the psychophysical task, the mean gap thresholds were 7.8 ms for the elderly participants and 6.4 ms for the young participants, not quite a significant difference (the 2-tailed $p = 0.09$), whereas in the passive MMN procedure, the gap thresholds were significantly different, 15 ms for the elderly and 9 ms for the young participant. The elderly listeners with a measurable MMN had smaller amplitudes and longer latencies than the young listeners, but four elderly subjects and one young subject had no measurable MMN for any gap duration.

Gap detection depends on high-frequency spectral components of the markers for the gap in humans (Snell et al. 1994) and the large failure rate in the MMN task found in these elderly subjects suggests the possibility that elderly listeners with high-frequency hearing loss can at least partially compensate for this loss with focused attention in the psychophysical task but not in passive MMN detection.

Tremblay et al. (2003) studied the effects of age on the P1, N1, and P2 AEP responses evoked by an ordered sequence of gap durations within speech syllables, with the AEP method accompanied by a psychophysical test. These seven syllables varied in voice-onset time (VOT) between /ba/ and /pa/, the extremes having a VOT of 0 and 60 ms, respectively. Three groups were tested, young listeners (mean = 26 years), older listeners (mean = 68 years) with preserved hearing within 10 dB of the younger group, and older listeners (mean = 72 years) with high-frequency hearing loss, their thresholds being 50 dB above those of the young at 8 kHz. The psychophysical ability of the subjects to discriminate the different VOT cues was best in the young group, next in the elderly group with preserved hearing sensitivity, and poorest in the elderly listeners with hearing impairment. For the AEP measures, the latency of P1 did not differ among the groups and neither did its amplitude, but the amplitude of N1 was increased with hearing loss in the elderly hearing-impaired group, another apparent example of central hyperreactivity associated with hearing loss. Both elderly groups showed longer N1 latencies for the more delayed VOT stimuli in comparison to the young group, and P2 peaks were delayed for both elderly groups across the entire range of VOT times. These results indicate that the neural encoding of VOT is related to age alone, whereas the encoding necessary for perception is affected by both age and hearing loss. Speech perception is critically dependent on VOT, and the delayed response in older adults with mild hearing loss may explain their difficulties in understanding speech in difficult listening situations.

Another temporal cue that has been studied in older adults is the interaural phase difference (IPD), a cue that contributes to the perception of sound location. Using magnetoencephalography (a procedure that measures an auditory evoked magnetic field generated in the brain, yielding an AEF rather than an AEP), Ross et al. (2007) measured the P1-N1-P2 complex to assess the effects of aging on the physiological capacity to detect interaural timing cues Fig. 4.5. The passive recording paradigm included stimuli being presented through insert earphones that included a change in IPDs within a succession of simple stimulus onsets and offsets while the participants watched a silent movie. Three age groups were tested in this experiment, young adults (mean = 26.8 years), a middle-aged group (mean = 50.8 years), and an elderly group (mean = 71.4 years). The middle-aged mean hearing thresholds were no more than 5 to 10 dB poorer than those of the young group, whereas the old group had a relative hearing loss of ~10 to 15 dB for the low frequencies up to 1,000 Hz, rising to 50 dB at 8 kHz. The stimuli were 40-Hz AM signals. For the first 2 seconds, the stimuli were diotic, i.e., the two ears received the same stimulus in phase, then for the last 2 seconds the presentation was dichotic, with the tone in one ear 180° out of phase with the other ear. The active psychophysical task consisted of listening to pairs of 1-second-long diotic and dichotic stimuli at 1 IPD presented in semirandom order and choosing which stimulus was separated between

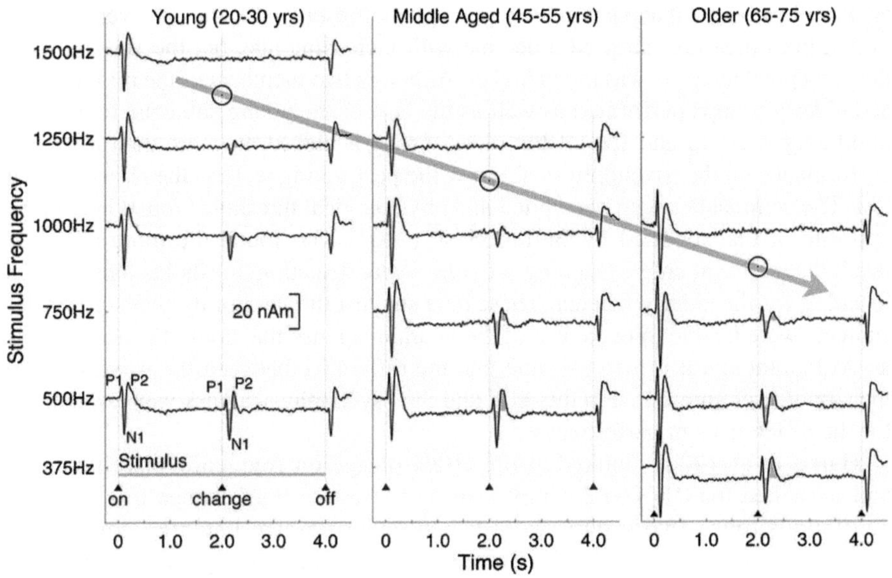

Fig. 4.5 Grand averaged auditory evoked fields (AEFs) for 3 age groups and for each test frequency. At the lowest frequency, all age groups provided an AEF to the phase shift that occurred at 2 s, but then the amplitude of the response diminished with increasing frequency and disappeared near each behavioral threshold frequency (circles connected by the arrow). (Reprinted from Ross et al. 2007, Fig. 3, with permission.)

the 2 ears (i.e., the dichotic presentation). This session began with the longest IPD and then adaptively converged on a threshold.

The AEF data are shown in Fig. 4.5. It is evident that clearly defined onset and offset responses were present and equal in all age groups and for all frequencies, suggesting that there were no age differences in the ability to detect the onset of the AM stimuli. The onset response was in fact largest in the oldest group, and this appears to be another example of hyperreactivity in elderly subjects. However, greater responsivity may also have resulted because these stimuli were presented at equal decibel SL for all groups: it is possible that (here and in some other studies) using SL as the reference level may result in a greater perceived loudness in the hearing-impaired group because of loudness recruitment. All age groups had their most prominent AEF amplitudes when the IPD was introduced for the lowest frequency. The amplitude of this response declined with an increase in tonal frequency, most rapidly in the oldest group with their threshold for responding between 750 and 1,000 Hz, more slowly in the middle-aged group, with their threshold between 1,000 and 1,250 Hz, and most slowly in the youngest group that overall was the most sensitive to these stimuli, with the group threshold between 1,250 and 1,500 Hz. In addition to these effects, old age had a much stronger impact on the latency measures for this IPD change condition, with P2 being especially delayed for the elderly group beyond the range of the younger groups.

The physiological and the median perceptual thresholds for the IPD were similar in that both measures showed a decline with increasing age, but the variability in the perceptual measure was much higher. Although two members of the middle-aged and elderly groups performed as well as the best of the young subjects, four of the middle-aged group and five of the oldest group achieved no more than a chance performance on the psychophysical test while performing well on the physiological test. The contrast between perceptual and physiological thresholds found here is the opposite of that reported by Bertoli et al. (2002) who found the more expected result that active attentive listening provides lower detection thresholds than passive listening for the elderly listener. These data suggest the possibility that some older subjects were unable to make use of the neural response that could be measured as an AEF, although it is also possible that the difference between the way in which the stimuli were presented in the AEF and the psychophysical task was responsible for the differences in performance.

Harris et al. (2008) looked at the effect of age on frequency discrimination and measured the P1-N1-P2 complex in response to a brief change in a standard continuous tone. Two groups were tested under passive listening conditions: a young group ranging from 18 to 30 years of age and an older group ranging from 65 to 80 years of age. There were two standard frequencies: 500 and 3,000 Hz. Pure-tone thresholds in the groups were nearly identical up to 3 kHz. As given by the AEP data, listeners in the younger group were able to detect smaller frequency excursions and, furthermore, the difference favoring the young group was greater for the 500-Hz standard than the 3,000-Hz standard: in the younger group, the relative threshold (i.e., $\Delta F/F$) was lower for the 500-Hz standard, whereas in the older group, the relative threshold was lower for the 3,000-Hz standard. This effect had been found previously in a psychophysical study from the same laboratory (He et al. 2007), where it had been reasonably interpreted as revealing a deficit in temporal processing in the elderly listeners who were less able to make use of changes in the small differences in the fine structure of the low-frequency standard. This interpretation is similar to that of Ross et al. (2007) for the effect of age on encoding an IPD. Similar age effects were evident as well in latency and amplitude measures, with longer latencies in the older group and also higher amplitudes at 500 Hz but lower amplitudes at 3,000 Hz.

4.5.3 Behavioral Studies of Complex Auditory Processing

Behavioral studies in old animals parallel these AEP studies in humans, including the effects of increased age on gap detection, on spatial location and spatial release from masking, and, finally, on frequency selectivity. Barsz et al. (2002) described a set of three comparable experiments concerned with gap detection that fits very well with the theme of this chapter, providing comparisons of psychophysical gap thresholds in young and old human listeners (young between 17 and 40 years of age and old between 61 and 82 years of age); behavioral gap thresholds in young and old CBA mice (young mice between 2 and 3 months compared with old mice between 24 and 25 months); and electrophysiological gap thresholds in phasic IC neurons from young CBA mice (between 2 and 4 months) and old CBA mice (between 25 and 30 months old). The details of the procedures are presented in Barsz et al. (2002): humans were tested in a standard adaptive procedure to determine thresholds, the mice were tested in a reflex modification procedure, and the mouse neurons were tested with brief gaps placed near the center of noise bursts. The mean gap thresholds for young vs. old humans were 2.6 vs. 3.7 ms; the thresholds for young vs. old mice were 2.9 vs. 4.9 ms; and the thresholds for young vs. old IC cells were 2.7 vs. 26 ms. All three of the gap threshold age comparisons were statistically significant. Absolute hearing thresholds were also significantly different between young and old subjects, both mice and humans, in each case better in the younger group. Hearing thresholds for the human listeners and response thresholds for the cells were not significantly correlated with gap thresholds, but for young and old mice combined, the correlation was significant, $r = 0.45$; i.e., the greater the hearing loss, then the higher was the gap threshold. A second regression analysis performed after the effect of absolute sensitivity was removed from the data revealed a significant effect of increased age on the gap threshold that was independent of hearing loss.

Turning from gap detection to sound localization, Heffner et al. (2001) reported that middle-aged C57BL/6 mice with high-frequency hearing loss were less able to discriminate between sounds from different locations in young mice; this is consistent with the observations of McFadden and Willott (1994a, b) that in contrast to young C57BL/6 mice, single cells in the IC of middle-aged C57BL/6 mice are less sensitive to differences in sound source along the azimuth and also show less benefit in signal detection from increasing the spatial separation between signals and maskers. In mice, sound location primarily depends on the interaural level differences that are provided only by high-frequency stimuli, and these are no longer available to the middle-aged C57BL mouse.

It is thus of interest to determine whether an old mouse with maintained high-frequency hearing would show these deficits. Ison and Agrawal (1998) examined the effect of having a signal and its masker presented from the same location or separated 180° apart in the free field as a function of signal frequency and level (4 and 25 kHz, presented at 10-dB intervals between 30 and 80 dB) and of age (4 vs. 20 months). The masker was a 1-octave narrow-band noise centered on either 4 or 25 kHz and presented at 50-dB SPL. The mice were the F_1 hybrid offspring of a CBA male and a C57BL/6 female, mice known to have less hearing loss with advancing age than even the CBA parent. Compared with the young mice, the older mice had a greater ABR hearing loss of just 5 dB at 24 kHz, ~12 dB at 4 kHz, and near 0 dB at 20 kHz. The effectiveness of the signals was measured in their inhibition of the startle reflex elicited by a noise burst. The effect of spatial separation depended on the stimulus and the masker frequency, so that the inhibitory effect of the 24-kHz stimulus was greater when this signal and its 24-kHz masker were separated in space, whereas the effect of the 4-kHz stimulus was not affected by the relative position of this stimulus from its 4-kHz masker. It is noteworthy that these effects were the same in young and old mice, indicating that in the absence of peripheral hearing loss, there was no loss of spatial localization ability. It is generally thought that this benefit of spatially separating the masker from high-frequency signals results because the subject, usually a human rather than a mouse, is able to selectively attend to the ear with the better signal-to-noise ratio, and the results of this experiment suggest that these older mice had maintained this ability. The data are similar to those provided by Gelfand et al. (1988) showing that aged human listeners with good high-frequency hearing benefit from the separation of noise from signal as much as the young do.

In contrast to the mouse, the rat has a more distinct and substantially larger medial superior olivary nucleus (Harrison and Irving 1966) and so may be expected to make better use of IPDs to locate low-frequency stimuli. Brown (1984) studied the trained performance of rats when they pressed bars to the right or left of a center orienting bar. Which bar was pressed depended on the location of a brief noise pulse from a speaker that was 1 meter to the left or right of the orienting bar. Brown used a longitudinal life-span experimental design in which he began training the rats at 3 months of age and continued testing for 5 times/week (save for school vacations) until they neared the end of their life span at 21 months of age. Their discrimination performance was stable from 10 to 14 or 15 months of age, averaging ~90% correct responses, and then steadily declined as they approached 21 months of age to an average below 70%. In contrast, performance was maintained over this time interval at 100% success in a visual discrimination task. The author noted that these rats would have little absolute threshold loss (which agrees with the more recent data of Stenqvist (2000) that were obtained in the same strain) and thus concluded that the performance decrement resulted from a decline in the ability to use binaural timing cues. He pointed out that this conclusion was consistent with the histopathology data in the auditory nerve, cochlear nucleus, and superior olivary complex observed in the rat by Feldman and his associates (see section 2.2 above) and consistent also with the results of a study of elderly human listeners by Herman et al. (1977).

May et al. (2006) provided a study of changes in the auditory processing of spectral cues in old mice that is also important because it is another of these rare longitudinal studies of "life-span" ARHL. These authors began training and testing a group of mice at ~1 month of age and then continued the experiment until the mice were close to 30 months, i.e., beyond the average life span of mice. The specific rationale was based on the possibility that age may increase the width of the spectral filter in mice as had been previously shown in human listeners by Patterson et al. (1982). This phenomenon seems particularly useful for understanding the problems of signal detection in noise (which is one of the signature complaints of the elderly listener) because a filter centered on a particular signal is also sensitive to the immediately surrounding noise, which will serve to mask the signal; thus a wider filter must allow greater masking of a central signal in the presence of broadband noise. But although auditory filters are typically conceived as a peripheral mechanism that, e.g., occupies a particular swath on the basilar membrane, filter bandwidth can also be affected by a central cholinergic-based mechanism as the previously described experiment of Pickles and Comis (1973) has shown. May et al. (2006)

measured the width of the auditory filter by looking at the degree of masking of a tone by a flat broadband noise and measured the relief from masking provided by inserting quiet notches into the noise centered on the signal. Tracking the loss of masking with the increasing width of the notch provides an estimate of the filter width, given well-established algorithms developed in the human psychophysical laboratory. Fig. 4.6 presents the data for masking and derived filter shape as a function of the gap width at 2 ages, the baseline taken when the mice were less than 12 months and the final data when the mice were over 24 months of age. For the 11.2-kHz test tone, masking increased by ~10 dB overall as a function of age, but for the 16-kHz test tone the initial 15-dB masking effect increased as the notch widened from near 0 to 50% of its center frequency. The effect seen at 11.2 kHz may be an indication that the efficiency of signal-to-noise processing has weakened in the old

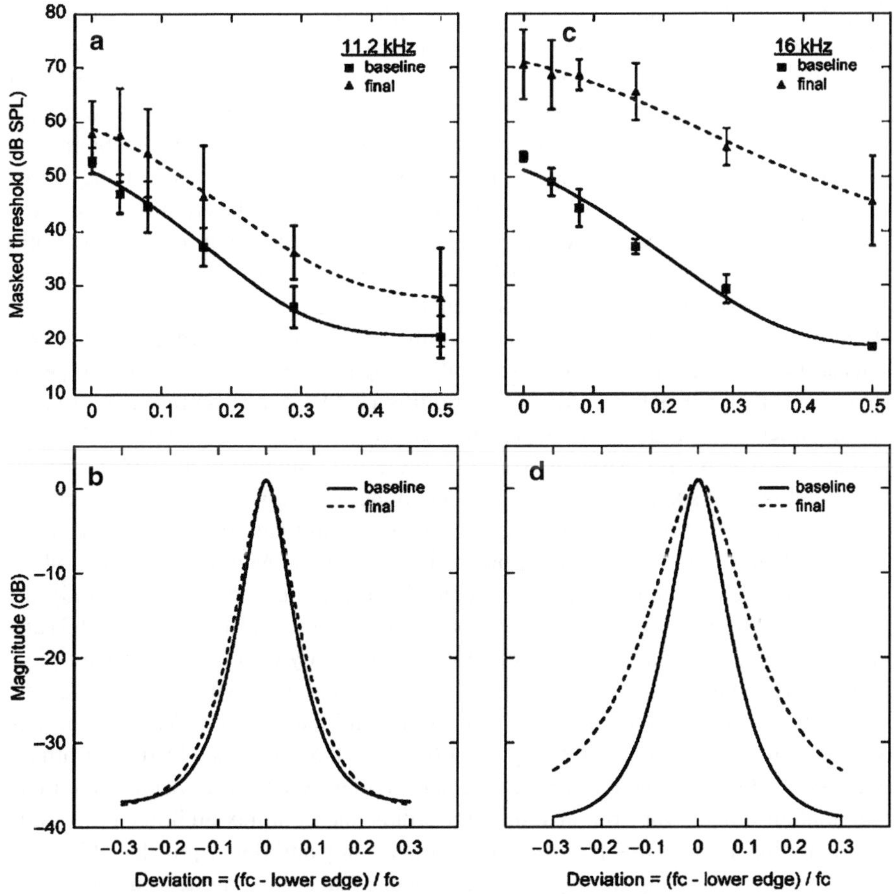

Fig. 4.6 Masked threshold in groups of mice tested when young (baseline) and near the end of their life span (final) at 11.2 (**a**) and 16 kHz (**c**). (**b** and **d**) Calculated filter shapes at these 2 ages. (Reprinted from May et al. 2006, Fig. 7, with permission.)

mouse, but the shape of the auditory filter has not (Patterson et al. 1982), whereas the effect at 16 kHz can be interpreted as showing both a loss of efficiency and a widening of the filter in the near-senescent mouse. It is of additional interest that changes in the filter shape did not correlate with the severity of absolute hearing loss, as the mouse that showed the largest degree of hearing loss at 16 kHz had a well-preserved filter shape for that frequency, whereas a mouse that had an extreme increase in filter width had only an average threshold change.

4.6 A Summary of Past Research and Its Implications for Moving Forward

Different animal species have been variously proposed as being the most appropriate animal model for human ARHL based on different criteria. Factored into these decisions may be a set of practical criteria favoring rodents with their relatively short life span that are reasonably inexpensive to acquire and maintain and that remain healthy in a vivarium setting. Another criterion is the similarity between the appearance of ARHL in the animal model and the common phenotypes obtained in human studies of ARHL, although there is some disagreement about the "true" phenotype of pure ARHL. As described in section 1.3 above, although many investigators would accept the pattern of steeply rising hearing loss for the high-frequency hearing seen in Figure 4.1A as representative of human ARHL, others point to the potential confound between chronological age and the cumulative effect of noise exposure, e.g., that may contribute to high-frequency hearing loss. This approach suggests that the less sharply rising profiles of hearing loss that were also present in the data of Allen and Eddins (2009) better represent biological aging. Other criteria reflect the similarity between the animal's genome and that of humans, in which case primates would be preferred (Bennett et al. 1983), or their having a long life span because some effects of aging might well differ with chronological rather than biological time; this favors an animal such as the cat or the chinchilla that have life spans approaching 20 years (McFadden et al. 1997). And most recently, new criteria to be considered are the opportunity to study the genetics of ARHL in inbred and hybrid strains of mice (e.g., Erway et al. 1993) and the relative ease of manipulating the mouse genome by genetic engineering (e.g., McCullough and Tempel 2004). This is no doubt the reason why the mouse model has in the last decade become the most preferred animal model of ARHL, although economy of upkeep and a relatively short life span have always added to its attractiveness. Fortunately, the basic phenomena of human ARHL have been found in all mammalian models for both absolute thresholds and complex auditory processing and including animals that provide both sharply rising and flatter profiles of hearing loss. This is a reassuring finding because it means that animal models can be chosen freely for their best fit to the needs of hypothesis testing; thus a hypothesis about the effect of high-frequency hair cell loss on central tonotopic reorganization could be tested in the C57BL/6 mouse (Willott 1986), a hypothesis about the interaction of age and noise exposure could be tested in the CBA mouse (Kujawa and Liberman 2006), and a hypothesis concerning the effects of age on sensitivity to interaural time differences and spatial localization could use the rat (Brown 1984) or the gerbil in an extension of the research on gerbils by Heffner and Heffner (1988).

Most impressive in both the animal and the human literature is that all of the functional data show substantial variation among individuals in hearing ability within an age group, perhaps not surprising in humans because of the great diversity in genetic background and experience in our species but also present in at least a reduced form between inbred animals of the same age that have been maintained in well-regulated environments (Ison et al. 2007). This pattern of variation suggests there may be significant effects on hearing of seemingly insignificant differences in the pre- or postnatal environment and raises as well the possibility of different patterns of epigenetic gene expression even between animals with seemingly identical genes. Not surprisingly, most experimental reports in hearing science focus on the significant differences between different age groups rather than the presence of individual differences within a group, but some researchers have shown that it is possible to use these individual differences to advantage. There are exemplary attempts to address the problem of individual differences in the animal laboratory by looking for their correlates in other neurobiological indices. For example, Tarnowski et al. (1991) examined the association between threshold measures at specific frequencies and counts of regional hair cells in a large group of old gerbils with varied degrees of hearing loss. These authors concluded that overall differences in thresholds among animals were correlated with their overall hair cell loss, and careful readers of their very informative figures will note also that the gerbils that were outliers in the degree of frequency-specific hearing loss appeared also to be outliers in their site-specific hair cell loss. Ison et al. (2007) were able to correlate individual differences in age-related low-frequency hyperreflexia in the aging inbred C57BL/6 mice with their degree of high-frequency hearing loss, suggesting that the degree of peripheral hearing loss was in part responsible for their individual differences in this abnormal behavior. Hequembourg and Liberman (2001) found that degeneration of fibrocytes in the spiral ligament was associated with a small decline in the ABR and predated the loss of hair cells in the C57BL/6 mice associated with the rapid progression of hearing loss, raising the interesting hypothesis that the composition of the cochlear endolymph may have some interactive role in the genetically determined progression of hair cell loss in this mouse strain. The lesson of these examples is that unique insights can be realized in multidisciplinary research projects that have a broad range of end points, including the correlations between and within functional and neurobiological measures that may then suggest causative hypotheses to be tested. Individual differences in neurobiological measures have only infrequently been noted, but it would be of great benefit to document their extent and their co-occurrence with individual differences in functional end points, as searched for in a study of human temporal bones by Nelson and Hinojosa (2006).

The major goal of this chapter was to provide the behavioral link between auditory psychophysics in aging humans and neurobiology research in aging animals. This link has been established most strongly for the deterioration of absolute thresholds in aging animals and aging human listeners, in some measure because many of the indices of cochlear pathology seen in animals are also open to investigation in human temporal bone specimens, and there are many studies of absolute thresholds in both humans and animals. The link has been less well developed between neurobiology and age-related deficits in complex auditory processing in humans and animals because the range of neurobiological observations available in humans is as yet much less developed than it is in animal models, whereas in contrast, the range of complex tasks completed in studies with human listeners is much greater than has been available for animal models. There are well-documented neurobiological indices of both peripheral and central deterioration with advancing age in animals in neurochemistry, connectivity, synaptic counts, and gene expression, but the evidence for similar effects of aging on the central auditory system in humans is at present limited and the further development of noninvasive methods for studying these end points in both animals and in humans would be very useful.

It is also clear that certain conditions accelerate ARHL (or perhaps mimic ARHL), e.g., exposure to loud noises and ototoxic drugs and systemic pharmacological manipulations and brain-lesion effects, all of which can also be demonstrated in animal models and observed in human research participants or, sometimes, patients. But it must be acknowledged that many of the neurobiological and functional studies of aging in animals and in humans have not been sufficiently well coordinated to foster integration across disciplinary lines or across species. A full accounting of presbycusis is best approached in a multidisciplinary and comparative research program in which data for multiple functional and multiple neurobiological end points are gathered not just in the same species or strains but in the same individual subjects to more directly establish the association between neurobiology and auditory processing. The search for an explanation of individual differences in the effects of age on auditory ability must rest on the assumption that their foundation is in individual differences in the neurobiological substrate.

Further advances may profit from borrowing a research design from functional programs that accumulate data for each participant on many auditory tests to study their pattern of intercorrelations (e.g., Humes 2005). At present, the results of "neurobiological tests" are typically reported separately for different end points and different neural sites, no doubt because each test result has required a major investment in time and resources. But it seems most sensible in the long-term plan to map out, e.g., how age-related neurobiological changes in one nucleus are correlated with changes in other nuclei in that same animal and whether any individual differences in these measures are correlated with individual changes in functional measures, again in the same animal. A program of this sort would depend on there being much more interaction between researchers in different disciplines than is common at present. But it is reasonable to think that translational research programs will advance most rapidly when multiple measures are taken in the same individual subjects, both animals and humans, to directly examine the empirical association between neurobiological and functional variables. A better understanding of individual differences in ARHL will help to suggest the types of interventions that will result in a beneficial functional outcome.

Acknowledgements This work was supported in part by National Institutes of Health Grants DC 007705 from the National Institute on Deafness and Communication Disorders to KLT and AG 09524 from the Institute on Aging and DC 05409 from the Institute on Deafness and Communication Disorders to JRI and PDA.

References

Allen PD, Eddins DA (2009) Cluster analysis reveals presbycusis phenotypes that group subjects by degree and configuration of hearing loss. Assoc Res Otolaryngol Abstr 32:140.

Allen PD, Bell J, Dargani N, Moore CA, Tyler CM, Ison JR (2003) kvcn1 knockout mice have a profound deficit in discriminating sound source location. Soc Neurosci Abstr 29:183

Barsz K, Ison JR, Snell KB, Walton JP (2002) Behavioral and neural measures of auditory temporal acuity in aging humans and mice. Neurobiol Aging 23:565–578.

Bennett CL, Davis RT, Miller JM (1983) Demonstration of presbycusis across repeated measures in a nonhuman primate species. Behav Neurosci 97:602–607.

Bergman M, Blumenfeld VG, Cascardo D, Dash B, Levitt H, Margulies MK (1976) Age-related decrement in hearing for speech. Sampling and longitudinal studies. J Gerontol 31:533–588.

Bertoli S, Smurzynski J, Probst R (2002) Temporal resolution in young and elderly subjects as measured by mismatch negativity and a psychoacoustic gap detection task. Clin Neurophys 113:396–406.

Birch LM, Warfield D, Rubin RJ, Mikaelian DO (1968) Behavioral measurements of pure tone thresholds in normal CBA-J mice. J Aud Res 8:459–468.

Boettcher FA (2002) Presbyacusis and the auditory brainstem response. J Speech Lang Hear Res 45:1249–1261.

Boettcher FA, Mills JH, Norton BL (1993) Age-related changes in auditory evoked potentials of gerbils. I. Response amplitudes. Hear Res 71:137–145.

Bohne BA, Gruner MM, Harding GW (1990) Morphological correlates of aging in the chinchilla cochlea. Hear Res 48:79–91.

Bowen GP, Taylor MK, Lin D, Ison JR (2003) Auditory cortex lesions impair both temporal acuity and intensity discrimination in the rat, suggesting a common mechanism for sensory processing. Cerebral Cort 13:815–822.

Brant LJ, Fozard JL (1990) Age changes in pure-tone hearing thresholds in a longitudinal study of normal human aging. J Acoust Soc Am 88:813–820.

Brown CH (1984) Directional hearing in aging rats. Exp Aging Res 10:35–38.

Buchtel HA, Stewart JD (1989) Auditory agnosia: apperceptive or associative disorder? Brain Lang 37:12–25.

Burkard RF, Sims D (2001) The human auditory brainstem response to high click rates: aging effects. Am J Audiol 10:53–61.

Caine ED, Weingartner H, Ludlow CL, Cudahy EA, Wehry S (1981) Qualitative analysis of scopolamine-induced amnesia. Psychopharmacology 74:74–80.

Casey MA (1990) The effects of aging on neuron number in the rat superior olivary complex. Neurobiol Aging 11:391–394.

Casey MA, Feldman ML (1985) Aging in the rat medial nucleus of the trapezoid body. III. Alterations in capillaries. Neurobiol Aging 6:39–46.

Cody AR, Russell IJ (1987) The response of hair cells in the basal turn of the guinea-pig cochlea to tones. J Physiol 383:551–569.

Dayal VS, Bhattacharyya TK (1986) Comparative study of age-related cochlear hair cell loss. Ann Otol Rhinol Laryngol 95:510–513.

Di Girolamo S, Quaranta N, Picciotti P, Torsello A, Wolf F (2001) Age-related histopathological changes of the stria vascularis: an experimental model. Audiology 40:322–326.

Einsiedel LJ, Luff AR (1992) Alterations in the contractile properties of motor units within the ageing rat medial gastrocnemius. J Neurol Sci 112:170–177.

Erway LC, Willott JF, Archer JR, Harrison DE (1993) Genetics of age-related hearing loss in mice:I. Inbred and F1 hybrid strains. Hear Res 65:125–132

Fay RR (1988) Hearing in Vertebrates: A Psychophysics Database. Winnetka, IL: Hill-Fay Associates.

Finlayson PG (2002) Paired-tone stimuli reveal reductions and alterations in temporal processing in inferior colliculus neurons of aged animals. J Assoc Res Otolaryngol 3:321–331.

Ford JM, Roth WT, Isaacks BG, White PM, Hood SH, Pfefferbaum A (1995) Elderly men and women are less responsive to startling noises: N1, P3 and blink evidence. Biol Psychol 39:57–80.

Gates GA, Couropmitree NN, Myers RH (1999) Genetic associations in age-related hearing thresholds. Arch Otolaryngol 125:654–659.

Gelfand SA, Ross L, Miller S (1988) Sentence reception in noise from one versus two sources: effects of aging and hearing loss. J Acoust Soc Am 83:248–256.

Gleich O, Hamann I, Klump GM, Kittel M, Strutz J (2003) Boosting GABA improves impaired auditory temporal resolution in the gerbil. NeuroReport 14:1877–1880.

Gordon-Salant S, Fitzgibbons PJ (1995) Recognition of multiply degraded speech by young and elderly listeners. J Speech Hear Res 38:1150–1156.

Gratton MA, Schulte BA (1995) Alterations in microvasculature are associated with atrophy of the stria vascularis in quiet-aged gerbils. Hear Res 82:44–52.

Guimaraes P, Zhu X, Cannon T, Kim S, Frisina RD (2004) Sex differences in distortion product otoacoustic emissions as a function of age in CBA mice. Hear Res 192:83–89.

Halling DC, Humes LE (2000) Factors affecting the recognition of reverberant speech by elderly listeners. J. Speech Lang. & Hear. Res. 43:414–431.

Harris KC, Mills JH, He N-J, Dubno JR (2008) Age-related differences in sensitivity to small changes in frequency assessed with cortical evoked potentials. Hear Res 243:47–56.

Harrison J, Buchwald J (1982) Auditory brainstem responses in the aged cat. Neurobiol Aging 3:163–171.

Harrison JM, Irving R (1966) Visual and nonvisual auditory systems in mammals. Anatomical evidence indicates two kinds of auditory pathways and suggests two kinds of hearing in mammals. Science 154:738–743.

Hawkins JEJ, Miller JM, Rouse RC, Davis JA, Rarey K (1985) Inner ear histopathology in aging rhesus monkeys (*Macaca mulatta*). In: Davis RT, Leathers CW (eds) Behavior and Pathology of Aging in Rhesus Monkeys. New York: Alan R. Liss, pp. 137–154.

He NJ, Mills JH, Dubno JR (2007) Frequency modulation detection: effects of age, psychophysical method, and modulation waveform. J Acoust Soc Am 122:467–477.

Heffner RS, Heffner HE (1988) Sound localization and use of binaural cues by the gerbil (Meriones unguiculatus). Beh Neurosci 102:422–428.

Heffner HE, Koay G, Heffner RS (2006) Behavioral assessment of hearing in mice - conditioned suppression. Curr Protoc Neurosci 8:Unit 8.21D.

Heffner RS, Koay G, Heffner HE (2001) Sound-localization acuity changes with age in C57BL/6J mice. In: Willott JF (ed) Handbook of Mouse Auditory Research: From Behavior to Molecular Biology. New York: CRC Press, pp. 31–35.

Helfert RH, Sommer TJ, Meeks J, Hofstetter P, Hughes LF (1999) Age-related synaptic changes in the central nucleus of the inferior colliculus of Fischer-344 rats. J Comp Neurol 406:285–298.

Helfert RH, Krenning J, Wilson TS, Hughes LF (2003) Age-related synaptic changes in the anteroventral cochlear nucleus of Fischer-344 rats. Hear Res 183:18–28.

Hellstrom LI, Schmiedt RA (1991) Rate/level functions of auditory-nerve fibers in young and quiet-aged gerbils. Hear Res 53:217–222.

Henry KR (2004) Males lose hearing earlier in mouse models of late-onset age-related hearing loss; females lose hearing earlier in mouse models of early-onset hearing loss. Hear Res 190:141–148.

Henry KR, McGinn M, Chole R (1980) Age-related auditory loss in the Mongolian gerbil. Arch Otorhinolaryngol 228:233–238.

Herman GE, Warren LR, Wagener JW (1977) Auditory lateralization: age differences in sensitivity to dichotic time and amplitude cues. J Gerontol 32:187–191.

Hequembourg S, Liberman MC (2001) Spiral ligament pathology: a major aspect of age-related cochlear degeneration in C57BL/6 mice. J Assoc Res Otolaryngol 2:118–129.

Hill K, Boesch C, Goodall J, Pusey A, Williams J, Wrangham R (2001) Mortality rates among wild chimpanzees. J Human Evol 40:437–450.

Hoffman HS, Ison JR (1992) Reflex modification and the analysis of sensory processing in developmental and comparative research. In: Campbell BA, Hayne H, Richardson R (eds) Attention and Information Processing in Infants and Adults: Perspectives from Human and Animal Research. Hillsdale, NJ: Lawrence Erlbaum Associates, pp. 83–111.

Humes LE (2005) Do 'auditory processing' tests measure auditory processing in the elderly? Ear Hear 26:109–119.

Humes LE, Watson BU, Christensen LA, Cokely CG, Halling DC, Lee L. (1994) Factors associated with individual differences in clinical measures of speech recognition among the elderly. J Speech Hear Res 37:465–474.

Hunter KP, Willott JF (1987) Aging and the auditory brainstem response in mice with severe or minimal presbycusis. Hear Res 30:207–218.

Ichimiya I, Suzuki M, Mogi G (2000) Age-related changes in the murine cochlear lateral wall. Hear Res 139:116–122.

Ingham NJ, Comis SD, Withington DJ (1999) Hair cell loss in the aged guinea pig cochlea. Acta Otolaryngol 119:42–47.

Ison JR, Bowen GP (2000) Scopolamine reduces sensitivity to auditory gaps in the rat, suggesting a cholinergic contribution to temporal acuity. Hear Res 145:169–176.

Ison JR, Agrawal P (1998). The effect of spatial separation of signal and noise on masking in the free field as a function of signal frequency and age in the mouse. J Acoust Soc Am 104:1689–1695.

Ison JR, Agrawal P, Pak J, Vaughn WJ (1998) Changes in temporal acuity with age and with hearing impairment in the mouse: a study of the acoustic startle reflex and its inhibition by brief decrements in noise level. J Acoust Soc Am 104:1696 –1704.

Ison, JR, Allen, PD (2007) Pre- but not post-menopausal female CBA/CaJ mice show less prepulse inhibition than male mice of the same age. Behavioral Brain Res. 185:76–81.

Ison JR, Allen PD, Rivoli PJ, Moore JA (2005) The behavioral response of mice to gaps in noise depends on its spectral components and its bandwidth. J Acoust Soc Am 117:3944–3951.

Ison JR, Allen PD, O'Neill WE (2007) Age-related hearing loss in C57BL/6J mice has both frequency-specific and non-frequency-specific components that produce a hyperacusis-like exaggeration of the acoustic startle reflex. J Assoc Res Otolaryngol 8:539–550.

Jerger J, Chmiel R, Stach B, Spretnjak M (1993) Gender affects audiometric shape in presbyacusis. J Am Acad Audiol 4:42–49.

Jero J, Coling DE, Lalwani AK (2001) The use of Preyer's reflex in evaluation of hearing in mice. Acta Otolaryngol 121:585–589.

Jewett DL, Romano MN, Williston JS (1970) Human auditory evoked potentials: possible brain stem components detected on the scalp. Science 167:1517–1518.

Jimenez AM, Stagner BB, Martin GK, Lonsbury-Martin BL (1999) Age-related loss of distortion product otoacoustic emissions in four mouse strains. Hear Res 138:91–105.

Johnsson LG, Hawkins JEJ (1972) Sensory and neural degeneration with aging, as seen in microdissections of the human inner ear. Ann Otol Rhinol Laryngol 81:179–193.

Joris PX, Carney LH, Smith PH, Yin TC (1994) Enhancement of neural synchronization in the anteroventral cochlear nucleus. I. Responses to tones at the characteristic frequency. J Neurophysiol 71:1022–1036.

Kazee AM, Han LY, Spongr VP, Walton JP, Salvi RJ, Flood DG (1995) Synaptic loss in the central nucleus of the inferior colliculus correlates with sensorineural hearing loss in the C57BL/6 mouse model of presbycusis. Hear Res 89:109–120.

Keithley EM, Feldman ML (1982) Hair cell counts in an age-graded series of rat cochleas. Hear Res 8:249–262.

Kelly JB, Masterton B (1977) The auditory sensitivity of the albino rat. J Comp Physiol Psychol 91:930–936.

Kemp DT (1978) Stimulated acoustic emissions from within the human auditory system. J Acoust Soc Am 64:1386–1391.

Kopp-Scheinpflug C, Fuchs K, Lippe WR, Tempel BL, Rubsamen R (2003) Decreased temporal precision of auditory signaling in *Kcna1*-null mice: an electrophysiological study in vivo. J Neurosci 23:9199–9207.

Krauter EE, Wallace JE, Campbell BA (1981) Sensory-motor function in the aging rat. Behav Neural Biol 31:367-392.

Kujawa SG, Liberman MC (2006) Acceleration of age-related hearing loss by early noise exposure: evidence of a misspent youth. J Neurosci 26:2115–2123.

Lang H, Schulte BA, Schmiedt RA (2002) Endocochlear potentials and compound action potential recovery: functions in the C57BL/6J mouse. Hear Res 172:118–126.

Lavoie BA, Mehta R, Thornton AR (2008) Linear and nonlinear changes in the auditory brainstem response of aging humans. Clin Neurophysiol 119:772–785.

Long GR (1994) Psychoacoustics. In: Fay RR, Popper AN (eds) Comparative Hearing: Mammals. New York: Springer-Verlag, pp. 18–56.

Lutz J, Hemminger F, Stahl R, Dietrich O, Hempel M, Reiser M, Jager L (2007) Evidence of subcortical and cortical aging of the acoustic pathway: a diffusion tensor imaging (DTI) study. Acad Radiol 14:692–700.

May BJ, Kimar S, Prosen CA (2006) Auditory filter shapes of CBA/CaJ mice: behavioral assessments. J Acoust Soc Am 120:321–330.

McCullough BJ, Tempel BL (2004) Haplo-insufficiency revealed in deafwaddler mice when tested for hearing loss and ataxia. Hear Res 195:90–102.

McFadden SL, Willott JF (1994a) Responses of inferior colliculus neurons in C57BL/6J mice with and without sensorineural hearing loss: effects of changing the azimuthal location of an unmasked pure-tone stimulus. Hear Res 78:115–131.

McFadden SL, Willott JF (1994b) Responses of inferior colliculus neurons in C57BL/6J mice with and without sensorineural hearing loss: effects of changing the azimuthal location of a continuous noise masker on responses to contralateral tones. Hear Res 78:132–148.

McFadden SL, Campo P, Quaranta N, Henderson D (1997) Age-related decline of auditory function in the chinchilla (*Chinchilla laniger*). Hear Res 111:114–126.

Mendelson JR, Ricketts C (2001) Age-related temporal processing speed deterioration in auditory cortex. Hear Res 157:84–94.

Mikaelian DO, Warfield D, Norris BA (1974) Genetic progressive hearing loss in the C57/b16 mouse: relation of behavioral responses to cochlear pathology. Acta Otolaryngol 77:327–334.

Mills JH, Schmiedt RA, Kulish LF (1990) Age-related changes in auditory potentials of Mongolian gerbil. Hear Res 46:201–210.

Mrak RE, Griffin ST, Graham DI (1997) Aging-associated changes in human brain. J Neuropathol Exp Neurol 56:1269–1275.

Nabelek AK, Robinson PK (1982) Monaural and binaural speech perception in reverberation for listeners of various ages. J Acoust Soc Am 71:1242–1248.

Nelson EG, Hinojosa R (2006) Presbycusis: a human temporal bone study of individuals with downward sloping audiometric patterns of hearing loss and review of the literature. Laryngoscope 116:1–12.

Oeken J, Lenk A, Bootz F (2000) Influence of age and presbyacusis on DPOAE. Acta Otolaryngol 120:396–403.

Ohlemiller KK, Lett JM, Gagnon PM (2006) Cellular correlates of age-related endocochlear potential reduction in a mouse model. Hear Res 220:10–26.

Patterson RD, Nimmo-Smith I, Weber DL, Milroy R (1982) The deterioration of hearing with age: frequency selectivity, the critical ratio, the audiogram, and speech threshold. J Acoust Soc Am 72:1788–1803.

Pearson JD, Morrell CH, Gordon-Salant S, Brant LJ, Metter EJ, Klein LL, Fozard JL (1995) Gender differences in a longitudinal study of age-associated hearing loss. J Acoust Soc Am 97:1196–1205.

Pickles JO, Comis SD (1973) The role of centrifugal pathways to the cochlear nucleus and the detection of signals in noise. J Neurophysiol 36:1131–1137.

Prosen CA, Moody DB (1991) Low-frequency detection and discrimination following apical hair cell destruction. Hear Res 57:142–152.

Rivoli P, Moore J, O'Neill W, Allen P, Ison J (2005) Colony-wide analysis of mouse auditory brainstem responses (II): maturational, gender and aging effects in C57BL/6J and CBA/CaJ mice. Assoc Res Otolaryngol Abstr 28:433.

Roosa DBStJ (1885) Presbykousis. Trans Am Otol Soc 3:449–460.

Ross B, Tremblay KL, Picton TW (2007) Aging in binaural hearing begins in mid-life: evidence from cortical auditory-evoked responses to changes in interaural phase. J Neurosci 27:11172–11178.

Ryan A (1976) Hearing sensitivity in the mongolian gerbil, *Meriones unguiculatis.* J Acoust Soc Am 59:1222–1228.

Schmiedt RA (1993) Cochlear potentials in quiet-aged gerbils: does the aging cochlea need a jump start? In: Verillo RT (ed) Sensory Research: Multimodel Perspectives Hillsdale, NJ: Erlbaum Associates, pp. 91–103.

Schulte BA, Schmiedt RA (1992) Lateral wall Na,K-ATPase and endocochlear potentials decline with age in quiet-reared gerbils. Hear Res 61:35–46.

Shimada A, Ebisu M, Morita T, Takeuchi T, Umemura T (1998) Age-related changes in the cochlea and cochlear nuclei of dogs. J Vet Med Sci 60:41–48.

Snell KB (1997) Age-related changes in temporal gap detection. J Acoust Soc Am 101:2214–2220.

Snell KB, Ison JR, Frisina DR (1994) The effects of signal frequency and absolute bandwidth on gap detection in noise. J Acoust Soc Am 96:1458–1464.

Spongr VP, Flood DG, Frisina RD, Salvi RJ (1997) Quantitative measures of hair cell loss in CBA and C57BL/6 mice throughout their life spans. J Acoust Soc Am 101:3546–3553.

Spruijt BM, van Hooff JA, Gispen WH (1992) Ethology and neurobiology of grooming behavior. Physiol Rev 72:825–852.

Stebbins WC (1990) Perception in animal behavior. In: Berkley MA, Stebbins WC (eds) Comparative Perception New York: John Wiley & Sons, pp 1–26.

Stenqvist M (2000) Age-related hearing changes and effects of exotoxin on inner ear function in aging rat. A frequency-specific auditory brainstem response study. ORL J Otorhinolaryngol Relat Spec 62:13–19.

Tarnowski BI, Schmiedt RA, Hellstrom LI, Lee FS, Adams JC (1991) Age-related changes in cochleas of mongolian gerbils. Hear Res 54:123–134.

Torre P, Fowler CG (2000) Age-related changes in auditory function of rhesus monkeys (*Macaca mulatta*). Hear Res 142:131–140.

Tremblay KL, Burkard RF (2007) The aging auditory system: confounding effects of hearing loss on AEPs In: Burkard R, Don M, Eggermont JJ (eds) Auditory Evoked Potentials: Basic Principles and Clinic Application. Baltimore, MD: Lippincott Williams and Wilkins, pp. 403–425.

Tremblay KL, Piskosz M, Souza P (2002) Aging alters the neural representation of speech cues. NeuroReport 13:1865–1870.

Tremblay KL, Piskosz M, Souza P (2003) Effects of age and age-related hearing loss on the neural representation of speech cues. Clin Neurophys 114:1332–1343.

Van Eyken E, Van Camp G, Van Laer L (2007) The complexity of age-related hearing impairment: Contributing environmental and genetic factors. Audiol. Neurotol. 12:345–358.

Vaughan DW (1977) Age-related deterioration of pyramidal cell basal dendrites in rat auditory cortex. J Comp Neurol 171:501–515.

Walton JP, Frisina RD, O'Neill WE (1998) Age-related alteration in processing of temporal sound features in the auditory midbrain of the CBA mouse. J Neurosci 18:2764–2776.
Walton JP, Simon H, Frisina RD (2002) Age-related alterations in the neural coding of envelope periodicities. J Neurophysiol 88:565–578.
Walton JP, Barsz K, Wilson WW (2008) Sensorineural hearing loss and neural correlates of temporal acuity in the inferior colliculus of the C57BL/6 mouse. J Assoc Res Otolaryngol 9:90–101.
Willott JF (1986) Effects of aging, hearing loss, and anatomical location on thresholds of inferior colliculus neurons in C57BL/6 and CBA mice. J Neurophysiol 56:391–408.
Willott JF (1996) Auditory system plasticity in the adult C57BL/6J mouse. In: Salvi RJ, Henderson D, Fiorino F, Colletti V (eds) Auditory system Plasticity and Regeneration New York: Thieme Medical Publishers, Inc, pp. 297–316.
Willott JF, Bross LS (1990) Morphology of the octopus cell area of the cochlear nucleus in young and aging C57BL/6J and CBA/J mice. J Comp Neurol 300:61–81.
Willott JF, Carlson S (1995) Modification of the acoustic startle response in hearing-impaired C57BL/6J mice: prepulse augmentation and prolongation of prepulse inhibition. Behav Neurosci 109: 396–403.
Willott JF, Jackson LM, Hunter KP (1987) Morphometric study of the anteroventral cochlear nucleus of two mouse models of presbycusis. J Comp Neurol 260: 472–480.
Willott JF, Aitkin LM, McFadden SL (1993) Plasticity of auditory cortex associated with sensorineural hearing loss in adult C57BL/6J mice. J Comp Neurol 329:402–411.
Yerkes RM (1905) The sense of hearing in frogs. J Comp Neurol Psychol 15:279–304.
Yerkes RM (1907) The Dancing Mouse. New York: The MacMillan Company.
Zettel ML, Zhu X, O'Neill WE, Frisina RD (2007) Age-related decline in Kv3.1b expression in the mouse auditory brainstem correlates with functional deficits in the medial olivocochlear efferent system. J Assoc Res Otolaryngol 8:280–293.

Chapter 5
Behavioral Studies With Aging Humans: Hearing Sensitivity and Psychoacoustics

Peter J. Fitzgibbons and Sandra Gordon-Salant

5.1 Introduction

Historically, the study of human aging and hearing has focused on problems of speech perception and comprehension experienced by older listeners. An earlier comprehensive review of these problems (Committee on Hearing, Bioacoustics, and Biomechanics [CHABA] 1988) outlined the scope of the age-related listening difficulties and pointed to several areas of interest where information was lacking and greater study was needed. Some of the research needs cited are specific to speech processing and aging (see Humes and Dubno, Chapter 8) and to general aspects of cognition and audition (see Schneider, Pichora-Fuller, and Daneman, Chapter 7). Equally important are questions about the extent to which aging of the auditory system compromises listeners' ability to process simple and complex nonspeech sounds. This topic is explored in the present chapter by a review of recent psychoacoustic studies that were conducted to identify those auditory abilities and processes that appear to be affected by aging. Where possible, age-related alterations in processing mechanisms associated with spectral, intensive, and temporal aspects of sound are linked to known anatomical and physiological changes with age in the auditory system (see Schmiedt, Chapter 2; Canlon, Illing, and Walton, Chapter 3). A related goal of the psychoacoustic investigations is to identify some of the possible contributing factors that underlie the speech understanding difficulties of older listeners.

P.J. Fitzgibbons (✉)
Department of Hearing, Speech, and Language Sciences, Gallaudet University, Washington, DC 20002
e-mail: peter.fitzgibbons@gallaudet.edu

S. Gordon-Salant
Department of Hearing and Speech Sciences, University of Maryland, MD 20742, College Park,
e-mail: sgordon@hesp.umd.edu

S. Gordon-Salant et al. (eds.), *The Aging Auditory System*,
Springer Handbook of Auditory Research 34,
DOI 10.1007/978-1-4419-0993-0_5,

5.1.1 Methodology

Most of the psychophysical experiments are designed to examine sensory and perceptual processing in audition by conducting behavioral measurements on human listeners. For experiments cited in the present chapter, the sound stimuli are usually delivered to listeners monaurally through a headphone in an acoustically isolated environment (for studies in binaural processing, see Eddins and Hall, Chapter 6). Some of the listening tasks involve detection of a sound presented in a quiet background or in the presence of other sounds, usually referred to as maskers. In either case, a stimulus threshold value (e.g., in decibels) is measured that corresponds to a given percent-correct detection performance by the listener. Other tasks require listeners to compare two or more sounds that differ along a specific acoustic dimension such as frequency or duration. These discrimination tasks typically seek to measure the smallest physical difference between sounds that allows some predetermined level of listener performance in terms of percentage of correct responses. This measured stimulus difference then becomes an estimate of the difference threshold, or "difference limen (DL)," for the stimulus attribute that is varied, such as frequency, intensity, or duration. DLs can be expressed either in absolute physical units (e.g., in hertz) or in terms of a relative difference in which the absolute DL is expressed as a proportion of the baseline stimulus value. Some of the psychophysical experiments also utilize a "method of constant stimuli," in which each of a preselected series of stimuli featuring different physical values are presented to listeners an equal number of times. The percentage of correct listener responses to each stimulus in the series is then recorded to derive a performance curve, known as a "psychometric function," which displays percent correct as a function of stimulus value.

The study of auditory processing and aging faces certain obstacles that need to be addressed in the design of experiments. One factor of concern is the prevalence of hearing loss among older listeners (see Cruickshanks, Zhan, and Zhong, Chapter 9). A related issue is the hearing sensitivity of individual older listeners that generally varies as a function of age, gender, and frequency, as described in Section 2. As such, it is not uncommon for psychoacoustic performance measures collected from older listeners to contain confounding influences of both hearing loss and age effects. Current methodologies in psychoacoustic research on aging attempt to unravel this potential confounding of variable effects in one of several ways. For example, some studies attempt to recruit older listeners with normal or near-normal hearing, defined audiometrically, and compare their listening performance with that of younger listeners with normal hearing. Other studies utilize multiple groups of listeners and compare the performance of younger and older listeners with normal hearing and also compare the performance of younger and older listeners who exhibit matched degrees of hearing loss. Also available are statistical analysis tools, such as analysis of covariance, that have been applied in some studies to factor out the influence of individual variables that might contribute to the performance measures collected from older listeners.

Another general issue of relevance to the study of aging and auditory processing concerns the appropriate definition of the elderly listener. At present, the chronological age of an individual is the primary selection criterion used to assign listeners to an older subject group for a given psychophysical task. Thus 60-65 years is frequently stated as the minimum age of listeners who comprise the older groups of subjects in many studies on aging. However, it is already evident from this research that the elderly, when defined primarily by age, constitute a rather heterogeneous group in terms of their listening performance and, quite likely, the degree and nature of age-related deterioration to sensory structures within the auditory nervous system. As research on aging advances, alternative definitions of auditory aging may emerge, perhaps ones that characterize listener groups according to different functional and diagnostic criteria. Toward that goal, various psychoacoustic investigations have been conducted to explore some of the basic auditory abilities of older listeners. Listener sensitivity to the changing characteristics of sound in terms of variables such as frequency, intensity, and duration is considered essential to the accurate processing and understanding of the complex time-varying sounds of human speech.

5.2 Basic Hearing Sensitivity

Pure-tone detection thresholds elevate with increasing age during the adult life span. Corso (1963) published a classic study of hearing sensitivity data for a large sample of men and women, aged 18 to 65 years, who were screened for otologic disease and history of noise exposure. The cross-sectional data for men indicated that hearing thresholds were poorer at each successive age decade above age 32 years, particularly in the higher audiometric frequencies of 2 and 4 kHz. Women showed a decline in hearing sensitivity above age 37 years, but thresholds did not vary widely across frequency. Significant differences in hearing thresholds between men and women were observed most notably at 3 kHz and above, with the men exhibiting poorer thresholds and greater variability than the women. However, women had poorer hearing and greater variability than men at frequencies below 1 kHz. The crossover point for this "gender reversal" was 1 kHz, a frequency where the thresholds of men and women were not significantly different.

5.2.1 Cross-Sectional Epidemiologic Studies

Subsequent investigations employed an epidemiologic approach and extended the age range of participants to those over 65 years (Moscicki et al. 1985; Gates et al. 1990; Cruickshanks et al. 1998). Participants in these studies included all individuals residing in a particular town for whom audiometric data were available;

there were no exclusions on the basis of otologic disease or noise history. Representative data from the Epidemiology of Hearing Loss Study are shown in Fig. 5.1 (Cruickshanks et al. 1998), although data from the Framingham Heart

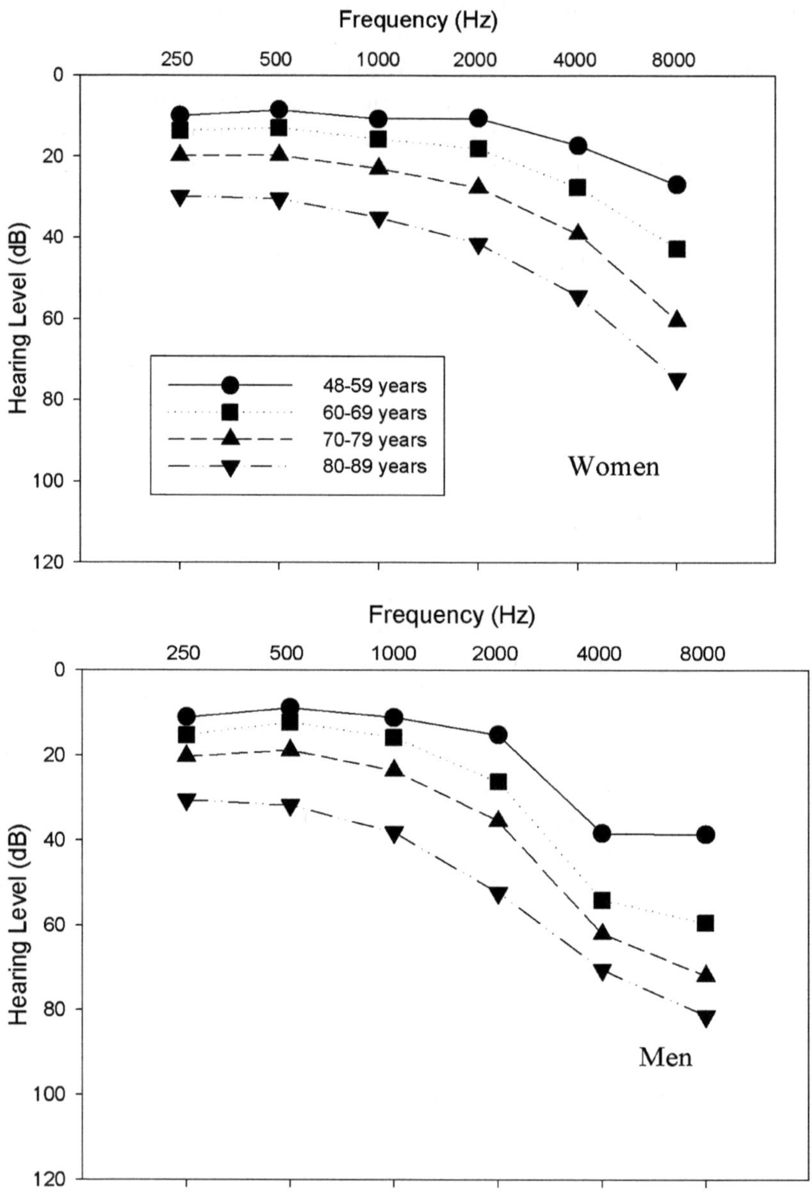

Fig. 5.1 Mean pure-tone air conduction thresholds (hearing level) in the right ear as a function of age, gender, and frequency from the Epidemiology of Hearing Loss Study. (Adapted from Cruickshanks et al. 1998.)

Cohort were highly similar (Moscicki et al. 1985; Gates et al. 1990). Each of these studies reported that hearing thresholds of both men and women were poorer in the higher frequencies than the lower frequencies, thresholds for men were poorer than those for age-matched women in the high frequencies, and hearing sensitivity declined with age at each frequency (Moscicki et al. 1985; Gates et al. 1990; Cruickshanks et al. 1998). The mean audiometric data of individuals over 65 years indicate a mild-to-moderately severe, sloping sensorineural hearing loss, which is more gradual in women than in men. However, Gates et al. (1990) reported that the audiometric configuration varied between individuals in the Framingham Heart Cohort. The most common audiometric configuration for men was sharply sloping (53.2%), followed by a gradually sloping loss (30.3%); in women, the most common configurations were gradually sloping (39.3%), followed by flat (38.6%). Notched audiometric configurations also were observed in a large proportion of individuals in these unscreened samples (Gates et al. 1990). In addition to gender, race appears to play a role in age-related hearing loss. One recent investigation of over 2,000 individuals, aged 73-84 years, reported that black men and women exhibited significantly lower (better) hearing thresholds than white men and women, particularly at 2 and 4 kHz (Helzner et al. 2005).

Hearing sensitivity in the frequency region above 8 kHz (sometimes referred to as extended high-frequency thresholds) is also affected by aging. Hearing thresholds in this extended frequency range are not tested routinely in the clinical setting but may provide early indications of damage to the cochlea. Indeed, samples of normal-hearing young and middle-age adults demonstrate deterioration in hearing thresholds at 14 kHz and higher above age 30 years and at all frequencies from 8 to 20 kHz above age 40 years (Stelmachowicz et al. 1989). Older listeners aged 60-79 years with normal hearing sensitivity in the standard audiometric range (0.25-8 kHz) also show significantly poorer hearing thresholds above 8 kHz compared with younger listeners with normal hearing (Matthews et al. 1997). Although women exhibit better hearing thresholds than men in the 8–to 12-kHz range, these gender differences are no longer observed at 14 kHz and above, in large part because thresholds in both groups exceed 100 dB SPL. Hearing thresholds in the extended high-frequency range generally increase with advancing age between 48 and 92 years (Wiley et al. 1998) at frequencies where hearing thresholds are measurable. Moreover, hearing sensitivity in the standard audiometric range is correlated with thresholds measured in the ultra-high frequency range (Wiley et al. 1998).

5.2.2 Longitudinal Data

The cross-sectional data described above compare hearing thresholds across different age groups at a fixed point in time but do not indicate the changes in hearing thresholds in individuals over time. Longitudinal studies track performance measures obtained from individuals over a period of time and may vary in the total time period of the measures, the frequency of testing, the number of repeated measures, the age of individuals at entrance to the study, and the number of individuals in a specified age category at any one period of time. One large-scale longitudinal study of aging examined hearing thresholds from 0.1 to 10 kHz biannually for 1,247 men and 588 women over a period of 23 and 13 years, respectively (Pearson et al. 1995). Individuals entered the study at any age, with the total sample ranging in age from 20 to 90 years. A screening protocol was used to eliminate data from participants who had significant otologic disease, history of noise exposure, or evidence of a notched audiogram either at the start of the study or at any subsequent visit. Threshold data from 0.5-8 kHz showed that the rate of change in hearing thresholds differs by gender and frequency. Fig. 5.2 shows the longitudinal change in hearing thresholds of the men and women. The longitudinal change in hearing accelerates above age 20-30 years in men, and above age 40-50 years in women. Thresholds in men decline twice as fast as those of women at most ages and frequencies, with the rate of change for both genders converging above age 60 years. At any age and frequency, however, men exhibit poorer thresholds than women at frequencies >1 kHz, but women exhibit poorer thresholds than men at frequencies <1 kHz. Considerable individual variability in the rates of change in hearing threshold also was reported. A subsequent investigation (Lee et al. 2005) reported longitudinal changes in hearing thresholds measured in the ultra-high audiometric frequencies among a sample of primarily older individuals (60-81 years at entrance to the study). Significant effects of age and gender were reported, with women showing faster rates of change (1.34-1.55 dB/year) than men (0.81-1.05 dB/year) at 6 to 12 kHz. Overall, the average rate of change in thresholds for both genders combined was greater at the high frequencies (1.2 dB/year at 8 and 12 kHz) than at the low frequencies (0.7 dB/year at 0.25 kHz). In general, both longitudinal studies found that older subjects had a faster rate of decline in hearing thresholds than younger subjects.

One potential confounding factor in all of these investigations is the effect of noise exposure on hearing thresholds. Even when methods are used to eliminate individuals with a significant history of noise exposure (such as in military service or hunting), most people in our industrialized society are exposed to noise or loud sounds during everyday activities (e.g., hairdryers, traffic noise, cell phones). Gates et al. (2000) examined longitudinal changes in hearing thresholds over a 15-year period in 203 men from the Framingham Cohort, aged 58-80 years, to track the natural history of notching in the 3-to 6-kHz region as well as at adjacent frequencies. Data from women were not examined because only 4% of the women developed a notch during the test period. Participants were assigned to notch groups, depending on the depth of the notch (no notch, >15-dB notch, >35-dB notch). The three groups differed in the rate of change in threshold in the 2-to 8-kHz region across the 15-year study period, and the pattern of group differences varied with frequency. At 2 kHz, the rate of change in hearing threshold increased with increasing notch group. However, at 3 and 4 kHz, the group with the deepest notch showed a decelerating pattern in which the rate of change was slower than that observed for the other two groups. Above 4 kHz, the two groups with notches showed an increase in the rate of change in hearing threshold, but the change was ~10 dB slower in the deeper notch group compared with the shallower notch group. The group with no evidence of notching generally

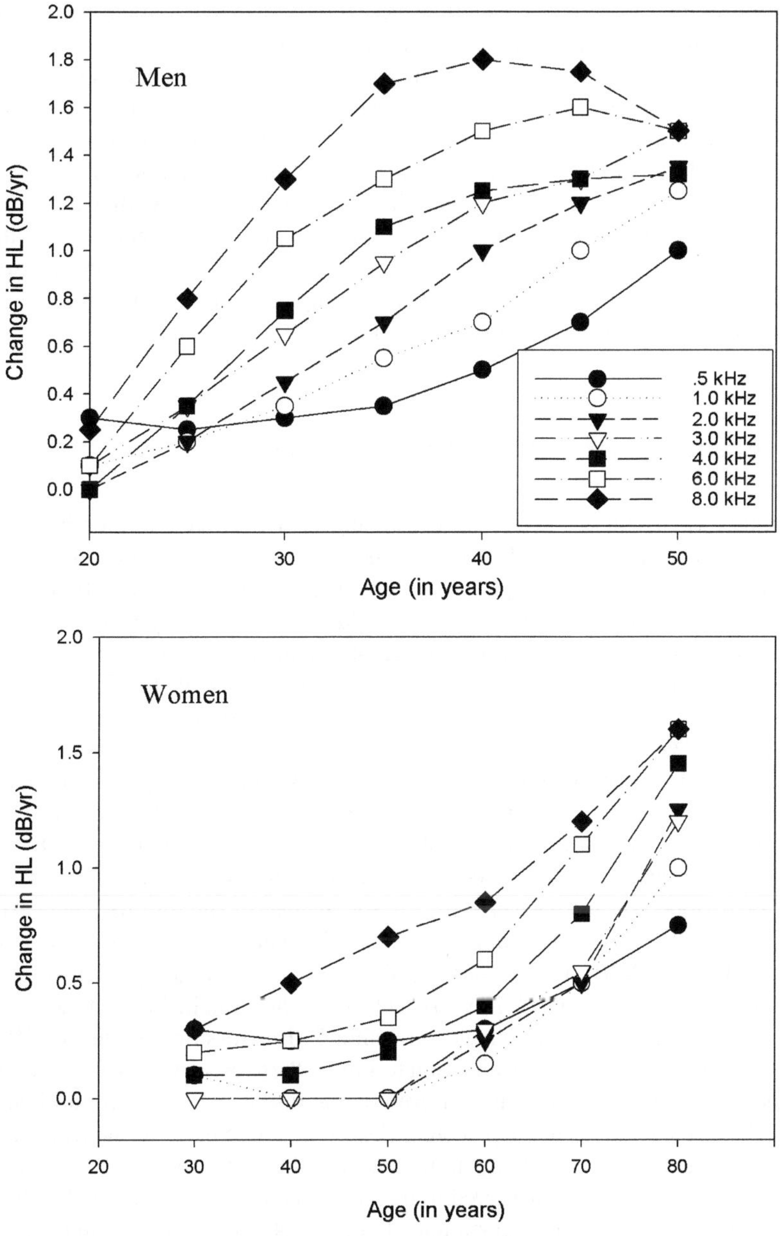

Fig. 5.2 Average 10-year changes in hearing levels (HL) for men and women from the Baltimore Longitudinal Study on Aging. (Adapted from Pearson et al. 1995.)

showed a more gradual longitudinal change in these high-frequency thresholds. Taken together, the findings indicate that the rate of change in hearing thresholds in the region of the notch decelerates over time compared with individuals without notching.

This longitudinal pattern suggests that the cochlear damage associated with aging is mediated by preexisting hair cell damage attributed to noise exposure.

In summary, hearing thresholds decline with increasing age across the adult life span for both men and women, and the greatest change in threshold occurs in the frequency region above 2 kHz. Although average audiograms of elderly individuals suggest a mild-to-moderately severe hearing loss, these mean audiograms do not necessarily reflect the hearing sensitivity of all individuals within a given age group. That is, some older people may exhibit relatively normal hearing sensitivity, better than that depicted in Fig. 5.1, whereas others may exhibit greater degrees of hearing loss. Indeed, wide individual variability in hearing thresholds between subjects is a common observation across studies.

5.3 Frequency/Intensity Discrimination

The ability of listeners to discriminate changes in the frequency or intensity of simple sounds has been explored in numerous studies, although relatively few experiments have examined age-related changes in these discrimination tasks. In addition, the potential for confounding influences of hearing loss is a concern because several investigations reveal that sensorineural hearing loss is a known factor that can diminish listener performance on various auditory discrimination tasks (Turner and Nelson 1982; Freyman and Nelson 1991; Florentine et al. 1993; Simon and Yund 1993).

Despite this situation, recent evidence from studies that controlled potential influences of hearing loss indicates that aging is an important factor that can act to diminish listeners' discrimination abilities. For example, He et al. (1998) used a same-different discrimination procedure with tonal stimuli of four different frequencies (500 to 4000 Hz) and measured the DL for changes of stimulus frequency (Δf Hz) and intensity (ΔL dB), with testing in each discrimination task conducted at stimulus levels of 40-and 80-dB SPL. Listeners in this study included groups of younger (age 20-33 years) and older (age 76-77 years) subjects, who exhibited normal hearing sensitivity and thresholds that were matched within 5 dB across frequencies of 250-4000 Hz. The findings from the discrimination testing in these experiments are summarized in Figs. 5.3 and 5.4. The frequency discrimination results are displayed in Figure 5.3, which shows the relative DLs in percent ($\Delta f/f$) for the group of aged listeners as a function of the standard test frequency at the two stimulus test levels. Figure 5.4 displays the intensity DLs (ΔL dB) measured at the two standard stimulus levels for the aged listeners at each of the four tonal stimulus frequencies. Shaded regions in both figures reflect the range of corresponding DLs reported in the literature for young listeners, including those tested by He et al. (1998).

The frequency discrimination results reported by He et al. (1998) indicated clearly that the DLs of the older listeners were larger than those of the younger listeners, with the magnitude of the age-related performance differences being greater at the lower stimulus frequencies; results were essentially the same at each stimulus level tested.

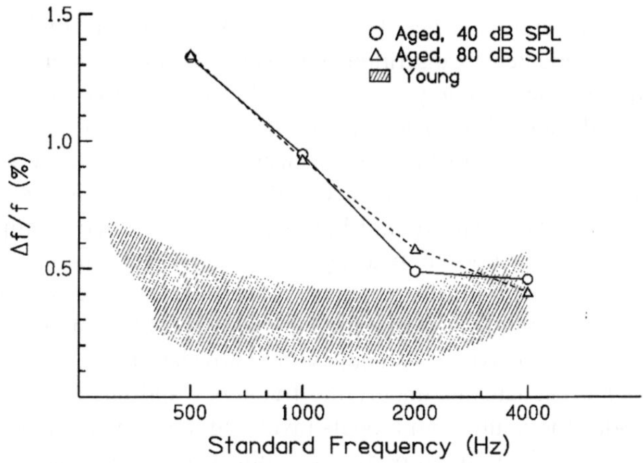

Fig. 5.3 Mean relative frequency difference limens (DLs; Δf/f) as a function of standard tone frequency for the aged listeners at stimulus levels of 40-and 80-dB SPL. Shaded region represents range of DLs for young listeners from He et al. (1998) and previous literature. (Adapted from He et al. 1998.)

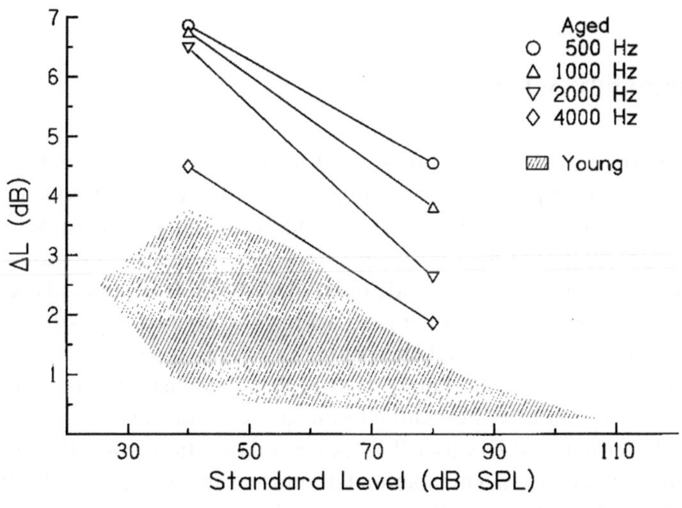

Fig. 5.4 Mean intensity DLs (ΔL) of the aged listeners as a function of standard stimulus level, with the parameter representing the stimulus frequency. Shaded region represents range of intensity DLs for young listeners from He et al. (1998) and previous literature. (Adapted from He et al. 1998.)

The intensity DLs of both younger and older listeners decreased as stimulus level increased. However, the intensity DLs of the older listeners were larger than those of the younger listeners, with the magnitude of the age-related differences being greatest at the lower test frequencies. Thus the age effects observed by He et al. (1998) for both frequency and intensity discrimination exhibit a strong frequency dependency. A similar frequency dependency was observed more recently by He et al. (2007) in a frequency-modulation (FM) detection task involving younger and older listeners. This experiment also revealed that FM detection among older listeners was diminished, with the largest age-related performance differences observed for stimuli with lower carrier frequencies. Moore and Peters (1992) also cited examples of older listeners with relatively normal hearing showing large frequency DLs at low stimulus frequencies. Similarly, intensity discrimination results reported by Florentine et al. (1993) indicated that the DLs for two older listeners in their task were larger than those of young listeners with equivalent hearing sensitivity.

The processing mechanisms underlying frequency and intensity discrimination are not fully understood, although models involving the role of peripheral auditory filtering and the phase-locking properties of auditory nerve fibers have been applied successfully to describe large bodies of discrimination data (e.g., Moore 1973; Sek and Moore 1995). However, the frequency-dependent effects showing large age-related discrimination differences in the lower frequency regions suggests that temporal mechanisms may be involved, perhaps as a result of age-related changes in the phase-locking properties of auditory nerve fibers (see Schmiedt, Chapter 2).

5.4 Frequency Selectivity

Another important aspect of auditory spectral processing concerns the ability of listeners to distinguish perceptually among sounds that differ in frequency. Indeed, a diminished ability among older listeners to resolve spectral aspects of sound stimuli is often suggested as a possible source of age-related difficulty in understanding speech in noisy backgrounds (CHABA 1988). It is generally known that active nonlinear processing mechanisms within the cochlea function to enhance sensitivity, spectral contrasts, and sharpening of neural and psychophysical tuning curves (Yates 1995). It is also known that healthy cochlear structures (e.g., outer hair cells, stria vascularis) are essential to the active nonlinear processes (Schmiedt et al. 1980; Liberman and Dodds 1984). A central question is whether aging alters these nonlinear processes in a manner that diminishes the tuning of auditory filtering or the suppression mechanisms that serve to enhance spectral contrasts.

5.4.1 Auditory Filtering

Behavioral estimates of frequency selectivity are derived primarily from masking experiments that measure either the sharpness of psychophysical tuning curves (PTCs) or the shape of peripheral auditory filters. As was the case for measures of frequency discrimination, the investigation of age effects on frequency resolution is hindered by potential independent influences associated with the presence of sensorineural hearing loss. For example, various studies report that listeners with sensorineural hearing loss exhibit reduced frequency selectivity (de Boer and Bouwmeester 1974; Pick et al. 1977; Glasberg and Moore 1986). Other masking studies that measured PTCs in listeners with hearing loss found abnormally broad tuning curves at each of several different test frequency regions (Wightman et al. 1977; Florentine et al. 1980; Tyler et al. 1982). Many of the listeners with hearing loss in these earlier studies were also older than the younger normal-hearing control subjects, although the specific contribution of age effects in the psychophysical measures were not usually examined. One study that did examine age effects in frequency selectivity was reported by Patterson et al. (1982), who conducted simultaneous masking experiments using a spectrally notched noise masker and tonal probe signal with a frequency that coincided with the center frequency of the masker spectral notch. In this procedure, probe-tone detection thresholds were measured as a function of the masker notch width, and the threshold data were used to derive auditory filter shapes according to procedures described by Patterson (1976) and later by Glasberg and Moore (1990). Results of Patterson et al. (1982) revealed that auditory filter widths increased progressively with the age of their listeners (23-75 years), indicating a progressive age-related loss in frequency selectivity. These results, however, may have been influenced in part by factors related to hearing loss in the listeners because sensitivity loss exhibited by the participating subjects was also correlated with listener age. A number of subsequent experiments that used similar notched-noise masking procedures to examine frequency selectivity have found no significant differences in the derived auditory filter bandwidths between younger and older listeners when tested at frequency regions with normal hearing sensitivity (e.g., Peters and Moore 1992; Sommers and Humes 1993; Sommers and Gehr 1998; Gifford and Bacon 2005). Thus the cumulative evidence to date indicates that in the absence of sensorineural hearing loss, aging alone does not appear to have a significant influence on cochlear mechanisms involved in auditory filtering.

5.4.2 Suppression

Suppression is another cochlear nonlinear process whereby stronger sensory and neural activity associated with one stimulus frequency can inhibit, or suppress, weaker activity associated with nearby frequency regions. Psychophysical evidence for suppression phenomena in hearing is documented in a number of past studies (Houtgast 1972; Shannon 1976; Terry and Moore 1977; Weber and Green 1978). Such evidence indicates that suppression plays an important role in determining the sharpness of psychophysical and neural tuning curves and auditory filter bandwidths (Moore and Vickers 1977) and may act to enhance spectral contrasts of certain phoneme features to aid listener processing of speech sounds. Although the mechanisms of suppression are not fully understood, there is clear evidence showing that subjects with a hearing loss of cochlear origin exhibit reduced suppression, as evidenced by poorer frequency selectivity depicted in PTCs and forward masking patterns (Wightman et al. 1977; Dubno and Ahlstrom 2001a). Additionally, normal-hearing listeners on a regimen of aspirin ingestion can also display abnormal nonlinear processes, as evidenced by psychophysical observations of broader auditory filtering and reduced two-tone suppression (Hicks and Bacon 1999). These findings revealed that even small temporary aspirin-induced shifts in hearing sensitivity are sufficient to disrupt the functioning of various nonlinear cochlear processes.

A small number of recent studies have explored the effects of listener age on psychophysical measures of auditory suppression. One of these (Sommers and Gehr 1998) tested younger and older listeners with normal hearing and compared auditory filter widths derived from data collected in simultaneous and forward masking tasks. Auditory filters derived from forward masking data typically reveal sharper tuning than seen with simultaneous masking, a result that is often attributed to suppression (Houtgast 1972; Moore and Glasberg 1981). That is, under forward masking, suppressive effects are thought to reduce the effective level of the noise masker but not the later occurring probe tone, whereas, in simultaneous masking, both the masker and probe signal are similarly suppressed. Thus the net effect of suppression operating in the two masking paradigms is to enhance the signal-to-masker ratio in forward masking, leading to improved probe detection and narrower filter-width estimates relative to those obtained from simultaneous masking. Sommers and Gehr (1998) found similar filter shapes for younger and older listeners with simultaneous masking but significant age-related differences in the degree of filter narrowing observed with forward masking; i.e., the narrowing of filter widths from simultaneous to forward masking was significantly greater for the younger listeners compared with the older listeners. These results are supportive of the conclusion that aging is associated with reduced suppression within the auditory system.

Subsequent investigation by Dubno and Ahlstrom (2001b) also indicated that auditory suppression may be reduced in older listeners with normal hearing. This psychophysical experiment examined suppression in younger and older listeners using a forward-masking procedure with noise maskers of different bandwidths and tonal probe signals centered at a lower-and higher-frequency region. In this procedure, suppression effects acting on the masker are expected to increase with masker bandwidth, thus reducing the effective masking, leading to lower (better) probe tone thresholds. A single estimate of suppression can be obtained by comparing the magnitude of change in probe threshold in decibels observed between conditions featuring the narrowest and widest noise masker bandwidths. Dubno and Ahlstrom observed that the magnitude of suppression exhibited by older listeners with normal hearing was reduced relative to that of younger normal-hearing listeners. A different conclusion about aging and auditory suppression was reported recently by Gifford and Bacon (2005). These investigators used essentially the same forward-masking procedure as Dubno and Ahlstrom (2001b) to measure probe-tone thresholds in younger and older listeners with normal hearing in conditions featuring different noise masker bandwidths. The masking results of Gifford and Bacon (2005) revealed no significant age-related differences in the magnitude of suppression shown by their young and older listeners. Clearly, these conflicting results indicate that questions remain concerning the influence of aging on the mechanisms of auditory suppression. It is possible that even small differences in hearing sensitivity or cochlear function among older normal-hearing listeners can produce different outcomes on psychophysical measures of auditory suppression.

5.5 Temporal Sensitivity: Simple Stimuli

A larger number of the psychophysical studies on age effects in audition have investigated temporal aspects of hearing. The reason for this research emphasis perhaps stems from the predominant complaints among older listeners regarding perceptual difficulties with temporally altered or accelerated speech. Additionally, various early research experiments that used temporally degraded speech stimuli (e.g., time compressed, reverberant, interrupted) demonstrated consistent age-related perception difficulties (Sticht and Gray 1969; Bergman et al. 1976; Konkle et al. 1977; Helfer and Wilbur 1990). The cumulative evidence from these and other speech studies suggested that aging was associated with a decline in time-dependent auditory-processing abilities. This hypothesis was also consistent with research findings from studies on cognition and aging, which postulated an age-related slowing of information processing (Salthouse 1985; Wingfield et al. 1985). Psychophysical studies on aging and temporal aspects of auditory processing are relatively recent and provide additional information about the specific components of the temporal processing that appear to undergo changes with aging. Some of the experiments used relatively simple stimuli and assessed the limits of auditory temporal acuity and discrimination capacity, whereas others used more complex stimulus patterns to examine the sequential processing abilities of older listeners. As with all psychoacoustic studies on aging, careful attention to subject selection for experiments is required to eliminate, or minimize, potential influences of listener hearing loss on the measure of temporal sensitivity under investigation.

5.5.1 Temporal Resolution

Estimates of temporal resolution in hearing have been measured with a variety of tasks and stimuli to determine the shortest time period in which listeners are capable of detecting a change in the waveform of a sound without the aid of spectral cues. Green (1971) reviewed much of the early research and reported that most estimates of auditory temporal acuity tend to converge on values of ~2-3 ms for the young trained listener with normal-hearing sensitivity. Acuity estimates of this general magnitude were also observed by Plomp (1964), who measured the smallest detectable silent interval (temporal gap) between a pair of broadband noise bursts, typically referred to as stimulus markers. This gap-detection procedure has proven to be a useful tool in assessing auditory temporal resolution, perhaps because it is relatively easy to administer, and has produced fairly consistent resolution estimates, at least for broadband stimulus markers presented to young normal-hearing listeners at clearly audible signal levels (Abel 1972; Penner 1977).

It is noteworthy, however, that estimates of temporal gap resolution can vary considerably from those observed by Plomp (1964) if the stimulus markers surrounding a gap differ from broadband noise bursts. For example, several studies since Plomp (1964) used a variety of filtered noise bursts as stimulus markers and observed that gap thresholds tend to decrease in magnitude with increases in noise bandwidth, center frequency, and, within limits, signal level (Fitzgibbons and Wightman 1982; Fitzgibbons 1983; Shailer and Moore 1983; Green and Forrest 1989). Additionally, with tonal stimulus markers, a systematic influence of tonal frequency on gap detection is observed in one study(Green and Forrest 1989) but not in others (Shailer and Moore 1987; Moore and Glasberg 1988). Finally, most reports agree that temporal resolution can become considerably more difficult for all listeners if stimulus markers surrounding a temporal gap differ substantially in frequency (Williams and Perrott 1972; Formby and Forrest 1991; Phillips et al. 1997).

Several studies have begun to investigate the possibility that aging may compromise listeners' temporal resolving capacity. One motivation for the studies stems from awareness among investigators that any reduction in temporal sensitivity of older listeners could hamper their ability to detect and discriminate many of the temporal properties of speech waveforms that provide perceptual cues to phoneme identity. Factors related to hearing loss are also important to consider in an assessment of any potential age effects in temporal resolution. This is the case because several studies report significant reductions in temporal gap resolution among listeners with sensorineural hearing loss (Fitzgibbons and Wightman 1982; Tyler et al. 1982; Florentine and Buus 1984; Glasberg et al. 1987). One study that focused on age effects in temporal resolution was conducted by Snell (1997), who compared temporal gap thresholds in groups of younger (age 17-40 years) and older (age 64-77 years) listeners with normal hearing and closely matched audiometric thresholds. Using various low-pass noise bursts as stimulus markers, Snell observed diminished gap-resolution abilities in about one-third of the older listeners. Another study by He et al. (1999) also used relatively broadband noise to measure gap thresholds in normal-hearing younger and older listeners with matched hearing sensitivity. This study also reported elevated gap thresholds among older listeners, but the age-related effects were apparent only for temporal gaps located near the onset or offset of a noise burst and not when gaps were centered within the noise burst.

Additional information about temporal resolution and aging has come from gap-detection experiments conducted with tonal stimulus markers. With tonal stimuli, introduction of a temporal gap will introduce a sudden spread of energy (spectral splatter), which could be used by listeners to cue the presence of a gap. To prevent this possibility, investigators frequently use a spectrally shaped background noise to eliminate audibility of the spectral splatter while preserving sufficient audibility at the signal frequency. Moore et al. (1992) used this approach with 400-ms tone bursts to compare gap thresholds in younger and older listeners with relatively normal hearing. Gap thresholds collected from the older listeners were generally observed to be larger than those of the younger listeners, but most of the age-related differences were attributed to the poor resolution performance of a few older listeners. Subsequent measurements of gap resolution using pairs of brief tone pips (Schneider et al 1994) or pairs of longer 200-ms tone bursts (Strouse et al. 1998) revealed that many older listeners with normal hearing do exhibit elevated gap thresholds. More recently, Schneider and Hamstra (1999) suggested that the observation of age-related effects in temporal gap resolution may depend on the duration of stimulus markers surrounding a gap. These investigators varied marker durations using pairs of Gaussian-shaped tone bursts to measure gap thresholds in groups of younger (mean age 21.9 years; $n = 20$) and older (mean age 72.4 years; $n = 20$) listeners with nearly equivalent degrees of normal hearing at the stimulus frequency of 2 kHz. The main findings of Schneider and Hamstra are shown in Fig. 5.5, which displays the mean gap thresholds measured for the younger and older adults as a function of tonal marker duration (2.5-500 ms). These results revealed significant age-related differences in gap thresholds, with the size of the age effect being a decreasing function of tone duration; resolution performance of the two listener groups was nearly equivalent at the longest tone duration of 500 ms.

In summary, the collective results of gap-detection studies reveal that many older listeners exhibit diminished temporal resolution, even when potential influences of hearing loss are absent or deemed negligible by data analysis. However, several of the reports cite large variability in the performance measures collected from older listeners and point to observations revealing that the resolution performance of some older listeners is quite similar to that of younger listeners.

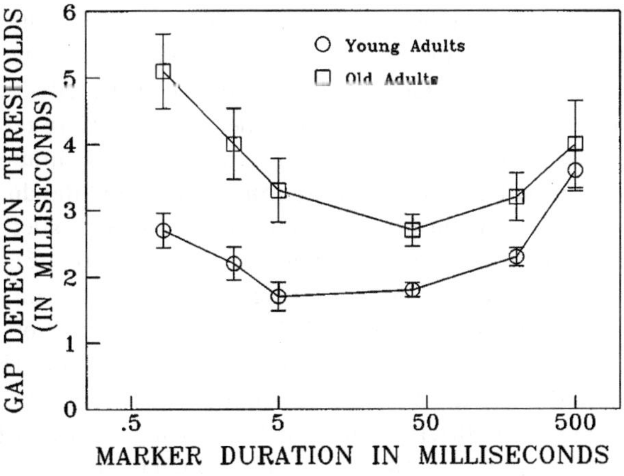

Fig. 5.5 Mean gap detection thresholds as a function of marker duration for the younger and older adults. (Adapted from Schneider and Hamstra 1999.)

Also, it appears that stimulus factors, such as marker duration and intensity; gap location within markers; and type of marker, noise or tones, can influence the magnitude of gap thresholds and age-related performance differences in any given resolution task. Finally, a gap threshold is simply one operational definition of auditory temporal resolution and does not indicate the underlying processes involved in detection. That is, the presence of a gap between stimulus markers reflects a waveform amplitude shift, and thus resolution of brief intensity decrements are needed for gap detection, as discussed and modeled by Plack and Moore (1990). The potential influence of nerve fiber adaptation and offset-onset response effects on gap detection is also cited (e.g., Schneider and Hamstra 1999). The influence of aging on some of these underlying processes is a focus of much ongoing research (see Schmiedt, Chapter 2).

5.5.2 Duration Discrimination

Another aspect of auditory temporal sensitivity concerns the ability of listeners to detect and discriminate changes in the duration of a sound. Stimuli used in these measurements have included samples of both shorter and longer noise or tone bursts, and some experiments measured listeners' discrimination of changes to duration of a silent interval bounded by a pair of stimulus markers. Stimuli for the these measurements are usually presented at intensity levels that are sufficient to ensure adequate signal audibility for the listener. Collective findings from early studies of duration discrimination with young listeners reveal that duration DL increases monotonically as a function of the reference stimulus duration, with relatively little effect of stimulus type over a broad range of reference durations exceeding ~10-20 ms (Creelman 1962; Small and Campbell 1962; Abel 1972; Divenyi and Danner 1977).

The investigation of age-related changes in listeners' sensitivity to changes in stimulus duration has been the focus of several studies in recent years. In one experiment, Fitzgibbons and Gordon-Salant (1994) measured duration DLs in younger and older listeners with and without hearing loss using reference stimuli consisting of 250-ms tones or silent intervals (gaps) of equal duration bounded by a pair of 250-ms tones. Results of this experiment indicated the duration DLs for the tones and gaps were about the same, but the DLs of the older listeners were significantly larger than those of the younger listeners; factors related to hearing loss and stimulus frequency did not have a significant influence on discrimination performance in any of the listener groups. Other measures of duration discrimination collected with band-limited noise (Abel et al. 1990) or tone bursts (Bergeson et al. 2001) also showed diminished temporal sensitivity among older listeners, with the magnitude of age-related discrimination deficits being greater when measured using shorter duration reference stimuli. An effect of the reference stimulus duration on duration discrimination was observed also in the results of a recent experiment (Fitzgibbons et al. 2007) that measured duration DLs for changes in the inter-onset interval separating a pair of successive 20-ms tone pulses separated by a silent interval. Fig. 5.6 shows the mean relative DLs in percent as a function of the reference tonal inter-onset interval for the groups of participating younger and older listeners with normal hearing. These results revealed that the mean relative DLs of older listeners were larger than those of the younger listeners for all values of the reference interval, but the magnitude of the age-related discrimination difference became progressively larger as the reference tonal inter-onset interval was reduced below ~200 ms.

Another potentially important influence on duration discrimination among older listeners is the spectral composition of stimulus markers surrounding a reference silent interval. That is, for many older listeners, difficulties in discriminating changes in the duration of a brief silent interval are observed to be exaggerated in conditions that feature a wide frequency disparity between leading and trailing stimulus markers (Lister et al. 2002; Pichora-Fuller et al. 2006). Other studies that used frequency-disparate stimulus markers also observed age-related differences in gap discrimination (Grose et al. 2006), with the influence of marker frequency disparity on discrimination being greater for brief reference gaps compared to longer reference intervals (e.g., 250 ms). Evidence from these gap discrimination studies also indicates that age effects in duration discrimination may extend to normal-hearing middle-aged listeners (e.g., 40-55 years) who were observed to perform more like older listeners than younger listeners (Grose et al. 2006). Generally, studies on duration discrimination and aging are conducted using clearly audible stimuli and report negligible effects of hearing loss on listener performance measures.

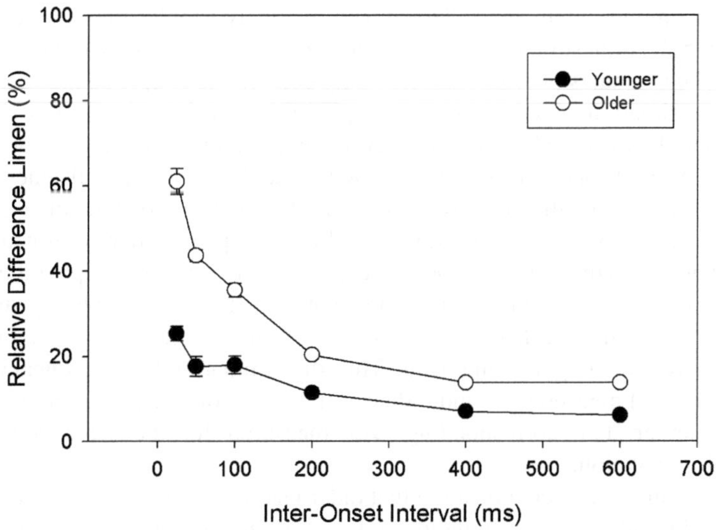

Fig. 5.6 Mean relative difference limens as a function of the tonal inter-onset interval for the younger listeners and older listeners with and without hearing loss. (Adapted from Fitzgibbons et al., 2007.)

Collective evidence from the various experiments indicates that older listeners are less sensitive to changes in stimulus duration than younger listeners. The processing mechanisms implicated in these effects of aging are not well specified and may include both peripheral and central auditory components. For example, most theoretical accounts of the processing of stimulus duration invoke the operation of central timing mechanisms that are modeled as counters that code duration by summing the neural firings elicited during stimulation. The models have been applied successfully to account for a large body of empirical data on duration discrimination collected from young trained listeners (Creelman 1962; Abel 1972; Diveny and Danner 1977). In this theoretical account, the accuracy of counting would depend also on the precision in sensory coding of stimulus onsets and offsets. Thus the observed decrements in duration discrimination for older listeners could be a consequence of diminished central processing mechanisms or of impoverished peripheral coding of stimulus onsets and offsets that might result from a reduced synchronous response of auditory nerve fibers. These more peripheral sensory effects would seem to be particularly relevant toward explaining some of the age-related deficits in discrimination seen for brief tones and noise stimuli or in brief temporal gaps bounded by markers featuring both similar and dissimilar spectral characteristics.

5.6 Temporal Sensitivity: Complex Stimuli

Most of the evidence for an age-related reduction in auditory temporal sensitivity has come from measurements collected with relatively simple stimuli, as described in Section 5.2. A question of particular interest to investigators concerns the extent to which the estimates of temporal sensitivity derived for simple sounds generalize to listening situations featuring more complex time-varying stimuli. For example, Watson and Foyle (1985) reviewed several experiments that used word-length sequences of brief tone pulses and examined the abilities of young listeners to discriminate changes in the properties of a single tonal component of a stimulus sequence. Results of the studies with the multitone sequences demonstrated clearly that listeners' discrimination performance depends not only on peripheral auditory sensitivity but also on a number of factors related to stimulus complexity, discrimination learning, and the degree of listener uncertainty regarding the number and location of variable target components. Thus discrimination DLs (frequency, intensity, or duration) measured for individual components of a sequence could be substantially larger than corresponding DLs measured for the same components presented in isolation.

There is some evidence indicating that older listeners also exhibit an exaggerated degree of difficulty in discrimination tasks involving complex sequential stimuli. Humes and Christopherson (1991) compared the auditory abilities of older and younger listeners using a battery of discrimination tests that featured both simple tonal stimuli and more complex sequences of tone bursts. Most of the age-related performance decrements were observed with the complex stimulus sequences that measured temporal aspects of hearing, such as duration discrimination, perception of sequence rhythm, or the discrimination of temporal order. Later, Fitzgibbons and Gordon-Salant (1995) measured duration DLs for a target tone or a silent interval that was embedded as a single target component in a sequence of five tones that differed in frequency. For older listeners participating in this study, the duration DLs measured for the embedded targets (tones or silent intervals) were substantially larger than the DLs measured for the same targets measured in isolation. The duration DLs for younger listeners in the study were significantly smaller than those of the older listeners, and for embedded tonal targets, discrimination performance was essentially the same as observed for isolated tones of equal reference duration. Thus these results indicate that stimulus complexity can have an important influence on the discrimination performance of older listeners.

Other aspects of aging and temporal processing can be examined by asking listeners to discriminate changes in the rhythm of sequential stimulus patterns. In one such study, Fitzgibbons and Gordon-Salant (2001) used isochronous sequences of five brief tones separated equally by silent intervals to examine the abilities of younger and older listeners to discriminate changes in the temporal spacing of tones in the sequence. Duration DLs for changes of the tonal onset-to-onset intervals were measured for each of several baseline sequence presentation rates. Results showed that older listeners were consistently poorer than younger listeners at discriminating changes in sequence rate over a broad range of baseline-sequence rates. Other conditions of this experiment revealed even greater age-related discrimination differences for localized changes of sequence rhythm imposed by modifications to a single tone interval within the stimulus sequence. None of the measured duration DLs with the stimulus sequences were influenced by factors related to hearing loss.

5.7 Temporal Order Perception

Listener sensitivity to the order of sounds in a sequence is considered a basic aspect of auditory processing and one that is essential to understanding a number of complex stimulus patterns, such as those found in speech or music. Additionally, the processing and recall of temporal order associated with sound sequences undoubtedly involves the contribution of various central processing mechanisms that may undergo changes with aging. Indeed, there is some evidence in the research literature indicating that older listeners do experience difficulty in discrimination and recognition tasks that require temporal order judgments (e.g., Trainor and Trehub 1989; Humes and Christopherson 1991). Moreover, the task demands and processing-time requirements associated with temporal order discrimination and recognition appear to be quite different. For example, in one study, Fitzgibbons and Gordon-Salant (1998) used contiguous three-tone stimulus sequences and observed that listeners could discriminate changes in tone order at sequence rates that permitted only chance-level performance for tone-order recognition. This study also compared the performance of younger and older listeners on the ordering tasks and found significantly reduced ordering abilities among the older listeners. Closer examination of stimulus factors in a subsequent experiment (Fitzgibbons and Gordon-Salant 2006) revealed that the age-related limitations in temporal order processing were determined primarily by stimulus presentation rate; i.e., older listeners required slower sequence rates than younger listeners to achieve equivalent levels of temporal order recognition.

5.8 Summary

The study of aging and auditory processing does not have a long history and is best viewed as being in its early stages. Nevertheless, progress has been made in understanding the processing difficulties experienced by older listeners. Some of these problems are undoubtedly attributed to the consequences of age-related sensorineural hearing loss, but some are not. In this chapter, an attempt was made to cite psychophysical studies that were specifically designed to separate hearing loss and age as independent variables that could influence a listener's auditory processing. A summary of the research findings indicates that some aspects of auditory processing appear to be minimally affected by age, whereas others show large effects. For example, there appears to be little influence of aging on frequency selectivity as measured by auditory filter bandwidths derived from simultaneous masking experiments. Also, the cochlear nonlinear processing mechanisms associated with auditory suppression appear to exhibit diminished function in some older listeners but not in other listeners. One aspect of spectral processing that does appear to undergo changes with aging is frequency discrimination but only for relatively low-frequency stimuli. Similarly, age effects are observed for intensity discrimination, again for low-frequency sounds. The restriction of these discrimination problems to lower-frequency regions has led investigators to suspect the involvement of temporal mechanisms, primarily those associated with phase-locked responses of auditory nerve fibers.

Age effects are more consistently observed in measures of auditory temporal sensitivity. However, for some temporal resolution measures, particularly gap detection, the observation of age-related performance differences seems to depend on parameters of the stimuli used in the measurements. Present evidence indicates that older listeners demonstrate the greatest difficulty in resolving temporal gaps surrounded by relatively brief stimulus markers, or markers that differ substantially in spectral characteristics. Temporal measures of duration discrimination show the most consistent effects of aging. This is the case for reference stimuli defined by tone or noise bursts and for empty intervals defined by temporal gaps. The age-related difficulties with duration discrimination have been observed for stimuli of various reference durations, although the largest age effects are evident for short-duration stimuli. Also, unlike many auditory measures, most estimates of duration discrimination appear to be relatively unaffected by hearing loss as long as the stimuli are clearly audible to the listener. This outcome points to a probable role of central auditory mechanisms in the processing of stimulus duration, although diminished sensory coding of stimulus boundaries may contribute to some of the age-related discrimination problems observed for short-duration sounds. Additionally, the discrimination of complex sequential stimuli and the processing of auditory temporal order seem particularly difficult for older listeners. Listener performance with these more complex stimulus patterns may be influenced by a combination of sensory and central factors that undergo changes with aging.

Much work remains to be done to advance understanding of the processing changes associated with aging. It is evident from the existing body of research that older listeners can be quite variable on several measures of auditory performance. This variability indicates a need for better ways to define auditory aging because chronological age appears to be an imperfect predictor of listening performance. It is clear also that many of the listening problems among older listeners are likely to involve central processing mechanisms and cognitive factors that influence performance measures in several of the psychophysical tasks, particularly those featuring complex sounds and greater task demands. Better understanding of these factors will emerge as the number and scope of investigations on aging and audition continue to grow.

References

Abel S (1972) Discrimination of temporal gaps. J Acoust Soc Am 52:519–524.

Abel S, Krever E, Alberti PW (1990) Auditory detection, discrimination, and speech processing in aging, noise-sensitive and hearing-impaired listeners. Scand Audiol 19: 43–54.

Bergeson TR, Schneider BA, Hamstra SJ (2001) Duration discrimination in younger and older adults. Can Acoust 29:3–9.

Bergman M, Blumenfeld MS, Cascardo D, Dash B, Levitt H, Margulies MK (1976). Age-related decrement in hearing for speech. J Gerontol 31:533–538.

Committee on Hearing, Bioacoustics, and Biomechanics (CHABA) (1988) Speech understanding and aging. J Acoust Soc Am 83:859–895.

Corso JF (1963) Age and sex differences in pure-tone thresholds. Arch Otolaryngol 77:385–405.

Creelman CD (1962) Human discrimination of auditory duration. J Acoust Soc Am 34:582–593.

Cruickshanks KJ, Wiley TL, Tweed TS, Klein BE, Klein R, Mares-Perlman JA, Nondahl DM (1998) Prevalence of hearing loss in older adults in Beaver Dam, Wisconsin: the epidemiology of hearing loss study. Am J Epidemiol 148:879–886.

de Boer E, Bouwmeester J (1974) Critical bands and sensorineural hearing loss. Audiology 13:236–259.

Divenyi PL, Danner WF (1977) Discrimination of time intervals marker by brief acoustic pulses of various intensities and spectra. Percept Psychophys 21:125–142.

Dubno JR, Ahlstrom JB (2001a) Forward-and simultaneous-masked thresholds in bandlimited maskers in subjects with normal hearing and cochlear hearing loss. J Acoust Soc Am 110:1049–1057.

Dubno JR, Ahlstrom JB (2001b) Psychophysical suppression measured with bandlimited noise extended below and/or above the signal: effects of age and hearing loss. J Acoust Soc Am 110:1058–1066.

Fitzgibbons PJ (1983) Temporal gap detection in noise as a function of frequency, bandwidth, and level. J Acoust Soc Am 74:67–72.
Fitzgibbons PJ, Gordon-Salant S (1994) Age effects on measures of auditory duration discrimination. J Speech Hear Res 37:662–670.
Fitzgibbons PJ, Gordon-Salant S (1995) Age effects on duration discrimination with simple and complex stimuli. J Acoust Soc Am 98:3140–3145.
Fitzgibbons PJ, Gordon-Salant S (1998) Auditory temporal order perception in younger and older adults. J Speech Hear Lang Res 41:1052–1060.
Fitzgibbons PJ, Gordon-Salant S (2001) Aging and temporal discrimination in auditory sequences. J Acoust Soc Am 109:2955–2963.
Fitzgibbons PJ, Gordon-Salant S (2006) Effects of age and sequence presentation rate on temporal order recognition. J Acoust Soc Am 120:991–999.
Fitzgibbons PJ, Wightman FL (1982) Gap detection in normal and hearing-impaired listeners. J Acoust Soc Am 72:761–765.
Fitzgibbons PJ, Gordon-Salant S, Barrett J (2007) Age-related differences in discrimination of an interval separating onsets of successive tone bursts as a function of interval duration. J Acoust Soc Am 122:458–466.
Florentine M, Buus S. (1984) Temporal gap detection in sensorineural and simulated hearing impairments. J Speech Hear Res 27:449–455.
Florentine M, Buus S, Scharf B, Zwicker E (1980) Frequency selectivity in normally-hearing and hearing-impaired observers. J Speech Hear Res 23:457–461.
Florentine M, Reed CM, Rabinowitz WM, Braida LD, Durlach NI, Buus S (1993) Intensity discrimination in listeners with sensorineural hearing loss. J Acoust Soc Am 94:2575–2586.
Formby C, Forrest TG (1991) Detection of silent temporal gaps in sinusoidal markers. J Acoust Soc Am 89:830–837.
Freyman RL, Nelson DA (1991) Frequency discrimination as a function of signal frequency and level in normal-hearing and hearing-impaired listeners. J Speech Hear Res 34:1371–1386.
Gates GA, Cooper JC, Kannel WB, Miller NF (1990) Hearing in the elderly: the Framingham cohort, 1983–1985. Ear Hear 11:247–256.
Gates GA, Schmid P, Kujawa SG, Nam B, d'Agostino R (2000) Longitudinal threshold changes in older men with audiometric notches. Hear Res 141:220–228.
Gifford RH, Bacon SP (2005) Psychophysical estimates of nonlinear cochlear processing in younger and older listeners. J Acoust Soc Am 118:3823–3833.
Glasberg BR, Moore BCJ (1986) Auditory filter shapes in subjects with unilateral and bilateral cochlear impairments. J Acoust Soc Am 79:1020–1033.
Glasberg BR, Moore BCJ (1990) Auditory filter shapes derived from notched-noise data. Hear Res 47:103–138.
Glasberg BR, Moore BCJ, Bacon SP (1987) Gap detection and masking in hearing impaired and normal hearing subjects. J. Acoust Soc Am 81:1546–1556.
Green DM (1971) Temporal auditory acuity. Psychol Rev 78:540–551.
Green DM, Forrest TG (1989) Temporal gaps in noise and sinusoids. J Acoust Soc Am 86:961–970.
Grose JH, Hall JW III, Buss E (2006) Temporal processing deficits in the pre-senescent auditory system. J Acoust Soc Am 119:2305–2315.
He N, Dubno JR, Mills JH (1998) Frequency and intensity discrimination measured in a maximum-likelihood procedure from young and aged normal-hearing subjects. J Acoust Soc Am 103:553–565.
He N, Horowitz AR, Dubno JR, Mills JH (1999) Psychometric functions for gap detection in noise measured from young and aged subjects. J Acoust Soc Am 106:966–978.
He N, Mills JH, Dubno JR (2007) Frequency modulation detection: Effects of age, psychophysical method, and modulation waveform. J Acoust Soc Am 122:467–477.
Helfer K, Wilbur L (1990) Hearing loss, aging, and speech perception in reverberation and noise. J Speech Hear Res 33:149–155.
Helzner EP, Cauley JA, Pratt SR, Wisniewski SR, Zmuda JM, Talbott EO, deRekeneire N, Harris TB, Rubin SM, Simonsick EM, Tylavsky FA, Newman AB (2005) Race and sex differences

in age-related hearing loss: the health, aging and body composition study. J Am Geriatr Soc 53:2119–2127.

Hicks ML, Bacon SP (1999) Effects of aspirin on psychophysical measures of frequency selectivity, two-tone suppression, and growth of masking. J Acoust Soc Am 106: 1436–1451.

Houtgast T (1972) Psychophysical evidence for lateral inhibition in hearing. J Acoust Soc Am 51:1885–1894.

Humes L, Christopherson L (1991) Speech identification difficulties of hearing-impaired elderly persons: the contributions of auditory processing deficits. J Speech Hear Res 34:686–693.

Konkle D, Beasley D, Bess F (1977) Intelligibility of time-altered speech in relation to chronological aging. J Speech Hear Res 20:108–115.

Lee F-S, Matthews LJ, Dubno JR, Mills JH (2005) Longitudinal study of pure-tone thresholds in older persons. Ear Hear 26:1–11.

Liberman MC, Dodds LW (1984) Single-neuron labeling and chronic cochlear pathology. III. Stereocilia damage and alterations of threshold tuning curves. Hear Res 16:55–74.

Lister J, Besing J, Koehnke J (2002) Effects of age and frequency disparity on gap discrimination. J Acoust Soc Am 111:2793–2800.

Matthews LJ, Lee F-S, Mills JH, Dubno JR (1997) Extended high-frequency thresholds in older adults. J Speech Lang Hear Res 40:208–214.

Moore BCJ (1973) Frequency difference limens for short-duration tones. J Acoust Soc Am 54:610–619.

Moore BCJ, Glasberg BR (1981) Auditory filter shapes derived in simultaneous and forward masking. J Acoust Soc Am 70:1003–1014.

Moore BCJ, Glasberg BR (1988) Gap detection with sinusoids and noise in normal, impaired, and electrically stimulated ears. J Acoust Soc Am 83:1093–1101.

Moore BCJ, Peters RW (1992) Pitch discrimination and phase sensitivity in young and elderly subjects and its relationship to frequency selectivity. J Acoust Soc Am 91: 2881–2893.

Moore BCJ, Vickers DA (1997) The role of spread of excitation and suppression in simultaneous masking. J Acoust Soc Am 102:2284–2290.

Moore BCJ, Peters RW, Glasberg BR (1992) Detection of temporal gaps in sinusoids by elderly subjects with and without hearing loss. J Acoust Soc Am 92:1923–1932.

Moscicki EK, Elkins EF, Baum HM, McNamara PM (1985) Hearing loss in the elderly: an epidemiologic study of the Framingham Heart Study Cohort. Ear Hear 6:184–190.

Patterson RD (1976) Auditory filter shapes derived with noise stimuli. J Acoust Soc Am 59:640–654.

Patterson RD, Nimmo-Smith I, Weber DL, Milroy R (1982) The deterioration of hearing with age: frequency selectivity, the critical ratio, the audiogram and speech threshold. J Acoust Soc Am 72:1788–1803.

Pearson JD, Morrell CH, Gordon-Salant S, Brant LJ, Metter EJ, Klein LL, Fozard JL (1995) Gender differences in a longitudinal study of age-associated hearing loss. J. Acoust Soc Am 97:1196–1205.

Penner MJ (1977) Detection of temporal gaps as a measure of the decay of auditory sensation. J Acoust Soc Am 61:552–557.

Peters RW, Moore BCJ (1992) Auditory filters shapes at low center frequencies in young and elderly hearing-impaired subjects. J Acoust Soc Am 91:256–266.

Phillips DP, Taylor TL, Hall SE, Carr MM, Mossop JE (1997) Detection of silent intervals between noises activating different perceptual channels: some properties of "central" auditory gap detection. J Acoust Soc Am 101:3694–3705.

Pichora-Fuller MK, Schneider BA, Benson NJ, Hamstra SJ, Storzer E (2006) Effect of age on detection of gaps in speech and nonspeech markers varying in duration and spectral symmetry. J Acoust Soc Am 119:1143–1155.

Pick GF, Evans EF, Wilson JP (1977) Frequency resolution in patients with hearing loss of cochlear origin. In: Evans EF, Wilson JP (eds) Psychophysics and Physiology of Hearing. London: Academic Press, pp. 273–282.

Plack CJ, Moore BCJ (1990) Temporal window shape as a function of frequency and level. J Acoust Soc Am 87:2178–2187.

Plomp R (1964) Rate of decay of auditory sensation. J Acoust Soc Am 36:277–282.
Salthouse TA (1985) A Theory of Cognitive Aging. New York: North-Holland.
Schmiedt RA, Zwislocki JJ, Hamernik RP (1980) Effects of hair cell lesions on responses of cochlear nerve fibers. I. Lesions, tuning curves, two-tone inhibition, and responses to trapezoidal-wave patterns. J Neurophysiol 43:1367–1389.
Schneider BA, Hamstra SJ (1999) Gap detection thresholds as function of tonal duration for younger and older listeners. J Acoust Soc Am 106:371–379.
Schneider BA, Pichora-Fuller MK, Kowalchuk B, Lamb M (1994) Gap detection and the precedence effect in young and old adults. J Acoust Soc Am 95:980–991.
Sek A, Moore BCJ (1995) Frequency discrimination as a function of frequency, measured in several ways. J Acoust Soc Am 97:2479–2486.
Shailer MJ, Moore BCJ (1983) Gap detection as function of frequency, bandwidth, and level. J Acoust Soc Am 74:467–473.
Shailer MJ, Moore BCJ (1987) Gap detection and the auditory filter: phase effects using sinusoidal stimuli. J Acoust Soc Am 81:1110–1117.
Shannon RV (1976) Two-tone unmasking and suppression in a forward-masking situation. J Acoust Soc Am 59:1460–1470.
Simon HJ, Yund EW (1993) Frequency discrimination in listeners with sensorineural hearing loss. Ear Hear 14:190–201.
Small AM, Campbell RA (1962) Temporal differential sensitivity for auditory stimuli. Am J Psychol 75:401–410.
Snell KB (1997) Age-related changes in temporal gap detection. J Acoust Soc Am 101: 2214–2220.
Sommers MS, Gehr SE (1998) Auditory suppression and frequency selectivity in older and younger adults. J Acoust Soc Am 103:1067–1074.
Sommers MS, Humes LE (1993) Auditory filter shapes in normal-hearing, noise-masked normal, and elderly listeners. J Acoust Soc Am 93:2903–2914.
Stelmachowicz PG, Beauchaine KA, Kalberer A, Jesteadt W (1989) Normative thresholds in the 8-to 20-kHz range as a function of age. J Acoust Soc Am 86:1384–1391.
Sticht T, Gray B (1969) The intelligibility of time-compressed words as function of age and hearing loss. J Speech Hear Res 12:443–448.
Strouse A, Ashmead DH, Ohde RN, Grantham DW (1998) Temporal processing in the aging auditory system. J Acoust Soc Am 104:2385–2399.
Terry M, Moore BCJ (1977) Suppression effects in forward masking. J Acoust Soc Am 62:781–784.
Trainor LJ, Trehub SE (1989) Aging and auditory temporal sequencing: ordering the elements of repeating tone patterns. Percept Psychophys 45:417–426.
Turner CW, Nelson DA (1982) Frequency discrimination in regions of normal and impaired sensitivity. J Speech Hear Res 25:34–41.
Tyler RS, Summerfield AQ, Wood EJ, Fernandes MA (1982) Psychological and phonetic temporal processing in normal and hearing impaired listeners. J Acoust Soc Am 73:740–752.
Watson CS, Foyle DC (1985) Central factors in the discrimination and identification of complex sounds. J Acoust Soc Am 78:375–380.
Weber DL, Green DM (1978) Temporal factors and suppression effects in backward and forward masking. J Acoust Soc Am 64:1392–1399.
Wightman FL, McGee T, Kramer M (1977) Factors influencing frequency selectivity in normal and hearing impaired listeners. In: Evans EF, Wilson JP (eds) Psychophysics and Physiology of Hearing. London: Academic Press, pp. 295–306.
Wiley TL, Cruickshanks KJ, Nondahl DM, Tweed TS, Klein R, Klein BEK (1998) Aging and high-frequency hearing sensitivity. J Speech Lang Hear Res 41:1061–1072.
Williams KN, Perrott DR (1972) Temporal resolution of tone pulses. J Acoust Soc Am 51:644–647.
Wingfield A, Poon LW, Lombardi L, Lowe D (1985) Speed of processing in normal aging: effects of speech rate, linguistic structure, and processing time. J Gerontol 40: 579–585.
Yates, GK (1995) Cochlear structure and function. In: Moore BCJ (ed) Hearing. San Diego: Academic Press, pp. 41–74.

Chapter 6
Binaural Processing and Auditory Asymmetries

David A. Eddins and Joseph W. Hall III

6.1 Introduction

Many aspects of peripheral and central auditory processing peak by adolescence (Maxon and Hochberg 1982; Hall et al. 2004, 2005) and a growing body of evidence indicates a substantial decline in audition in presenescent adults (Gates et al. 1990; Lee et al. 2005; Grose et al. 2006) that continues to deteriorate with advancing age (for reviews, see Schneider 1997; Divenyi and Simon 1999; Chisholm et al. 2003). The functional significance of this decline is often gauged in terms of the accurate perception of speech in complex acoustic backgrounds (e.g., Committee on Hearing, Bioacoustics, and Biomechanics [CHABA] 1988; van Rooij and Plomp 1990; Humes et al. 1994; Pichora-Fuller 1997; Divenyi et al. 2005), which, in turn, relies on both monaural and binaural auditory processing. Declines with age in monaural and binaural auditory processing are often compounded by concomitant peripheral hearing loss, resulting in reduced audiometric sensitivity, reduced frequency selectivity, and increased linearity of coding of intensity.

For young listeners with normal hearing, the accurate perception of many sounds can be accomplished on the basis of the input from a single ear. Monaural coding of intensity, frequency, and time leads to an acute perception of changes in loudness, pitch, timbre, and sound duration as well as the recognition of patterns of intensity changes across time and frequency. Such monaural processing supports the perception of speech, music, and environmental sounds and the ability to integrate related acoustic features and parse other acoustic features. With identical

D.A. Eddins (✉)
Department of Otolaryngology, University of Rochester, 2365 S. Clinton Avenue, Suite 200, Rochester, NY 14618,
e-mail: david_eddins@urmc.rochester.edu

J.W. Hall III
Department of Otolaryngology, University of North Carolina, Chapel Hill, 170 Manning Drive, CB: 7070, Chapel Hill, NC 27599-7070,
e-mail: joseph_hall@med.unc.edu

S. Gordon-Salant et al. (eds.), *The Aging Auditory System*,
Springer Handbook of Auditory Research 34,
DOI 10.1007/978-1-4419-0993-0_6, © Springer Science+Business Media, LLC 2010

input from both ears (diotic listening), modest improvements in signal detection and discrimination may be observed. The most important benefits of binaural hearing, however, result from the comparison of disparate inputs from the two ears that are analyzed centrally. Binaural hearing greatly improves our ability to locate sound sources in space, judge source distance, focus on the momentarily desirable sounds in the environment to the exclusion of other simultaneously occurring sounds (i.e., signal versus background), and to quickly develop an awareness of the physical characteristics of the listening environment. Age-related deficits in monaural processing (as detailed in Fitzgibbons and Gordon-Salant, Chapter 5) along with deficits in binaural processing combine to gradually reduce the advantages of binaural hearing with increasing age.

Investigations of binaural processing span a wide range of psychophysical tasks. Sound localization involves the identification of the position of the sound source in space relative to the position of the listener and is measured in the free field. The importance of sound source localization is obvious and includes enhancements in safety, environmental awareness, and communication. When listening with headphones or insert phones, sounds may appear as diffuse images within the head, may be lateralized to a specific location somewhere within the head, or may be perceived as emanating from the center of the head. Dichotic presentation may provide an internalized spatialization of the sound stage having important counterparts to the externalized spatialization that occurs for music in the free field. Auditory scientists realized many years ago that the presentation of stimuli via headphones permits very precise control of interaural parameters, including parameters that cannot occur in free-field listening, allowing systematic investigation of those parameters to better understand binaural hearing and underlying mechanisms (e.g., Jeffress 1948; Klump and Eady 1956). Thus much of what we know about binaural hearing has emerged from a variety of binaural listening tasks performed under headphones.

Although many studies of binaural hearing have involved the localization, lateralization, detection, or discrimination of contrived laboratory stimuli to explore the nature of the binaural system, an equally large body of research has focused on the perception of speech sounds in various binaural conditions under headphones and in the free field to better understand the nature of speech perception in natural environments.

The potential effects of aging per se on binaural hearing have received much less attention to date than the effects of hearing loss on binaural hearing. Furthermore, because hearing sensitivity declines with increasing age, separating the effects of aging and hearing loss on binaural processing suffers from the same problems discussed by Fitzgibbons and Gordon-Salant (Chapter 5) in the context of monaural processing. The most common experimental approaches to age-related changes in hearing include evaluating (1) younger and older listeners with normal hearing, (2) younger and older listeners with hearing loss, (3) older listeners with hearing loss and younger listeners with simulated hearing loss, or (4) unselected older listeners followed by statistical analyses used to separate the effects of age and hearing loss. The following overview of age-related changes in binaural processing includes examples of each of these approaches. Where data are sparse or not available, consideration will be given to known age-related changes in binaural processing and other aspects of the aging auditory system to provide insight into potential age effects.

6.2 Perception of Auditory Space in the Free Field

6.2.1 Duplex Theory

Sound originating from a point source in an anechoic space may differ in terms of the time of arrival at the two ears, resulting in interaural time differences (ITDs), and may differ in the intensity at the two ears, resulting in interaural intensity differences (IIDs). If one considers the head to be a simple sphere, then ITDs are dependent on the speed of sound in air, the distance between the ears, and the angle of incidence of the sound source relative to the head. The perception of the ITDs is dependent on the time resolution of the binaural system and the frequency of the sound. For low-frequency sounds with relatively long wavelengths, the delay between ears translates to a direction-dependent phase shift within a single cycle of the waveform. For high-frequency sounds with shorter wavelengths, the phase shift can span more than one cycle and thus provide ambiguous directional information.

IIDs are dependent on the angle of incidence, the size of the head, and the resulting acoustic shadow cast by the head as well as the frequency of the incoming sound. The wavelengths of high-frequency sounds are small relative to the diameter of the head and are diffracted to a greater extent, producing larger IIDs than low-frequency sounds, where the wavelengths are large relative to the diameter of the head. Lord Rayleigh (Strutt 1907) championed the notion of a duplex theory of sound localization, surmising that high-frequency sounds would be localized on the basis of IID cues and low-frequency sounds on the basis of ITD cues.

The position of a sound source is generally specified relative to horizontal and vertical planes that both bisect the head and cross in the center of the head. As shown in Figure 6.1, locations in the horizontal plane are specified as the difference in angle (azimuth) between a given location and the median plane. Locations in the vertical plane are specified as the difference in angle (elevation) between a given location and the horizontal plane. For sources positioned at 0° elevation, horizontal localization is largely dependent on ITD and IID cues, whereas for sources located in the median plane (0° azimuth), vertical localization primarily depends on spectral cues. The localization of sources off the 0° azimuth and 0° elevation references depends on a combination of ITD, IID, and spectral cues. Despite the strong empirical and theoretical evidence supporting the basic tenets of duplex theory of sound localization, it should be emphasized that although the sound from a single source may produce different inputs at the two ears, the binaural system integrates those inputs into a single percept.

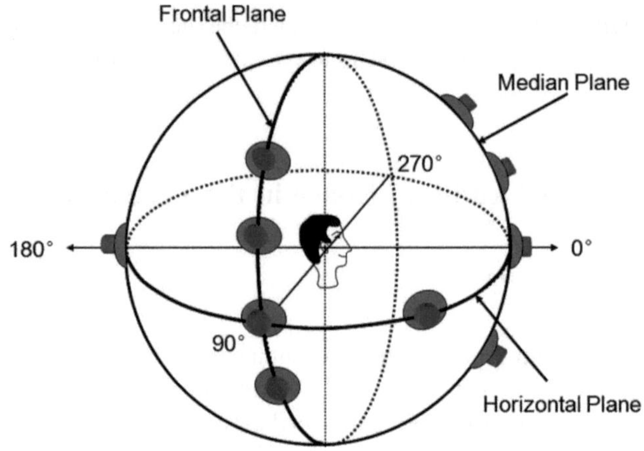

Fig. 6.1 Coordinate system for localizing sounds in free field. (Adapted from Blauert 1983 with permission.)

6.2.2 Spectral Cues in Sound Localization

Monaural spectral cues arise from the direction-dependent filtering functions of the pinnae. Such cues are a by-product of the complex delay-and-add network produced by the individual ear-specific convolutions of the pinnae and are colored somewhat by the direction-dependent features of the head, neck, and torso. To optimally use the spectral changes imposed on the sound source by the pinnae, the spectrum of the original sound source must be known by the listener. In laboratory settings, this is straightforward, but in the real world, the source spectrum is uncertain. The use of spectral cues, therefore, relies on a sort of template matching and most likely a knowledge base of the spectral features of common sounds. For a given source location, the position-dependent filtering consists of a series of intensity peaks and valleys across audiofrequency. Changes in source elevation result in systematic changes in the location of those peaks and valleys of the filter function (e.g., Mehrgart and Mellert 1977).

Spectral cues are known to be important for elevation localization, both in the median plane where spectral cues dominate and off the median and horizontal planes where ITD and IID cues define azimuthal position but not elevation (Roffler and Butler 1968; Blauert 1969; Gardner and Gardner 1973; Wright et al. 1974; Butler and Planert 1976). The importance of pinna cues in sound localization was highlighted in a pinna occlusion experiment (Oldfield and Parker 1984), in which the insertion of a mold in the curves of the pinna produced poorer localization of sound elevation but no corresponding change in azimuthal localization. Although sound localization is generally considered in the context of binaural hearing, listeners with unilateral deafness have limited sound localization abilities, even in the absence of head movements, which has been attributed to the use of monaural spectral cues (e.g., Oldfield and Parker 1986; Butler et al. 1990). Furthermore, studies of both monaural and binaural spectral cues (Gardner 1973; Butler and Planert 1976) have shown more accurate binaural than monaural sound localization in the median plane, which implies the use of combined binaural spectral cues by the auditory system.

6.2.3 *Effects of Age and Hearing Loss on Sound Localization*

6.2.3.1 Horizontal Sound Localization

Despite the fact that rigorous investigations of sound localization have been ongoing since the pioneering work of Stevens and Newman (1936), remarkably few investigations have addressed the possible changes in sound localization ability with advancing age. Several studies have included younger and older listeners in the context of investigating the relationship between sound localization ability and sensorineural hearing loss, of which presbycusis is the most common etiology. One of the few studies that directly addressed aging and sound localization was conducted in the context of assessing the effects of hearing protection devices on localization in groups ($n = 24$) of younger and older listeners with clinically normal hearing and older listeners with sensorineural hearing loss (Abel and Hay 1996). Listeners identified the apparent location of the sound source by button press on a response box, with buttons positioned in a circle that represented the array of 6 speakers separated by 60° (30°, 90°, 150°, 210°, 270°, and 330°). Percent correct identification in the forced-choice task was computed for one-third octave noise bands centered at 500 and 4,000 Hz. In their quiet condition without any hearing protection device, localization accuracy in the 500- and 4,000-Hz conditions was significantly worse, by 8% for older (41 to 58 years) listeners with normal-hearing than for younger (18-38 years) listeners with normal hearing. Performance was significantly worse for the hearing-impaired than for the older normal-hearing group. Age was not a significant factor in left/right discrimination ability. For front/back discrimination in the quiet, unoccluded condition, localization accuracy was better for the 4,000- than for the 500-Hz condition and was significantly worse, by 10% for older listeners with normal hearing than the younger group. Poorer performance by the older than the younger normal-hearing group at both frequencies is consistent with potential age-related deficits in the use of both ITD (i.e., 500 Hz) and IID (i.e., 4,000 Hz) cues.

Perhaps the most comprehensive study to date investigating the effects of aging on sound localization was an extension of the work of Abel and Hay (1996) by Abel et al. (2000), who focused exclusively on localization in the horizontal plane. In that study, listeners heard broadband or one-third octave band noise bursts from one of an array of loudspeakers surrounding the listener in the horizontal plane using the same task as described above. Experimental conditions varied in terms of the number of speakers in the array, the separation between adjacent speakers (in azimuthal angle), and the stimulus type. A total of 112 listeners were separated by age in decades into 7 groups of 16 listeners ranging from 10-19 to 60-69 years, with the final group ranging from 70 to 81 years. Inclusion criteria permitted mild high-frequency hearing loss, primarily in the older groups. Collapsing across experimental conditions, localization accuracy declined 15% from the youngest to the oldest age range and nearly all of that decline was attributed to listeners 40 years and older. To determine the extent to which age rather than hearing sensitivity contributed to the apparent age-dependent decline in localization accuracy, a series of multiple linear regressions was reported, and in all but one condition, age accounted for a significant proportion of the variance above and beyond the contributions of hearing thresholds. Subsequent analyses revealed that accuracy did not decline significantly between the first and the third age groups but did decline significantly from the third to the seventh age groups. Significant age-related declines in localization accuracy were reported for the broadband noise as well as the one-third octave bands centered at 500 and 4,000 Hz; however, there were differences across the three noise conditions. The age effects observed for the broadband and 4,000-Hz noises were largely attributable to errors in the localization of sources behind the listener, whereas the age effects for the 500-Hz condition indicated an increased number of errors for stimuli located on the right than on the left side. The primary acoustic difference between sources located in front and in back of the listener is the spectral shape resulting from pinna filtering. Because front/back confusions increased with age, one possibility is that the ability to process spectral-shape cues declines with age. Such a conclusion based on localization data is supported by the work of Weinrich (1982) who showed that moving a broadband source from the front to the rear results in the introduction of a low-frequency spectral notch (~1,200 Hz) and a broad midfrequency peak (~5.000 Hz) to the sound reaching the ear canal. When considering front versus back classifications, Abel et al. (2000) noted that substantially more front/back errors occurred for stimuli on the right than on the left side. They suggested that, consistent with the observations of Butler (1994), these results might point to an advantage for processing spectral-shape information in the right versus left cerebral hemifield. A similar right hemisphere advantage was revealed by positron emission tomography (PET) imaging during an active spectral-shape classification task under conditions of minimal spectral uncertainty (Eddins et al. 2001).

Noble et al. (1994) showed that older listeners with moderate flat versus moderate, gradually sloping sensorineural hearing loss differed in their ability to localize sources in the lateral horizontal plane and suggested that this might reflect a reduced ability to use spectral cues associated with such source positions. Because the two-subject subgroups differed in their pure-tone thresholds in the 5,000-Hz but not in the 1,000-Hz region, the role of mid- to high-frequency spectral cues is emphasized in lateral horizontal localization. Interestingly, Noble et al. (1994) concluded that although older listeners with hearing loss clearly show declines in localization performance relative to younger listeners with normal hearing, performance differences are not well predicted by the degree or configuration of hearing loss alone, allowing for the possibility that age-related factors may contribute substantially to individual differences. Unfortunately, age was not considered in their data analysis.

6.2.3.2 Vertical Sound Localization

The ability to perceive changes in source elevation is a fundamental component of sound localization and binaural processing in general. To date, there are no published studies on vertical sound localization that have specifically addressed potential changes with advancing age separate from hearing loss. Here, several studies that have explored vertical sound localization in older listeners with hearing loss are discussed. It has been known since the pioneering work of Batteau (1967), Butler (1969), and Blauert (1969) that relatively high-frequency (>6,000 Hz) spectral information in broadband sounds provides the primary cue for vertical sound localization. The logical extension that high-frequency hearing loss might lead to poorer vertical localization was confirmed by Butler (1970). Because typical presbycusic hearing loss is characterized by a gradually sloping high-frequency sensorineural hearing loss, it stands to reason that such listeners should have poorer than normal vertical localization ability.

The older listeners with sensorineural hearing loss who participated in the Noble et al. (1994) study demonstrated poor sound localization in the median vertical plane that was correlated with high-frequency (6,000 and 8,000 Hz) pure-tone thresholds. Their localization in the lateral vertical plane also was poor and was correlated with pure-tone thresholds from 4,000 to 8,000 Hz. Analysis of the same two subgroups of listeners mentioned above revealed that localization accuracy of the subgroup with moderate, flat sensorineural hearing loss was worse than that in the normal-hearing group in the median vertical plane but was clearly above chance levels, whereas the accuracy of listeners with moderate, sloping sensorineural hearing loss was close to chance. The latter group apparently mapped sources in the median vertical plane to positions in the horizontal plane. Localization of sources in the lateral vertical plane by the same two subgroups was poorer than for localization of sources in the median vertical plane; however, the subgroup with flat hearing losses again was more accurate in their localization than the group with sloping hearing loss. These results are consistent with later work by Noble et al. (1997) and in combination serve to confirm the importance of high-frequency hearing sensitivity to localization in the vertical plane. Noble and colleagues concluded that although high-frequency hearing sensitivity played a major role in sound localization in the vertical plane, knowledge of the audiogram alone was insufficient to predict localization performance in either the frontal or lateral horizontal or vertical planes. Unfortunately, the relationship between age and sound localization among the 144 middle-aged to older listeners was not considered.

Rakerd et al. (1998) also investigated median plane sound localization in 25 presbycusic listeners ranging in age from 66 to 88 years with sloping high-frequency sensorineural hearing losses of varying degrees. These investigators examined three localization tasks, each involving only three distinct sound source positions. Figure 6.2 shows the results of their elevation task in which speakers were located at elevations of −15, 0, and 15°. The data for each subject are labeled with a letter from A to Y and are ordered from the least to the greatest hearing loss based on a pure-tone average, progressing from left to right and top to bottom.

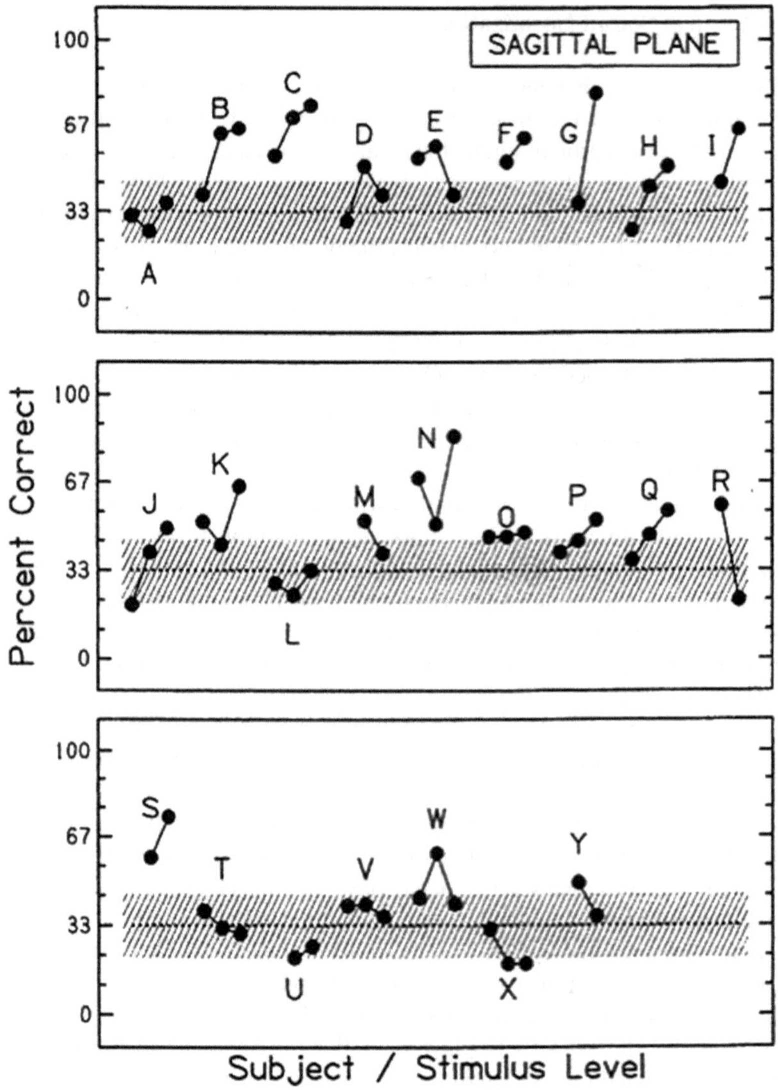

Fig. 6.2 Sound localization in the median plane as a function of degree of hearing loss. Percent correct location identification is shown for 25 subjects labeled A to Y. For each subject, localization was measured at three stimulus levels, shown by connected symbols. Note that the best performance was not always at the highest level. The important feature to note is that subjects are ordered by degree of hearing loss based on a pure-tone average from subject A (least loss) to Y (most loss). The fact that localization performance is not so ordered is indicative of the weak relationship between degree of hearing loss and localization performance. (From Rakerd et al. 1998 with permission.)

For each subject, there are 3 data points reflecting test levels of 48-, 66-, and 84-dB SPL. The shaded area indicates the 5% confidence interval above and below chance performance. The best performance corresponded to different levels for different listeners. More importantly, note that the percent correct performance was not a monotonic function of hearing level, providing strong evidence that something in addition to degree of hearing loss contributes substantially to sound localization. Given that the average performance (data not shown) for their normal-hearing group (mean age = 23.5 years) ranged from 78 to 84% correct across the 3 test levels, it is clear that the hearing loss associated with presbycusis results in a severely impaired vertical localization.

6.2.4 Minimum Audible Angles

In addition to evaluating sound localization accuracy and associated errors, a common measure of binaural hearing is the minimum audible angle (MAA) or the smallest change in perceived location along a single dimension (e.g., azimuth or elevation). As noted by Wightman and Kistler (1993), determination of the just-noticeable difference in the angular separation between speaker positions is useful in the context of understanding and characterizing binaural hearing but may provide little useful information about the apparent position of a sound source. In a study of the relationship between binaural processing and hearing impairment, Hausler et al. (1983) investigated the MAA for broadband noise presented from speakers located at multiple reference locations in the horizontal and vertical planes. In the horizontal plane, the listeners with sensorineural hearing loss had MAAs that ranged from well within the range of MAAs for normal-hearing listeners to MAAs that were considerably worse than normal. Interestingly, the listeners with MAAs within the normal range also had relatively good speech perception abilities, whereas the listeners with worse than normal MAAs generally had poorer speech perception abilities. When measuring MAAs in the vertical plane, listeners wore ear plugs in the ear contralateral to the test ear, permitting an evaluation of the use of monaural spectral-shape cues. Again, one subset of listeners with sensorineural hearing loss had MAAs that were within the range of MAAs obtained by the normal-hearing listeners as well as relatively good speech perception abilities. For other listeners with sensorineural hearing loss, MAAs could not be measured and those listeners generally had relatively poor speech perception abilities. Because listeners with hearing loss spanned a wide age range, the apparent relationship between performance on the MAA and speech perception tasks raises the interesting possibility of a complex relationship among etiology, age, and underlying central auditory mechanisms required for the two tasks.

The potential effect of aging on the discrimination of MAAs in the horizontal plane was investigated by Chandler and Grantham (1991) in listeners with normal or near-normal hearing ranging in age from 60 to 80 years. These results can be compared with the performance of younger listeners with normal hearing reported by Chandler and Grantham (1990). Measurements were made in a darkened anechoic chamber to minimize visual cues and the potential effects of reverberation. MAAs for narrowband noise bursts of different bandwidths and center frequencies were larger for the older than for the younger listeners, reflecting an age-related decline in binaural processing.

6.2.5 Distance Perception

The perception of sound source distance is dependent on a complex set of monaural and binaural acoustic cues. To our knowledge, investigations of the potential declines in distance perception with increasing age have not been published. As summarized by Grantham (1995), the primary cues for distance perception are sound level, the ratio of direct to reverberant energy, spectral shape, and binaural (IID and ITD) cues. Based on these cues alone, one may speculate as to possible changes in distanced perception with advancing age. For example, He et al. (1998) showed that intensity discrimination by listeners with clinically normal hearing was worse for older than for younger listeners and that the difference between age groups increased for stimulus frequencies from 500 to 4,000 Hz. Poorer intensity discrimination would be consistent with reduced ability to use sound level cues to distance perception. A growing body of evidence indicates that perception in reverberant conditions declines with age (e.g., Nabelek and Robinson 1982; see Section 4.1). If the ability to code or process the ratio of direct to reverberant energy declines with age in a similar manner, then one would expect this aspect of distance perception to be negatively affected by age as well. The dependence of distance perception on spectral cues results from the fact that attenuation over distance is greater for high- than for low-frequency sounds. Thus changes in relative spectral tilt are interpreted as changes in source distance. The sound localization experiments reviewed above as well as others in the literature clearly indicate that the use of spectral cues in sound localization is worse for older than for younger listeners, although the relative contributions of hearing loss and age per se are not well understood. Interestingly, Eddins et al. (2006) showed that older listeners with clinically normal hearing had thresholds that were comparable to young listeners with clinically normal hearing on a spectral-modulation detection task with a noise carrier ranging from 400 to 3,200 Hz. Spectral-shape perception for higher-frequency ranges in older adult listeners with good high-frequency hearing have not been reported. Finally, as reviewed in Section 2.1, age-related declines in the use of interaural cues would also lead to reductions in distance perception. Combining these results, it may be expected that distance perception declines substantially with increasing age, particularly when hearing loss is present.

6.3 Binaural Processing Under Headphones

6.3.1 ITDs

The human auditory system is acutely sensitive to ITDs of sounds presented over headphones, with the just-noticeable difference being on the order of 10 μs (Klump and Eady 1956; Zwislocki and Feldman 1956) when performance is based on the coding of the temporal fine structure of relatively low-frequency (<1,200 to 1,500 Hz) stimuli. The subjective percept is a lateralization toward the ear associated with the temporally leading sound. The auditory system is insensitive to interaural differences in the ongoing temporal fine structure for stimulus frequencies above 1,200 to 1,500 Hz. However, if a high-frequency stimulus has a pronounced low-frequency envelope (e.g., a high-frequency carrier amplitude modulated by a low-frequency tone), ITDs can be discriminated on the basis of interaural differences in the temporal envelope (Henning 1974; McFadden and Pasanen 1976, 1978; Bernstein and Trahiotis 2002). It has been hypothesized that the processing of interaural fine structure and interaural envelope differences are processed by essentially the same binaural neural mechanisms (Colburn and Esquissaud 1976; Bernstein and Trahiotis 2002).

The study of sensitivity to ITDs is of interest from the perspective of human aging for several reasons. First, because the coding of the temporal cues underlying sensitivity to ITDs is so acute in young adults, the ITD paradigm offers a highly sensitive measure of age-related decline. It is thought that neural mechanisms from the auditory nerve to the brainstem must function with a high degree of temporal precision to account for the sensitivity to ITDs that is observed in young adults (e.g., Jeffress 1948; Goldberg and Brown 1969). A second reason that sensitivity to ITDs is of particular interest is that the cues for performance on the task are restricted specifically to the timing domain (and, for low-frequency pure tones, restricted further to temporal fine structure). Although many other tasks (e.g., some aspects of frequency discrimination and speech understanding) may depend strongly on the coding of temporal fine structure, few, if any, tasks depend as clearly and exclusively on temporal coding as the ITD task. The ITD task, therefore, has the potential to provide insights about the effects of aging on a particular type of important auditory cue. Third, ITDs provide the dominant cue for sound localization in the free field (Wightman and Kistler 1992) and interaural time discrimination may therefore provide important insights about the ability of binaural cues to provide benefit to older listeners in natural environments.

Although the literature pertaining to the effect of age on sensitivity to ITDs is not large, the available data are consistent with some age-related decline in this binaural ability (Matzker and Springborn 1958; Kirikae 1969; Herman et al. 1977; Strouse et al. 1998; Babkoff et al. 2002; Ross et al. 2007). Although some studies have not controlled the factor of hearing loss carefully, clear age effects appear to occur even when criteria are adopted to match audiometric sensitivity between age groups (Herman et al. 1977; Strouse et al. 1998). For example, Strouse et al. found that the threshold ITD for a 100-Hz pulse train was approximately twice as large for a group of older listeners (mean age of ~71 years) in comparison to a group of young adults. In addition, the Strouse et al. study reported an interaction between age and the intensity at which the stimulus was presented; compared with the young adults, the older listeners showed particularly poor performance for stimuli presented at a low sensation level. In addition to older listeners being generally less sensitive to ITD cues, the results of Babkoff et al. (2002) indicate that the ITD necessary to lateralize a sound one subjective unit increases with increasing age of the listener.

A study by Ross et al. (2007) provides some further insight about ITD processing with aging, particularly in terms of stimulus frequency and the age range over which effects are likely to occur. This study examined cortical evoked potentials to changes in interaural phase from diotic to binaurally antiphasic amplitude modulated tones as a function of the frequency of the sound. Responses to this phase change occurred up to an average frequency of 1,225 Hz in young listeners but occurred up to a frequency of only 760 Hz in older listeners (~71 years old). Interestingly, the response occurred in listeners of middle age (~51 years old) up to a frequency of only 940 Hz. Behavioral data on a related task indicated the same general trend but with more variability. Ross et al. suggested that a decline in the ability to use ITD cues for sound lateralization may begin well before what is typically viewed as old age.

Overall, the available results indicate a strong likelihood that sensitivity to ITDs decreases in older people due to factors beyond those related to audiometric sensitivity. Areas of interest for further research include possible frequency effects over the range for which listeners are sensitive to interaural differences in fine structure timing and possible age effects related to fine structure versus envelope cues for ITDs. It is also of interest to determine the underlying reason for poor ITD coding in older listeners. For example, poor sensitivity to ITDs in older listeners might arise due to a smaller number of neural elements contributing timing information, imprecise coding of timing within peripheral neural elements, or some other factors or combination of factors.

6.3.2 IIDs

Another stimulus feature that can change the lateralization of a sound presented binaurally over headphones is an IID of the sound. In cases where the level of stimulation is unequal between the ears, the sound is lateralized toward the side with the higher intensity. In a listener with normal hearing, a stimulus is lateralized fully to the ear associated with the more intense sound when the IID is ~10 dB. In young adults, sensitivity to the IID is ~0.5 dB. Although only a few studies have investigated sensitivity to IIDs in older listeners, there is some indication that sensitivity to IIDs may not show much degradation with increasing age. Whereas the results of studies by Herman et al. (1977) and Babkoff et al. (2002) were consistent with an interpretation of poor sensitivity to ITDs in older listeners (see Section 2.1), neither of these studies found an age effect for the processing of IIDs in their listeners.

Another clue to the effect of age on IID processing versus ITD processing may be evident in results reported by Kirikae (1969). One of the factors addressed in that study was a phenomenon that is referred to as time/intensity trading. In the time/intensity trading paradigm, lateralization due to a time lead in one ear can be approximately offset by an intensity increase in the opposite ear (Yost et al. 1971; Hafter and Carrier 1972). The Kirikae study (1969) reported that compared with young adults, older listeners appeared to weight IIDs more than ITDs in the time/intensity trading paradigm. This result is consistent with a broad interpretation that deleterious effects related to advancing age may be greater for ITD processing than for IID processing. Such an interpretation would have significance for the idea that ITDs and IIDs are ultimately coded by the same auditory process. For example, it has been speculated that the effects of IIDs in binaural lateralization could be driven by the decrease in latency of neural responses for increases in stimulus level (Jeffress 1948; David et al. 1958) and, therefore, that both IIDs and ITDs are coded in terms of interaural differences in the latency of neural firing. The finding that there is an age effect for ITDs but not for IIDs would not support such a hypothesis and would be consistent with psychophysical data indicating that time and intensity sometimes do not trade completely in lateralization paradigms (Hafter and Carrier 1972).

An issue that should be pointed out is that in tests of IID sensitivity, it is important to rule out the possibility that performance is influenced by the processing of monaural intensity difference cues. For example, if the intensity is increased in one ear and decreased in the other to create an IID, it is possible that the listener could track the level change in a single ear and base performance on that ear alone. This has not been addressed carefully in previous research on aging and greater insight into the question of possible aging effects related to the processing of IID cues might be obtained in future research that takes this factor into account.

6.3.3 Binaural Masking-Level Difference

The binaural masking-level difference (BMLD) is a masking release paradigm in which the threshold of a signal presented in noise is improved by the availability of binaural difference cues that result when the signal is added to the masker (Hirsh 1948; Licklider 1948; Webster 1951). In the most common BMLD paradigm, the reference condition consists of a diotic noise that masks a diotic pure-tone signal and the masking release condition consists of a diotic noise that masks a pure-tone having an interaural phase shift of 180° (binaurally out of phase). Conventional shorthand terms for these reference and masking release conditions are N_0S_0 and N_0S_π, respectively, where N refers to the noise, S refers to the signal, and the subscript refers to the interaural phase difference expressed in radians. The magnitude of the masking release (N_0S_0 threshold minus the N_0S_π threshold) depends on many factors but is ~15 dB for a 500-Hz tone presented in a moderately intense broadband masking noise in listeners with normal hearing. Although the BMLD usually refers to a detection advantage, it can also refer to a recognition advantage for speech where the speech recognition threshold (SRT) is better for N_0S_π than for N_0S_0 presentation (e.g., Levitt and Rabiner 1967).

One factor that strongly affects the magnitude of the BMLD for a wideband noise masker is the frequency of the signal. The BMLD is ~15 dB for low-frequency signals, decreases to ~6 dB at 1,000 Hz, and asymptotes near ~3 dB for frequencies above ~1,500 Hz (Hirsh 1948). The reduction in BMLD magnitude with increasing signal frequency is generally believed to result from a loss in the ability to encode fine temporal information at higher frequencies. However, it is important to bear in

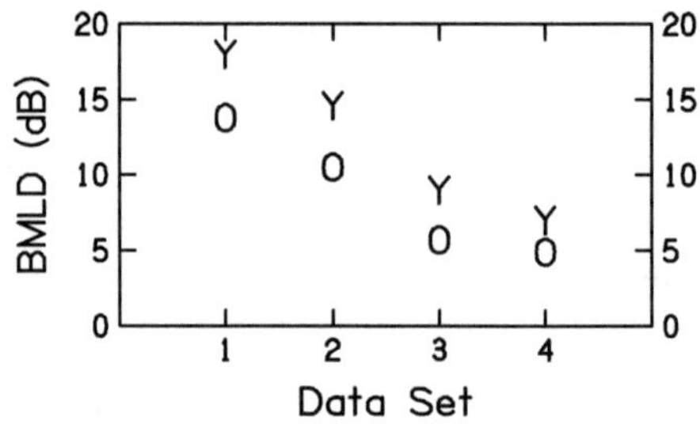

Fig. 6.3 Average binaural masking-level differences (BMLDs) for young (Y) and older (O) listeners for continuous noise maskers. Data sets 1 and 2 show data for tonal signals and data sets 3 and 4 show data for speech signals. Data set 1 shows data from Grose et al. (1994) where the signal was a 500-Hz pure tone and the masker was a 100-Hz-wide noise band centered on 500 Hz. Data set 2 shows data from Pichora-Fuller and Schneider (1991) where the signal was a 500-Hz pure tone and the masker was a band of noise from 100 to 5,000 Hz. Data set 3 shows data from Grose et al. (1994) where the signals were spondaic words and the masker was a speech-shaped noise. Data set 4 shows data from Strouse et al. (1998) where the signals were spondaic words and the masker was a speech-shaped noise.

mind that most BMLD paradigms are associated with the availability of both interaural time and interaural amplitude cues and that performance may be affected by each cue or by a combination of the two cues.

The BMLD is of interest from the perspective of aging for several reasons. One reason is that the BMLD may be viewed as an indicator of the advantage that binaural hearing may provide in natural environments for the processing of signals in noise. The BMLD may therefore be informative about the effect of age on this important auditory ability. Furthermore, because large BMLDs in many paradigms probably depend on sensitivity to interaural timing cues and because sensitivity to such cues may well be reduced in older listeners, it follows that the BMLD may be reduced in older listeners. Although the data on this question are limited, the available findings are consistent with some reduction in the BMLD in older listeners for pure-tone detection and speech reception (Pichora-Fuller and Schneider 1991; Grose et al. 1994; Strouse et al. 1998). Figure 6.3 summarizes this basic BMLD finding across the tonal and speech signal paradigms. Note that in these studies, the BMLD was reduced in older listeners due to an elevated N_0S_π threshold (N_0S_0 thresholds are approximately the same in young and older listeners). It should also be noted that not all studies have found reduced BMLDs with increasing age. Both Wilson and Weakley (2005) and Dubno et al. (2008) found trends for generally higher masked thresholds in elderly listeners, but no age effect for the BMLD.

Because the magnitude of the BMLD is often reduced in listeners with sensorineural hearing loss (Olsen et al. 1976; Hall et al. 1984; Jerger et al. 1984; Gabriel et al. 1992), it is informative to determine whether reduced BMLDs that are sometimes found in older listeners are associated with an age factor that is separate from factors related to hearing loss. Importantly, two of the studies that found significant reductions in the BMLD in older listeners (Grose et al. 1994; Strouse et al. 1998) matched audiometric sensitivity between age groups relatively closely. Although the mechanisms responsible for reduced BMLDs that are sometimes found in older listeners have not been identified, one hypothesis is that poorer performance in older listeners may be related to increased temporal jitter in the neural elements that underlie sensitivity to ITD cues (Pichora-Fuller and Schneider 1991).

6.4 Additional Temporal Aspects of Binaural Perception

The age-related decline in the perception of ITDs (Sections 3.1 and 3.3) is one manifestation of a general decline in auditory temporal processing in older listeners (see Fitzgibbons and Gordon-Salant, Chapter 5). Although the physiological mechanisms underlying such a decline are not clear, there is evidence that temporal-processing deficits in older listeners extend to other aspects of binaural processing, including perception in reverberant environments, the phenomenon known as the precedence effect, and possibly to the perception of motion.

6.4.1 Effects of Reverberation

Reverberation resulting from reflective surfaces in the environment colors the acoustic signal with fluctuations in intensity over time. In many cases, reverberation goes unnoticed, yet the distortion that results from reverberation may alter the perception of source location and the spectrotemporal details of the stimulus. Reverberation differentially affects auditory perception in younger and older listeners (Bergman et al. 1976; Duquesnoy and Plomp 1980; Nabelek and Robinson 1982; Gordon-Salant and Fitzgibbons 1993,1999; Roberts and Lister 2004) and marked effects can been seen in presenescent listeners (Nabelek and Robinson 1982). The effects of reverberation are more pronounced in listeners with hearing loss (Helfer 1992; Halling and Humes 2000). However, the relationships among the acoustic and perceptual effects of reverberation, spectrotemporal processing, and the underlying mechanisms of aging are as yet unknown.

Effects of reverberation are typically explored either by making audio recordings in real rooms having various degrees of reverberation or by simulating the acoustical effects of reverberation using mathematical algorithms (e.g., Moorer 1979; Schroeder 1962, 1970; Smith 2006) and then replaying stimuli to listeners via headphones. To gauge the effect of reverberation on speech perception in quiet across the life span, Nabelek and Robinson (1982) measured performance on the modified rhyme test (Bell et al. 1972) as a function of level, degree of reverberation, monaural versus binaural presentation, and age with 6 groups of 10 listeners ranging in mean age from 10 to 72 years. As is true in many studies, average pure-tone thresholds, particularly at moderate-to-high audiofrequencies, increased gradually with increasing age. Figure 6.4 shows the relationship between age, word identification

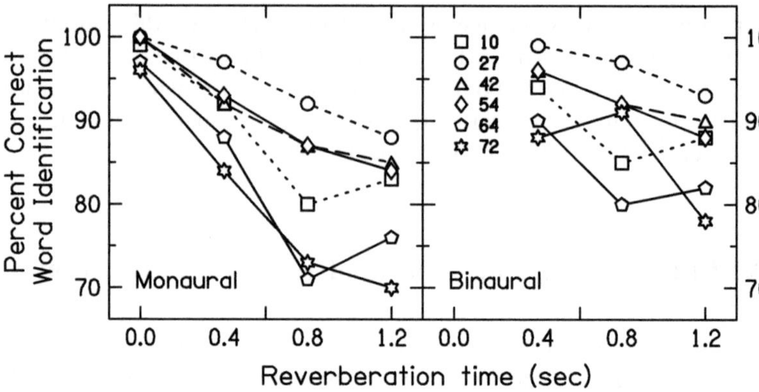

Fig. 6.4 Word identification score as a function of age (numbers in years) from Table III of Nabelek and Robinson (1982). Monaural word identification (left) is shown for three degrees of reverberation (0.4-, 0.8-, and 1.2-s delay) and without reverberation. Binaural word identification (right) is shown for three levels of reverberation.

score, and reverberation time constant for speech presented at an overall level of 70-dB SPL in quiet. Clearly, monaural word identification declines with reverberation time and varies with age, gradually decreasing from 27 to 72 years. Furthermore, the perception of reverberant speech was better for binaural than for monaural listening and this binaural advantage was present for all age groups. After arcsine transformation and averaging across reverberation times, Nabelek and Robinson (1982) reported binaural advantages that were slightly higher for older (72 years; 6.7 percentage points) than for younger (10 to 45 years; 4.0 percentage points) listeners. Although reverberation has a greater impact on older than on younger listeners in both monaural and binaural listening conditions, it appears that older listeners may benefit more from the information gained via binaural listening.

Subsequent research (Nabelek 1988; Gordon-Salant and Fitzgibbons 1993) has sought to provide a clearer delineation between the relationship of age and hearing loss on the perceptual effects of reverberation. Nabelek (1988) investigated vowel identification in quiet, with reverberation (reverberation time = 1.2 s), or in the presence of multitalker babble with four listener groups including a younger group with normal hearing and three older groups with either slight, mild, or moderate high-frequency hearing losses. Although partial correlations revealed significant effects of age and hearing loss for the noise and reverberant conditions, subsequent multiple comparisons failed to indicate a significant difference between the three better hearing groups in the quiet and reverberant conditions despite substantial differences in group mean performance. In a similar study, Gordon-Salant and Fitzgibbons (1993) investigated speech recognition using sentences with limited contextual cues, yielding low predictability (R-SPIN, Bilger et al. 1984) in four listener groups including younger and older listeners with normal hearing or mild-to-moderate sloping hearing losses. Using analyses of covariance to control for differences in performance across the four groups in the condition without reverberation, they reported significant effects of both age and hearing loss on speech perception in reverberation. Subsequent canonical correlations indicated that age and gap duration discrimination contributed significantly to the variance associated with recognition of speech in reverberation. Taken together, these studies provide some evidence of age-related declines in the perception of speech in quiet in the presence of reverberation.

Communication in realistic situations often involves a combination of reverberation as well as background competition. Thus age-related declines in speech perception in noise, as discussed by Humes and Dubno (Chapter 6), might be exacerbated by the presence of reverberation. Helfer and Wilber (1990) explored this possibility in four groups of listeners, including groups of young listeners with and without hearing loss and older listeners with minimal hearing loss and mild hearing loss that, on average, was less severe than in the younger group with hearing loss. They found marked decreases in performance on the nonsense syllable test (NST; Resnick et al. 1975) as a function of both age and hearing loss. In quiet, performance decrements were largely dominated by the degree of hearing loss. In noise, however, the performance by the younger and older listeners with hearing loss was nearly the same despite the fact that the younger listeners had substantially greater hearing loss than the older group. The results indicated a significant correlation between age and the ability to understand speech under conditions of noise and reverberation, a relationship that remained significant even when the effects of hearing loss were statistically controlled.

In an effort to experimentally control for age-related change in hearing sensitivity, Gordon-Salant and Fitzgibbons (1999) investigated age-related declines in the perception of reverberant speech in quiet and in the presence of multitalker babble in groups of younger and older listeners with either normal hearing or a mild-to-moderate sloping hearing loss. Speech perception in quiet and with minimal reverberation (0.4 s) did not differ significantly across age groups, whereas speech perception for the older listeners was significantly worse than for the younger listeners for moderate reverberation in quiet and mild (0.4 s) and moderate (0.6 s) reverberation in multitalker babble.

Gordon-Salant and Fitzgibbons (1999) suggested that the age-related decline in speech perception associated with reverberant conditions might be more closely related to susceptibility to temporal masking (e.g., Zwicker and Schorn 1982) rather than declines in temporal acuity per se. To explore this possibility, Halling and Humes (2000) measured modulation preservation and speech perception in noise under anechoic, reverberant, and intermediate conditions. Modulation preservation was estimated using a masking paradigm in which the detection of a pure-tone was measured in the presence of an amplitude-modulated noise. Significant effects of aging were reported for both types of listening task; however, analyses using a modified power law that models hearing loss as an independent noise source were able to account for the age effects and may be interpreted as an indication that slight differences in pure-tone threshold between the younger and older listeners with normal hearing could account for the differences in both speech perception under reverberant conditions and modulation preservation.

The overview provided here supports the conclusion that there are age-related declines in the perception of speech in reverberant conditions. Reverberation functions as a delay-and-add network that produces changes in both the temporal and spectral characteristics of the speech signal. The negative effects of reverberation on speech perception are greater when listening monaurally than binaurally; however, it appears that binaural listening rather than monaural listening may be more beneficial for older than for younger listeners. The perceptual and physiological mechanisms underlying reduced speech perception that occurs with increased age and reverberation have not been explored in depth; however, it is likely that audibility resulting from age-related changes in hearing sensitivity as well as age-related changes in the central auditory pathway combine to make it more difficult for older listeners to process sound accurately under reverberant conditions.

6.4.2 Precedence Effect

In a typical acoustic environment, the sound from a single sound source arrives at the ears of the listener as a combination of direct waves from the source combined with reflections from the surfaces in the room. In large rooms (e.g., auditoria), the reflections are heard as echoes, but in smaller rooms, the echoes are present but are not perceived. Suppression of the later arriving reflections is normal and is governed by the coding of interaural time and level cues by the central auditory system. In a common laboratory study of such binaural processing, sounds from two speakers are presented in an anechoic space. If the two sound sources are positioned in the horizontal plane and equidistant from the median plane, then the perceived location depends critically on the relative delay between the sounds from the two speakers. When the interspeaker delay is close to 0 ms, the percept is a single sound source at the midline. For brief clicks, as the delay is increased from 0 ms to ~0.7 ms, the percept is a single sound source, the location of which is shifted toward the leading speaker. Blauert (1983) refers to this as "summing localization," reflecting the fact that the perceived source position is a joint function of the position of the two speakers and the delay. For longer delays (e.g., 1 to 6 ms), the source position is perceived to correspond to the leading speaker. The latter has often been referred to as the "precedence effect" because the position of the first speaker takes precedence over the second (e.g., Wallach et al. 1949). This phenomenon is also referred to as the law of the first wave front and the Haas effect after the work by Haas (1951). For long interspeaker delays, the fusion between the two sounds begins to fail and the perception evolves into two distinct sound sources. Summing localization and the precedence effect are governed by factors such as stimulus type and timing and possibly by different neural processes (e.g., Blauert 1983). It has been suggested that summing localization under conditions of equal-level sounds from the two sources reflects interaural timing cues most likely coded at the level of the brainstem or midbrain. The precedence effect, on the other hand, is much more susceptible

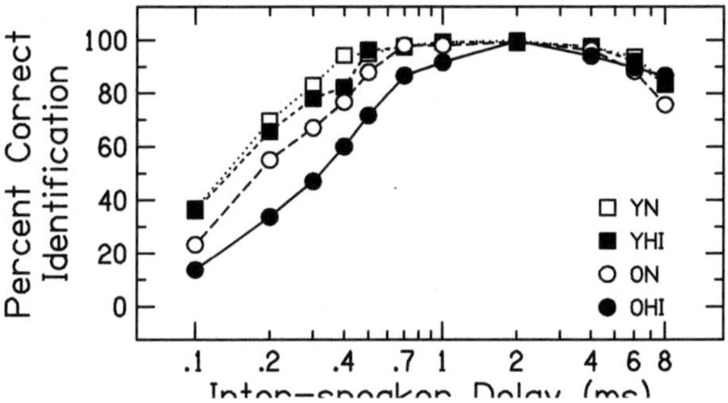

Fig. 6.5 Leading speaker identification as a function of interspeaker delay in four different listener groups: young normal hearing (YN); young hearing impaired (YHI); older normal hearing (ON); older hearing impaired (OHI). Individual data points represent the average across left-leading and right-leading speaker conditions. (Adapted from Cranford et al. 1993 with permission.)

to stimulus manipulations as well as to practice effects and such "echo suppression" may reflect processing at the level of the auditory cortex.

Taking into consideration age-related declines in temporal processing (see Fitzgibbons and Gordon-Salant, Chapter 5), including the perception of interaural time cues (see Section 2.1), it is reasonable to expect some age-related differences in summing localization and/or the precedence effect. In perhaps the most comprehensive study of this effect, Cranford et al. (1993) studied apparent source location as a function of interspeaker delay for groups of younger and older listeners with clinically normal hearing or sensorineural hearing loss. Figure 6.5 summarizes the primary results for the four listener groups averaged across left- and right-leading conditions. Note that the younger groups with and without hearing loss judged apparent source location in a similar delay-dependent manner, reflecting the gradual onset of the precedence effect from a midline percept at 0-ms delay to accurate identification of the leading speaker over interspeaker delays from 0.1 to 0.5 ms. The onset of the precedence effect was slower for older listeners without hearing loss than for their younger counterparts and that onset was slower still for older listeners with hearing loss. The fact that the precedence effect was present but delayed for the older group with clinically normal-hearing sensitivity was interpreted by Cranford et al. (1993) as reflecting a loss of auditory temporal acuity rather than a deficit in cortical processing. These results are similar to the earlier work of Cranford et al. (1990) showing an age effect for younger listeners with clinically normal hearing and older listeners with slight-to-mild, presumably age-related hearing loss.

In the context of a more comprehensive study of auditory temporal processing in younger and older listeners, Roberts and Lister (2004) measured fusion perception using 4-ms noise bursts presented bilaterally with variable interaural delays. There was no effect of age or hearing loss on their "lag-burst" thresholds in conditions without reverberation. Interestingly, in reverberant conditions, the older listeners with normal hearing performed worse than younger listeners with normal hearing and worse than older listeners with moderate sensorineural hearing loss. These results, combined with the lack of an age effect for fusion perception using trains of Gaussian-shaped tone bursts (Schneider et al. 1994), may be indicative of a fundamental difference between the fusion perception and the classic click-based studies of summing localization and the precedence effect. Although neither paradigm is fully understood, it is clear that the former relies on trial-by-trial labeling of the number of perceived sound sources, whereas the latter relies on the identification of an apparent source location. Such perceptual tasks are quite different and may interact with any age-related change in binaural processing.

6.4.3 Binaural Perception of Dynamic Stimuli

Many sound sources change position over time, leading to the perception of motion based on the acoustic signal even in the absence of reliable visual cues. The perception of auditory motion has been studied using a variety of different stimulus parameters and measurement techniques, including the use of actual moving sound sources as well as simulated motion via manipulation of interaural cues. A common measure is the minimum audible movement angle (MAMA) in which listeners discriminate a moving from a stationary source to determine the smallest discriminable angular change between the two (e.g., Harris and Sargeant 1971). One may also measure the smallest detectible change in the velocity of a moving standard stimulus (e.g., Grantham 1986). Mechanisms underlying motion detection are not well understood, and there is evidence to suggest that multiple mechanisms likely combine to yield the perception of auditory motion (e.g., Chandler and Grantham 1992; Grantham 1995; Saberi and Hafter 1997; Griffiths et al. 1996; Lutfi and Wang 1999). Furthermore, some measures of sensitivity to dynamically changing binaural cues indicate poorer temporal resolution than that associated with monaural hearing, a result that has been termed "binaural sluggishness" (e.g., Grantham and Wightman 1979).

Because motion perception can depend on underlying timing cues, it might be speculated that increased age should be associated with deficits in auditory motion perception. To date, no behavioral studies of auditory motion perception have directly addressed potential age-related changes. The one published study on aging and motion perception in humans used event-related auditory evoked potentials (ERPs) to compare simulated auditory motion across the life span (Jerger and Estes 2002). The investigation included three listener groups: children (9-12 years), young adults (18-34 years), and older adults (65-80 years). The two younger groups had clinically normal hearing, whereas, on average, the older listeners had typical presbycusic-sloping, high-frequency sensorineural hearing loss. The ERP occurring in the 500- to 700-ms latency range was used as the index of motion sensitivity. Based on minimal group-average differences in the amplitude of the earlier auditory evoked potentials, the authors reasoned that hearing loss per se was not likely to be a major factor in the ERP of interest. Examination of group-average waveforms revealed a gradual decline in the amplitude of the ERP response with age and no change with age in ERP latency. Importantly, however, the multiple-electrode recording system used by Jerger and Estes allowed the revelation of hemispheric differences in brain activity across the age groups. Movement to the left or right evoked greater activity in the right than in the left hemisphere for the youngest and oldest groups and slightly greater activity in the left than in the right hemisphere in the young-adult group. It was suggested that the hemispheric differences as a function of age might reflect the use of different brain mechanisms for the perception of motion in the different age groups, with the young-adult group employing a "velocity change" rather than a "positional change" strategy for detecting the presence of auditory motion. Clearly, much more work is needed to determine and understand possible age-related changes in the perception of auditory motion.

6.5 Perception of Masked Speech in the Free Field

An essential high-level binaural process for communication in natural environments is sound source segregation. This allows the listener to focus on a desired sound and deemphasize less desirable sounds. As discussed by Schneider, Pichora-Fuller, and Daneman (Chapter 7) and Humes and Dubno (Chapter 8), sound source segregation is easier when the target (desired sound) and masker (less desired sound) are spatially separated than when colocated in space. Segregation based on spatial separateness depends in large measure on binaural processing of the input sounds.

Measures of speech perception in background competition are frequently employed in clinical and laboratory settings and can be used to gauge the ability to perform sound source segregation. Speech perception in the presence of background sounds depends on the acoustic characteristics of the background sounds and improves when the sources of the target and background sounds are separated in space (e.g., Freyman et al. 2001). This improvement, termed a spatial release from masking, results from at least three factors, two of which are reasonably well understood (i.e., improvements in signal-to-noise ratio as a result of the head shadow effect and benefits gained from binaural hearing via the processing of interaural time and level cues; e.g., Zurek 1993). The third factor is more complex and concerns non-energetic masking that may arise from the failure of central hearing mechanisms to differentiate auditory sources on the basis of a variety of sound segregation cues.

Kim et al. (2006) used the hearing in noise test (HINT; Nilsson et al. 1994) to evaluate the use of spatial hearing cues in the free field in young and older adult listeners with normal hearing below 4,000 Hz. In this test, SRTs were measured in broadband noise filtered to match the long-term spectrum of speech. The masking noise was either colocated with the speech or was spatially separated by 90° (speech in front, noise to the side of the listener). Older adults had higher SRTs for colocated speech and noise and demonstrated less spatial release from masking when the speech and noise were separated. Dubno et al. (2002) have also reported a smaller benefit arising from spatially separated speech and speech-shaped noise in older listeners with relatively good hearing thresholds. Dubno et al. (2008) also investigated a "binaural squelch" phenomenon that can be characterized as the benefit conferred by the ear with the poorer signal-to-noise ratio. In this testing, the speech signal had an orientation of 0° and the noise had an orientation of 90° with respect to the listener. To determine the magnitude of the squelch effect, the SRT was determined with the ear on the same side as the noise plugged or unplugged. The outcome of interest was whether the benefit due to the binaural difference cues in the unplugged condition outweighed any benefit of plugging the ear with the poorer signal-to-noise ratio. The results indicated that the addition of the ear with the poorer signal-to-noise ratio resulted in improved performance (binaural squelch) for young but not for older adult listeners with normal hearing below 4,000 Hz. Overall, the available results for free-field listening appear to suggest that the elderly may have a reduced benefit related to binaural difference cues for speech recognition in speech-shaped noise. Such an impairment may help account for the fact that older individuals frequently complain that speech is difficult to understand in noisy environments.

The masking effects in the free-field speech studies considered up to this point are interpretable in terms of the overlap of the energy of the signal and the background noise and can be conceptualized in terms of "energetic masking." However, the results of several studies have demonstrated that factors other than energetic masking can sometimes result in significant degradations in the perception of sound. Such masking is sometimes referred to as "informational masking" and can be conceptualized as a failure of central auditory analysis to segregate target and background acoustical components (e.g., Watson 1987; Kidd et al. 1994). Several studies have shown that in complex speech masking studies, informational masking can occur wherein the identification of a target word is hampered by the presence of nontarget speech (Brungart 2001; Freyman et al. 2001, 2004; Arbogast et al. 2002). Although some of the masking in such studies arises from energetic factors (where it is assumed that the signal and masker stimulate an overlapping set of peripheral sensorineural elements), much of the masking results from an inability of the central auditory system to separate peripheral activity that is associated with the target speech from peripheral activity that is associated with the nontarget speech masker. It has been shown that substantial release from this kind of informational masking can occur when binaural cues are provided that result in a perceived spatial separation between the target and interfering material (e.g., Freyman et al. 2001; Johnstone and Litovsky 2003).

It is possible that spatial release from informational masking can provide insights into an important binaural hearing benefit that occurs in acoustically complex natural environments. A recent study by Helfer and Freyman (2008) found that although older listeners were likely to demonstrate a reduced release from informational masking when the target and masker were derived from talkers of the opposite sex, older and young listeners did not seem to differ greatly in terms of informational masking release based on spatial cues. These results are similar to reports by Li et al. (2004) and Marietta et al. (2007). Thus although the results of some studies indicate that older listeners are likely to have reduced sensitivity for some binaural cues, such a finding may not necessarily translate into a reduced benefit to use spatial cues for informational masking release in complex listening situations involving multiple talkers. The issue of spatial release from masking is discussed at length by Schneider, Pichora-Fuller, and Daneman (Chapter 7) and Humes and Dubno (Chapter 8).

6.6 Effects of Laterality

Many paradigms indicate a right ear advantage for the understanding of words under some monaural masking conditions where performance is compared for right ear versus left ear presentation or conditions where different words are presented simultaneously to the two ears and one or both words must be reported (Kimura 1961; Broadbent and Gregory 1964; Gerber and Goldman 1971). These laterality effects are typically interpreted in terms of cortical hemispheric asymmetries in speech processing, with the analysis of speech hinging critically on processing in the left cortical hemisphere. Although some results have indicated little effect of age on related asymmetries (e.g., Gelfand et al. 1980), the results of several studies have been interpreted in terms of an increasing right ear advantage in older listeners due to increasingly poor processing for the left ear with advancing age (e.g., Jerger and Jordan 1992; Alden et al. 1997; Divenyi and Haupt 1997a; Hallgren et al. 2001; Divenyi et al. 2005; Roup et al. 2006). There is some evidence that the right ear advantage (or left ear disadvantage) in older listeners may be particularly likely to occur under listening conditions that are relatively complex and conditions where the listener is asked to focus attention on a particular ear (Alden et al. 1997; Hallgren et al. 2001). Figure 6.6 shows data adapted from Alden et al. (1997) and Hallgren et al. (2001) that support such an interpretation. Hallgren et al. also suggested

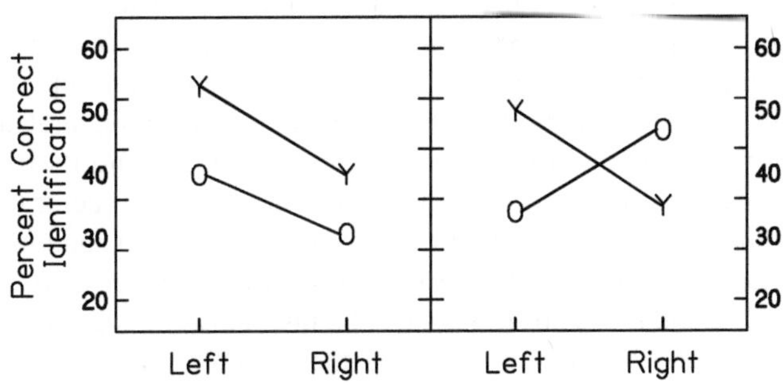

Fig. 6.6 Dichotic speech perception. In these conditions, two different consonant-vowels (CVs) were presented simultaneously (one to each of the two ears) and observers were directed to focus attention on the left ear. Average percent correct CVs for young (Y) and older (O) listeners are shown as a function of ear. (Data on left adapted from Hallgren et al. 2001 and data on right adapted from Alden et al. 1997.)

that decline in left ear performance was likely to be particularly notable in older listeners showing cognitive decline. The factors underlying an increased right ear advantage (or a left ear disadvantage) with advancing age have not been well accounted for, and it is not clear whether they might be conceptualized better in terms of changes in anatomy, auditory processing, general memory, or some other factors. As discussed by Humes and Dubno (Chapter 8) and noted above, the older listeners in many studies of age-related hearing loss, including dichotic speech perception, often have greater high-frequency hearing loss than their younger counterparts, and this factor often accounts for some proportion of the performance difference between younger and older listeners. It seems likely that the laterality effects demonstrated with advancing age are driven by central rather than peripheral factors (e.g., Jerger et al. 1991; Divenyi and Haupt 1997b; Hallgren et al. 2001).

In passing, it should be noted that a peripheral asymmetry effect, the slight right ear advantage in terms of a higher amplitude of otoacoustic emissions (Bilger et al. 1990; McFadden et al. 1996), also appears to be affected by age. Tadros et al. (2005) found that older listeners had a reversal of this asymmetry, with otoacoustic emissions having higher amplitude in the left ear.

6.7 Practical Considerations

As described above, there are several indications that age, independent from a loss of audiometric sensitivity, can have a deleterious effect on the ability of human listeners to take advantage of binaural cues both under headphones and in the free field. Because hearing loss can also have negative effects on the ability to benefit from binaural cues (e.g., Olsen et al. 1976; Hall et al. 1984; Jerger et al. 1984; Gabriel et al. 1992) and because advanced age is often associated with some degree of hearing loss, there is good reason to consider what the practical considerations may be with regard to binaural hearing abilities for older listeners. Unfortunately, this is a very challenging task. Binaural hearing benefits in everyday environments depend on multiple cues (and also the integration of those cues), and just because a listener may have a deficit in one or more particular aspects of binaural analysis does not mean that the listener will not, overall, reap important benefits from binaural hearing in many natural listening situations. One important factor to consider in this regard is the head shadow. Benefits derived from the head shadow can be substantial and depend simply on the possession of a head rather than on any fancy neural circuitry inside the head. It may also be instructive to consider the results of Helfer and Freyman (2008) in considering the possible practical implications of binaural deficits for older listeners. Although, as indicated in Sections 6.2.1, 6.3.2.1, and 6.4.2, there is evidence that older listeners may often have abnormal performance on specific tasks such as localization, the use of interaural timing cues, and the precedence effect, the results of Helfer and Freyman (2008) indicate that older listeners may perform similarly to younger listeners with normal hearing on complex tasks where spatial cues are used to help separate a signal from a complex informational masker. In light of these studies, one possible interpretation is that some specific binaural hearing abilities are impaired in older listeners, whereas the remaining capabilities may sometimes be sufficient to support very important aspects of binaural benefit.

Not withstanding the relatively optimistic point of view just presented, there is probably sufficient evidence to indicate that global binaural hearing abilities in real environments may be at least somewhat diminished in older listeners due to reduced sensitivity to one or more types of binaural cues. One implication of this observation is that it might be reasonable to counsel older hearing-impaired listeners that the benefit of bilateral and binaural hearing aids might be at least somewhat limited, not only due to sensorineural hearing loss but also due to factors related to aging. Investigators have even suggested that in some cases, the increased right ear advantage (or left ear disadvantage) that is sometimes seen with advancing age may make bilateral hearing aids inferior to monaural amplification (Chmiel et al. 1997; Henkin et al. 2007; see also Schnieder, Pichora-Fuller, and Daneman, Chapter 7). Likewise, independent automatic gain control in bilateral hearing aids may reduce the natural head-shadow effect, diminishing the cues for binaural hearing. Audiologists working with older patients should be aware of these possibilities. A final practical consideration is the possibility that amplification processing might mitigate the consequences of poor binaural processing in older listeners. For example, it is possible that older listeners who have difficulty separating spatially discrete signals might get particular benefit from directional hearing aids. Although some older (and perhaps younger) hearing aid wearers may not prefer to manually switch to a directional from an omnidirectional mode (e.g., Walden et al. 2000; Cord et al. 2002) and may not notice benefit from the directional mode in real-world listening (e.g., Cord et al. 2004), the degree to which older adults benefit from or prefer directional microphone settings in automatic-switching instruments is not known (see also Humes and Dubno, Chapter 8). It is possible that more advanced directional features may prove to be more beneficial and, in paired comparisons, preferable to full-time omnidirectional mode, even in older listeners. Such features may include context-specific automatic switching from omnidirectional to directional mode, automatic frequency-specific directionality, and binaural instruments in which decisions about the moment-by-moment settings of various automatic features, including microphone directionality, are linked between the two hearing aids. In this regard, microphone array strategies aimed at spatial filtering (e.g., Saunders and Kates 1997) could prove to be valuable to older listeners with reduced binaural hearing capacities.

References

Abel SM, Hay VH (1996) Sound localization. The interaction of aging, hearing loss and hearing protection. Scand Audiol 25:3–12.

Abel SM, Giguere C, Consoli A, Papsin BC (2000) The effect of aging on horizontal plane sound localization. J Acoust Soc Am 108:743–752.

Alden JD, Harrison DW, Snyder KA, Everhart DE (1997) Age differences in intention to left and right hemispace using a dichotic listening paradigm. Neuropsychiatry Neuropsychol Behav Neurol 10:239–242.
Arbogast TL, Mason CR, Kidd G Jr. (2002) The effect of spatial separation on informational and energetic masking of speech. J Acoust Soc Am 112:2086–2098.
Babkoff H, Muchnik C, Ben-David N, Furst M, Even-Zohar S, Hildesheimer M (2002) Mapping lateralization of click trains in younger and older populations. Hear Res 165:117–127.
Batteau DW (1967) The role of the pinna in human localization. Proc R Soc Lond Ser B 168:158–180.
Bell DW, Kruel EJ, Nixon JC (1972) Reliability of the modified rhyme test for hearing. J Speech Hear Res 15:287–295.
Bergman M, Blumenfeld VG, Cascardo D, Dash B, Levitt H, Margulies MK (1976) Age-related decrement in hearing for speech: sampling and longitudinal studies. J Gerontol 31:533–538.
Bernstein LR, Trahiotis C (2002) Enhancing sensitivity to interaural delays at high frequencies by using "transposed stimuli." J Acoust Soc Am 112:1026–1036.
Bilger RC, Nuetzel JM, Rabinowitz WM, Rzeczkowski C (1984) Standardization of a test of speech perception in noise. J Speech Hear Res 27:32–48.
Bilger RC, Matthies ML, Hammel DR, Demorest ME (1990) Genetic implications of gender differences in the prevalence of spontaneous otoacoustic emissions. J Speech Hear Res 33:418–432.
Blauert J (1969) Sound localization in the median plane. Acustica 22:205–213.
Blauert J (1983) Spatial Hearing: The Psychophysics of Human Sound Localization. Cambridge, MA: MIT Press.
Broadbent DE, Gregory M (1964) Accuracy of recognition for speech presented to the right and left ears. Q J Exp Psychol 16:359–360.
Brungart DS (2001) Informational and energetic masking effects in the perception of two simultaneous talkers. J Acoust Soc Am 109:1101–1109.
Butler RA (1969) Monaural and binaural localization of noise bursts vertically in the median sagittal plane. J Aud Res 3:230–235.
Butler RA (1970) The effect of hearing impairment on locating sound in the vertical plane. Int J Audiol 9:117–126.
Butler RA (1994) Asymmetric performances in monaural localization of sound in space. Neuropsychology 32:221–229.
Butler RA, Planert N (1976) The influence of stimulus bandwidth on localization of sound in space. Percept Psychophys 19:103–108.
Butler RA, Humanski RA, Musicant AD (1990) Binaural and monaural localization of sound in two-dimensional space. Perception 19: 241–256.
Chandler DW, Grantham DW (1990) Minimum audible movement angle in the horizontal plane as a function of stimulus frequency and bandwidth, source azimuth and velocity, and number of ears. J Acoust Soc Am 87:S64.
Chandler DW, Grantham DW (1991) Effects of age on auditory spatial resolution in the horizontal plane. J Acoust Soc Am 89:1994.
Chandler DW, Grantham DW (1992) Minimum audible movement angle in the horizontal plane as a function of stimulus frequency and bandwidth, source azimuth, and velocity. J Acoust Soc Am 91:1624–1636.
Chisolm TH, Willott JF, Lister JL (2003) The aging auditory system: anatomic and physiologic changes and implications for rehabilitation. Int J Audiol 42:2S2–2S10.
Chmiel R, Jerger J, Murphy E, Pirozzolo F, Tooley-Young C (1997) Unsuccessful use of binaural amplification by an elderly person. J Am Acad Audiol 8:1–10.
Colburn HS, Esquissaud P (1976) An auditory nerve model for interaural time discrimination of high-frequency complex stimuli. J Acoust Soc Am 59:S23.
Committee for Hearing, Bioacoustics and Biomechanics (CHABA) (1988) Speech understanding and aging. J Acoust Soc Am 83:859–895.
Cord MT, Surr RK, Walden BE, Olsen L (2002) Performance of directional microphone hearing aids in everyday life. J Am Acad Audiol 13:308–322.

Cord MT, Surr RK, Walden BE, Dyrlund O (2004). Relationship between laboratory measures of directional advantage and everyday success with directional microphone hearing aids. J Am Acad Audiol 15:353–364.

Cranford JL, Boose M, Moore CA (1990) Effects of aging on the precedence effect in sound localization. J Speech Hear Res 33:654–659.

Cranford JL, Morgan M, Scudder R, Moore C (1993) Tracking of "moving" fused auditory images by children. J Speech Hear Res 36:424–430.

David EE, Guttman N, van Bergeijk WA (1958) On the mechanism of binaural fusion. J Acoust Soc Am 30:801–802.

Divenyi PL, Haupt KM (1997a) Audiological correlates of speech understanding deficits in elderly listeners with mild-to-moderate hearing loss. I. Age and lateral asymmetry effects. Ear Hear 18:42–61.

Divenyi PL, Haupt KM (1997b) Audiological correlates of speech understanding deficits in elderly listeners with mild-to-moderate hearing loss. II. Correlational analysis. Ear Hear 18:100–113.

Divenyi PL, Simon HJ (1999) Hearing in aging: issues old and new. Curr Opin Otolaryngol 7:282–289.

Divenyi PL, Stark PB, Haupt KM (2005) Decline of speech understanding and auditory thresholds in the elderly. J Acoust Soc Am 118:1089–1100.

Dubno JR, Ahlstrom JB, Horwitz AR (2002) Spectral contributions to the benefit from spatial separation of speech and noise. J Speech Lang Hear Res 45:1297–1310.

Dubno JR, Ahlstrom JB, Horwitz AR (2008) Binaural advantage for younger and older adults with normal hearing. J Speech Lang Hear Res 51:539–556.

Duquesnoy AJ, Plomp R (1980) Effect of reverberation and noise on the intelligibility of sentences in cases of presbyacusis. J Acoust Soc Am 68:537–544.

Eddins AC, Eddins DA, Coad ML, Lockwood AH (2001) Cognitive and sensory influence on the perception of complex auditory signals. J Acoust Soc Am 190:2309–2310.

Eddins DA, Frisina RD, Mapes FM, Guimaraes P (2006) Spectral modulation detection as a function of age and degree of hearing loss. Assoc Res Otolaryngol Abstr 29:328.

Freyman RL, Balakrishnan U, Helfer KS (2001) Spatial release from informational masking in speech recognition. J Acoust Soc Am 109:2112–2122.

Freyman RL, Balakrishnan U, Helfer KS (2004) Effect of number of masking talkers and auditory priming on informational masking in speech recognition. J Acoust Soc Am 115:2246–2256.

Gabriel KJ, Koehnke J, Colburn HS (1992) Frequency dependence of binaural performance in listeners with impaired binaural hearing. J Acoust Soc Am 91:336–347.

Gardner MB (1973) Some monaural and binaural facets of median plane localization. J Acoust Soc Am 54:1489–1495.

Gardner MB, Gardner RS (1973) Problem of localization in the median plane: effect of pinnae cavity occlusion. J Acoust Am Soc 53:400–408.

Gates GA, Cooper JC, Kannel WB, Miller NJ (1990) Hearing in the elderly: the Framingham cohort, 1983–1985. Part I. Basic audiometric test results. Ear Hear 11:247–256.

Gelfand SA, Hoffman S, Waltzman SB, Piper N (1980) Dichotic CV recognition at various interaural temporal onset asynchronies: effect of age. J Acoust Soc Am 68:1258–1261.

Gerber SE, Goldman P (1971) Ear preference for dichotically presented verbal stimuli as a function of report strategies. J Acoust Soc Am 49:1163.

Goldberg JM, Brown, B (1969) Response of binaural neurons of dog superior olivary complex to dichotic tonal stimuli: some physiological mechanisms of sound localization. J Neurophysiol 32:613–636.

Gordon-Salant S, Fitzgibbons PJ (1993) Temporal factors and speech recognition performance in young and elderly listeners. J Speech Hear Res 36:1276–1285.

Gordon-Salant S, Fitzgibbons PJ (1999) Profile of auditory temporal processing in older listeners. J Speech Lang Hear Res 42:300–311.

Grantham DW (1986) Detection and discrimination of simulated motion of auditory targets in the horizontal plane. J Acoust Soc Am 79:1939–1949.

Grantham DW (1995) Spatial hearing and related phenomena. In: Moore BCJ (ed) Hearing: Handbook of Perception and Cognition. New York: Academic Press, pp. 297–345.
Grantham DW, Wightman FL (1979) Detectability of a pulsed tone in the presence of a masker with time-varying interaural correlation. J Acoust Soc Am 65:1509–1517.
Griffiths TD, Rees A, Witton C, Shakir RA, Henning GB, Green GG (1996) Evidence for a sound movement area in the human cerebral cortex. Nature 383:425–427.
Grose JH, Poth EA, Peters RW (1994) Masking level differences for tones and speech in elderly listeners with relatively normal audiograms. J Speech Hear Res 37:422–428.
Grose JH, Hall JW, Buss E (2006) Temporal processing deficits in the pre-senescent auditory system. J Acoust Soc Am 119:2305–2315.
Haas H (1951) On the influence of a single echo on the intelligibility of speech. Acustica 1:48–58.
Hafter ER, Carrier SC (1972) Binaural interaction in low-frequency stimuli: the inability to trade time and intensity completely. J Acoust Soc Am 51:1852–1862.
Hall JW, Tyler RS, Fernandes MA (1984) Factors influencing the masking level difference in cochlear hearing-impaired and normal-hearing listeners. J Speech Hear Res 27:145–154.
Hall JW, Buss E, Grose JH, Dev MB (2004) Developmental effects in the masking-level difference. J Speech Lang Hear Res 47:13–20.
Hall JW, Buss E, Grose JH (2005) Informational masking release in children and adults. J Acoust Soc Am 118:1605–1613.
Hallgren M, Larsby B, Lyxell B, Arlinger S (2001) Cognitive effects in dichotic speech testing in elderly persons. Ear Hear 22:120–129.
Halling DC, Humes LE (2000) Factors affecting the recognition of reverberant speech by elderly listeners. J Speech Lang Hear Res 43:414–431.
Harris JD, Sergeant RL (1971) Monaural/binaural minimum audible angles for a moving sound source. J Speech Hear Res 14:618–629.
Hausler R, Colburn S, Marr E (1983) Sound localization in subjects with impaired hearing. Spatial-discrimination and interaural-discrimination tests. Acta Otolaryngol Suppl 400:1–62.
He N, Dubno JR, Mills JH (1998) Frequency and intensity discrimination measured in a maximum-likelihood procedure from young and aged normal-hearing subjects. J Acoust Soc Am 103:553–565.
Helfer KS (1992) Aging and the binaural advantage in reverberation and noise. J Speech Hear Res 35:1394–1401.
Helfer KS, Freyman RL (2008) Aging and speech-on-speech masking. Ear Hear 29:87–98.
Helfer KS, Wilber LA (1990) Hearing loss, aging, and speech perception in reverberation and noise. J Speech Hear Res 33:149–155.
Henkin Y, Waldman A, Kishon-Rabin L (2007) The benefits of bilateral versus unilateral amplification for the elderly: are two always better than one? J Basic Clin Physiol Pharmacol 18:201–216.
Henning GB (1974) Detectability of interaural delay in high-frequency complex waveforms. J Acoust Soc Am 55:84–90.
Herman GE, Warren LR, Wagener JW (1977) Auditory lateralization - age differences in sensitivity to dichotic time and amplitude cues. J Gerontol 32:187–191.
Hirsh IJ (1948) Influence of interaural phase on interaural summation and inhibition. J Acoust Soc Am 20:536–544.
Humes LE, Watson BU, Christensen LA, Cokely CG, Halling DC, Lee L (1994) Factors associated with individual differences in clinical measures of speech recognition among the elderly. J Speech Hear Res 37:465–474.
Jeffress LA (1948) A place theory of sound localization. J Comp Physiol Psychol 41:35–39.
Jerger J, Estes R (2002) Asymmetry in event-related potentials to simulated auditory motion in children, young adults, and seniors. J Am Acad Audiol 13:1–13.
Jerger J, Jordan C (1992) Age-related asymmetry on a cued-listening task. Ear Hear 13:272–277.
Jerger J, Brown D, Smith S (1984) Effect of peripheral hearing loss on the MLD. Arch Otolaryngol 110:290–296.

Jerger J, Jerger S, Pirozzolo F (1991) Correlational analysis of speech audiometric scores, hearing loss, age, and cognitive abilities in the elderly. Ear Hear 12:103–109.

Johnstone P, Litovsky RY (2003) Speech intelligibility and spatial release from masking in children and adults for various types of interfering sounds. Assoc Res Otolaryngol Abstr 26:214.

Kidd G, Mason CR, Deliwala PS, Woods WS, Colburn HS (1994) Reducing informational masking by sound segregation, J Acoust Soc Am 95:3475–3480.

Kim S-H, Frisina RD, Mapes FM, Hickman ED, Frisina DR (2006) Effect of age on binaural speech intelligibility in normal hearing adults. Speech Commun 48:591–597

Kimura D (1961) Cerebral dominance and the perception of verbal stimuli. Can J Psychol 15:166–171.

Kirikae I (1969) Auditory function in advanced age with reference to histological changes in the central auditory system. Int J Audiol 8:221–230.

Klump RG, Eady HR (1956) Some measurements of interaural time difference thresholds. J Acoust Soc Am 28:859–860.

Lee FS, Matthews LJ, Dubno JR, Mills JH (2005) Longitudinal study of pure-tone thresholds in older persons. Ear Hear 26:1–11.

Levitt H, Rabiner LR (1967) Binaural release from masking for speech and gain intelligibility. J Acoust Soc Am 42:601–608.

Li L, Daneman M, Qi J, Schneider BA (2004) Does the information content of an irrelevant source differentially affect spoken word recognition in younger and older adults? J Exp Psychol Hum Percept Perform 30:1077–1091.

Licklider JCR (1948) The influence of interaural phase relations upon the masking of speech by white noise. J Acoust Soc Am 20:150–159.

Lutfi RA, Wang W (1999) Correlational analysis of acoustic cues for the discrimination of auditory motion. J Acoust Soc Am 106:919–928.

Marietta C, Nutt RC, Frisina DR, Eddins DA (2007) The effects of age on the spatial release from informational and energetic masking. Assoc Res Otolaryngol Abstr 30:222.

Matzker VJ, Springborn E (1958) Richtungshoren und lebensalter. Z Laryngol Rhinol Otol Grenzgeb 37:739–745.

Maxon AB, Hochberg I (1982) Development of psychoacoustic behavior: sensitivity and discrimination. Ear Hear 3:301–308.

McFadden D, Pasanen EG (1976) Lateralization at high frequencies based on interaural time differences. J Acoust Soc Am 59:634–639.

McFadden D, Pasanen EG (1978) Binaural detection at high frequencies with time-delayed waveforms. J Acoust Soc Am 63:1120–1131.

McFadden D, Loehlin JC, Pasanen EG (1996) Additional findings on heritability and prenatal masculinization of cochlear mechanisms: click-evoked otoacoustic emissions. Hear Res 97:102–119.

Mehrgardt S, Mellert V (1977) Transformation characteristics of the external human ear. J Acoust Soc Am 61:1567–1576.

Moorer JA (1979) About this reverberation business. Comput Music J 3:13–28.

Nabelek AK (1988) Identification of vowels in quiet, noise, and reverberation: relationships with age and hearing loss. J Acoust Soc Am 84:476–484.

Nabelek AK, Robinson PK (1982) Monaural and binaural speech perception in reverberation for listeners of various ages. J Acoust Soc Am 71:1242–1248.

Nilsson M, Soli SD, Sullivan JA (1994) Development of the Hearing in Noise Test for the measurement of speech reception thresholds in quiet and in noise. J Acoust Soc Am 95: 1085–99.

Noble W, Byrne D, Lepage B (1994) Effects on sound localization of configuration and type of hearing impairment. J Acoust Soc Am 95:992–1005.

Noble W, Byrne D, Ter-Horst K (1997) Auditory localization, detection of spatial separateness, and speech hearing in noise by hearing impaired listeners. J Acoust Soc Am 102:2343–2352.

Oldfield S, Parker S (1984) Acuity of sound localization: a topography of auditory space. II. Pinna cues absent. Perception 13:601–617.

Olsen W, Noffsinger D, Carhart R (1976) Masking-level differences encountered in clinical populations. Audiology 15:287–301.
Pichora-Fuller MK (1997) Language comprehension in older listeners. J Speech Lang Hear Res 21:125–142.
Pichora-Fuller MK, Schneider BA (1991) Masking level differences in the elderly: a comparison of antiphasic and time-delay dichotic conditions. J Speech Hear Res 34:1410–1422.
Rakerd B, Vander Velde TJ, Hartmann WM (1998) Sound localization in the median sagittal plane by listeners with presbyacusis. J Am Acad Audiol 9:466–479.
Resnick SB, Dubno JR, Hoffnung S, Levitt H (1975) Phoneme errors on a nonsense syllable test. J Acoust Soc Am 58:114.
Roberts RA, Lister JJ (2004) Effects of age and hearing loss on gap detection and the precedence effect: broadband stimuli. J Speech Lang Hear Res 47:965–978.
Roffler SK, Butler RA (1968) Factors that influence the localization of sound in the vertical plane. J Acoust Soc Am 43:1255–1259.
Ross B, Fujioka T, Tremblay KL, Picton TW (2007) Aging in binaural hearing begins in mid-life: evidence from cortical auditory-evoked responses to changes in interaural phase. J Neurosci 27:11172–11178.
Roup CM, Wiley TL, Wilson RH (2006) Dichotic word recognition in young and older adults. J Am Acad Audiol 17:230–240.
Saberi K, Hafter ER (1997) Experiments on auditory motion discrimination. In: Gilkey RH, Anderson TR (eds) Binaural and Spatial Hearing in Real and Virtual Environments. Mahwah, NJ: Erlbaum, pp. 315–327.
Saunders GH, Kates JM (1997) Speech intelligibility enhancement using hearing-aid array processing. J Acoust Soc Am 102:1827–1837.
Schneider BA (1997) Psychoacoustics and Aging: Implications for Everyday Listening. J Spch-Lang Audiol 21:111–124.
Schneider BA, Pichora-Fuller MK, Kowalchuk D, Lamb M (1994) Gap detection and the precedence effect in young and old adults. J Acoust Soc Am 95:980–991.
Schroeder MA (1962) Natural sounding artificial reverberation. J Audio Eng Soc 10: 219.
Schroeder MA (1970) Digital simulation of sound transmission in reverberant spaces. J Acoust Soc Am 47:424–431.
Smith J (2006) Physical Audio Signal Processing for Virtual Musical Instruments and Digital Audio Effects, August 2006 Edition. Center for Computer Research in Music and Acoustics (CCRMA), Stanford University, Stanford, CA. Online book at http://ccrma.stanford.edu/~jos/pasp06/.
Stevens SS, Newman EB (1936) The localization of actual sources of sound. Am J Psychol 48:297–306.
Strouse A, Ashmead DH, Ohde RN, Grantham DW (1998) Temporal processing in the aging auditory system. J Acoust Soc Am 104:2385–2399.
Strutt JW, Lord Rayleigh (1907) On our perception of sound direction. Philos Mag 13: 214–232.
Tadros SE, Frisina ST, Mapes F, Kim S, Frisina DR, Frisina RD (2005) Loss of peripheral right-ear advanatage in age-related hearing loss. Audiol Neurootol 10:44–52.
van Rooij JC, Plomp R (1990) Auditive and cognitive factors in speech perception by elderly listeners. II: Multivariate analyses. J Acoust Soc Am 88:2611–2624.
Walden BE, Surr RK, Cord MT, Edwards B, Olson L (2000) Comparisons of benefits provided by different hearing aid technologies. J Am Acad Audiol 11:540–560.
Wallach H, Newman EB, Rosenzweig MR (1949) The precedence effect in sound localization. Am J Psychol LXII:315–336.
Watson CS (1987) Uncertainty, informational masking and the capacity of immediate auditory memory. In: Yost WA, Watson CS (eds) Auditory Processing of Complex Sounds. Hillsdale, NJ: Erlbaum, pp. 267–277.
Webster FA (1951) The influence of interaural phase on masked thresholds: the role of interaural time deviation. J Acoust Soc Am 23:452–462.
Weinrich S (1982) The problem of front-back localization in binaural hearing. Scand Audiol Suppl 15:135–145.

Wightman FL, Kistler DJ (1992) The dominant role of low-frequency interaural time differences in sound localization. J Acoust Soc Am 91:1648–1661.

Wightman FL, Kistler DJ (1993) Sound localization. In: Yost WA, Fay RR, Popper AN (eds) Handbook of Auditory Research: Human Psychophysics. New York: Springer-Verlag, pp. 155–192.

Wilson RH, Weakley DG. (2005) The 500 Hz masking-level difference and word recognition in multitalker babble for 40- to 89-year-old listeners with symmetrical sensorineural hearing loss. J Am Acad Audiol. 16:367–82.

Wright D, Hebrank JH, Wilson B (1974) Pinna reflections as cues for localization. J Acoust Soc Am 56:957–962.

Yost W, Wightman F, Green D (1971) Lateralization of filtered clicks. J Acoust Soc Am 50:1526–1530.

Zurek PM (1993) A note on onset effects in binaural hearing. J Acoust Soc Am 93:1200–1201.

Zwicker E, Schorn K (1982) Temporal resolution in hard-of-hearing patients. Audiology 21:474–492.

Zwislocki J, Feldman RS (1956) Just noticeable differences in dichotic phase. J Acoust Soc Am 28:860–864.

[illegible]

Chapter 7
Effects of Senescent Changes in Audition and Cognition on Spoken Language Comprehension

Bruce A. Schneider, Kathy Pichora-Fuller, and Meredyth Daneman

7.1 Introduction

Older individuals often find it difficult to communicate, especially in group situations, because they are unable to keep up with the flow of conversation or are too slow in comprehending what they are hearing. These communication difficulties are often exacerbated by negative stereotypes held by their communication partners who often perceive older adults as less competent than they actually are (Ryan et al. 1986). Sometimes, older adults' communication problems motivate them, often at the prompting of their family and friends, to seek help from hearing specialists (O'Mahoney et al. 1996). Quite often, however, older adults and/or their family members wonder if these comprehension difficulties are a sign of cognitive decline. Such uncertainty on the part of both older adults and their family members with respect to the source of communication difficulties is understandable given that age-related changes in the comprehension of spoken language could be due to age-related changes in hearing, to age-related declines in cognitive functioning, or to interactions between these two levels of processing. To participate effectively in a multitalker conversation, listeners need to do more than simply recognize and repeat speech. They have to keep track of who said what, extract the meaning of each utterance, store it in memory for future use, integrate the incoming information with what each conversational participant has said in the past, and draw on the listener's own knowledge of the topic under consideration to extract general themes and formulate responses. In other words, effective communication requires not only an intact auditory system but also an intact cognitive system.

B.A. Schneider (✉), K. Pichora-Fuller and M. Daneman
Department of Psychology, University of Toronto Mississauga, Mississauga, Ontario, Canada L5L 1C6
e-mail: bruce.schneider@utoronto.ca; k.pichora.fuller@utoronto.ca; daneman@psych.utoronto.ca

S. Gordon-Salant et al. (eds.), *The Aging Auditory System*, Springer Handbook of Auditory Research 34,
DOI 10.1007/978-1-4419-0993-0_7, © Springer Science+Business Media, LLC 2010

Previous chapters have identified a number of age-related changes in cochlear, retrocochlear, and central auditory processing that could interfere with a listener's ability to understand, memorize, integrate, and recall heard information (Schmiedt, Chapter 2; Canlon, Illing, and Walton, Chapter 3; Ison, Tremblay, and Allen, Chapter 4; Fitzgibbons and Gordon-Salant, Chapter 5). To comprehend spoken language in complex listening situations, a person needs to overcome peripheral (energetic) masking (Humes and Dubno, Chapter 8), parse the auditory scene into different sources of information (e.g., different talkers) to be able to keep track of who said what, focus attention on the target talker, suppress the processing of irrelevant information, and, when appropriate, switch attention from one talker to another. Clearly, a person's ability to carry out these functions in complex environments will depend not only on the status of that person's auditory system but also on the status of their cognitive system.

Cochlear pathology could reduce the audibility of speech sounds and increase a listener's susceptibility to energetic masking, leading to errors in speech perception (see Humes and Dubno, Chapter 8). These errors, in turn, could cascade upward, making it more difficult for listeners to use higher order processes to extract meaning, store information in memory, or perform any of the other necessary cognitive operations. Central auditory deficits (e.g., declines in binaural processing or declines in the synchrony of neural firing at various levels of the auditory system, discussed in Schmiedt, Chapter 2; Canlon, Illing, and Walton, Chapter 3; Eddins and Hall, Chapter 6) will interfere with scene analysis to make it more difficult for the listener to keep track of different auditory sources and to separate streams of information for subsequent processing. At the cognitive level, age-related declines in speed of processing, working memory capacity, and the ability to suppress irrelevant information might make it more difficult for the listener to handle multiple streams of information, rapidly switch attention from one talker to another, and comprehend and store information extracted from speech for later recall.

In other words, to acquire and use the information contained in spoken language requires the smooth and rapid functioning of an integrated system of perceptual and cognitive processes. This chapter begins by introducing some concepts important to an integrated systems approach in studying spoken language comprehension. It then reviews how this integrated system is affected by (1) age-related auditory declines, (2) age-related cognitive declines, and (3) the interaction between age-related auditory and cognitive declines. Finally, it ends by considering how the system functions in different clinical populations (e.g., users of hearing aids and those with some form of dementia).

7.2 An Integrated Approach to Investigating Spoken Language Comprehension

Taking an integrated approach to investigating sources of age-related declines in spoken language comprehension requires the introduction of a number of concepts that although well known to cognitive scientists, may be less familiar to those in the hearing sciences. These include the notion of "levels of processing" as well as the interrelated concepts of "executive control," "limited working memory and attentional resources," and "processing speed."

7.2.1 *Levels of Processing*

In their seminal paper, Craik and Lockhart (1972) argued that cognitive functions such as memory flow from a perceptual analysis of the stimulus but that the degree or elaborateness of information processing depends on the task demands. This approach, rather than viewing perception and cognition as separate modules (or boxes in a flow chart), treats the sensory and cognitive systems as an integrated whole in which those processes one calls sensory or perceptual occur relatively early in the processing sequence, whereas those that are labeled cognitive could be considered as elaborations of these early processes. Importantly, it assumes that the depth to which a stimulus is processed will depend on the task demands. To repeat a sentence does not require the same depth of processing that is required, e.g., to decide whether a statement is true or false or to extract the theme from a short lecture. To repeat a sentence, the listener need only process the acoustic stream to the point at which he or she is able to identify and repeat the words, i.e., up to the lexical stage of processing. Understanding and remembering a short lecture requires more than lexical processing; it requires that the listener integrate the successively encountered words with one another and with world knowledge as well as storing a coherent representation of the lecture in memory for later use. Hence, processing of the information carried in the stimulus when the task is to understand and remember what has been heard is likely to be much more elaborate than it would be when the task is simply to repeat a word.

7.2.2 *Executive Control*

Implicit in the levels of processing approach is the notion of executive control (see Baddeley 1993; Shallice and Burgess 1993; West 1996 for reviews). If tasks require different kinds or different levels of processing, there must be an executive function that organizes and controls how the acoustic stimulus is processed. For example, at a more cognitive level, the executive would identify what the main themes are, separate relevant from irrelevant information, exclude the latter from further processing, decide what should be stored in memory, and marshal and organize the required resources to accomplish these tasks. At a more perceptual level, it may decide that it is has to focus attention on the talker to the left of the listener rather than the one on the right. At an even lower level of processing, it might decide that the important information is coming through the auditory filters serving the coding of low-frequency components of the signal. If the manner in which an acoustic stimulus is processed has such flexibility, there must be executive control over stimulus processing. If so, one interesting question is how extensive is this control; i.e., does executive control extend all the way down to very early perceptual processes? A second question of interest is how aging affects executive control at the various levels of information processing during spoken language comprehension.

There is evidence to suggest that aging is associated with declines in the control processes involved in coordinating distinct tasks (see McDowd and Shaw 2000; Verhaeghen et al. 2003 for recent reviews) and switching between tasks (Mayr et al. 2001; Verhaeghen et al. 2005), although the extent to which such age deficits contribute to age declines in spoken language comprehension has been relatively unexplored. There is also evidence to suggest that aging is associated with declines in the inhibitory control mechanisms that ordinarily prevent irrelevant information from interfering with the processing of relevant information (Hasher and Zacks 1988; Zacks and Hasher 1994). Research that has addressed the extent to which an inhibition deficit can account for age-related declines in spoken language comprehension is discussed in Section 7.5.2.

7.2.3 Limited Working Memory Resources

If processing and storage resources were unlimited, then one could imagine an executive function that would simply assemble all of the resources required for any task no matter how difficult the task and how many different resources were required. Moreover, an unlimited resource model would permit several tasks to be conducted in parallel without any performance decrements in any of the tasks. The more likely scenario is that processing and storage resources are limited (e.g., Craik and Byrd 1982; Baddeley 1986). As a result, one might expect that performance on one or more aspects of a task would deteriorate as (1) the complexity of the acoustic scene is increased (e.g., through the addition of competing sound sources), (2) the semantic or syntactic difficulty of the speech material is increased (e.g., switching from a narrative story to a lecture on non-Euclidean geometry), or (3) the task demands are increased (e.g., attempting to answer e-mail while carrying on a phone conversation).

Current information processing models use the term "working memory" to refer to the limited-capacity system that is responsible for the processing and temporary storage of task-relevant information during the performance of everyday cognitive tasks such as language comprehension (Daneman and Carpenter 1980; Baddeley 1986; Daneman and Merikle 1996; Miyake and Shah 1999). In some models, the executive control functions described earlier are part and parcel of the working memory system, see, e.g., "the central executive" component of Baddeley's (1986) model of working memory. There is considerable evidence to suggest that aging is associated with reductions in working memory resources (Van der Linden et al.1994, 1999; Bopp and Verhaeghen 2005). The extent to which a working memory deficit can account for age-related declines in spoken language comprehension and how hearing deficits can alter the operation of the working memory system are discussed in Section 7.5.1.

7.2.4 Speed of Processing

Older adults are 1.5 to 2 times slower than younger adults at performing even the simplest of tasks, such as pressing a button in response to a tone or deciding whether two stimuli are perceptually alike (Cerella 1990). Given the ubiquity of age-related slowing, it is not surprising that one of the most dominant theories among aging researchers is that a generalized slowing in brain function with age is associated with most, if not all, of the age-related declines in performance on complex cognitive tasks such as problem solving, reasoning, and language comprehension (Salthouse 1996; for a meta-analysis, see Verhaeghen and Salthouse 1997). According to this theory, older adults would find it difficult to understand someone who is talking rapidly or to follow a conversation when there are multiple overlapping talkers because the rate of flow of information approaches or exceeds the maximum rate that can be accommodated by the cognitive processes involved in language comprehension (Wingfield 1996). The extent to which processing speed deficits can account for age-related declines in spoken language comprehension is discussed in Section 7.5.3.

7.2.5 Evaluating How Age Affects the Comprehension of Spoken Language

An integrated model of how listeners process speech and other complex acoustic stimuli is one in which the labels "sensory" and "cognitive" are simply convenient ways of referring to the earlier and later stages of a processing system in which the level or depth of processing depends on task demands (for a related model, see Wingfield and Tun 2007). Moreover, it is assumed that there is an executive function capable of marshalling and organizing the different resources required when a listener processes speech and that these resources are limited. In taking an integrated approach to understanding how adult aging affects spoken language comprehension, the following questions will be considered. How do age-related changes in sensory processes affect comprehension and memory? How do age-related changes in cognitive mechanisms such as working memory resources, inhibitory control, and processing speed affect comprehension and memory? How do sensory and cognitive processes interact in the context of spoken language comprehension? Section 3 begins by reviewing how age-related changes in the early (sensory and perceptual) processes might affect the comprehension and memory of complex auditory signals.

7.3 The Effects of Age-Related Changes in Sensory Processes on the Comprehension of Spoken Language

7.3.1 Listening in Quiet

As earlier chapters have indicated, aging is associated with elevated thresholds (especially in the high-frequency region, see Fitzgibbons and Gordon-Salant, Chapter 5), losses in spectral and temporal acuity (see Fitzgibbons and Gordon-Salant, Chapter 5), and possible losses of neural synchrony in the auditory pathways (see Schmiedt, Chapter 2; Canlon, Illing, and Walton, Chapter 3). Provided that these losses are not too severe and that the signal level is adequate, they have little or no effect on simple speech recognition tasks in quiet. However, even though word recognition accuracy measures may be at ceiling in such situations, there is no guarantee that all of the individual speech sounds or words are easily identified or that listening is effortless. For example, high-frequency hearing losses will lead to errors in identifying isolated phonemes in quiet (e.g., van Rooij and Plomp 1992; Humes 1996). However, when these phonemes are embedded in a sentence, the listener is able to make use of sentential context to correct such errors. Hence, if the listener did not clearly hear the final phoneme (was the last word "risk" or "wrist"), she or he could use semantic context to eliminate the ambiguity. In other words, the listener can use his or her knowledge of the language to enhance phoneme or word identification. Moreover, there is some evidence that older adults are at least as good, if not better, than younger adults at using sentential context to reduce ambiguity (Pichora-Fuller et al. 1995; Dubno et al. 2000; Sheldon et al. 2008b; Pichora-Fuller, 2008).

Such use of sentence context would be an example of how top-down cognitive-level resources can be deployed to enhance or support lower-level perceptual processes. However, within our current model, the allocation of higher-level processes to support phoneme or word identification could reduce the pool of resources available for higher-order tasks. Hence, although older adults in the early stages of presbycusis might be able to perform as well as younger adults when there is sufficient contextual support and minimal task demands, age-related losses in hearing could potentially lead to age differences in performance on more demanding tasks. For example, older listeners are able to identify 100% of the high-context sentence-final words of the revised speech perception in noise (R-SPIN) test (Pichora-Fuller et al. 1995) at moderate-to-high signal-to-noise ratios (SNRs) but are unable to remember detailed information or deduce and report themes as well as younger adults when listening to a short lecture in a quiet background (Schneider et al. 2000). Part of the reason why performance might be poorer in the latter than in the former situation may be that older adults, because they are more likely to suffer from peripheral auditory processing deficits, need to engage higher-order cognitive processes for word recognition more frequently than younger adults, thereby depleting the pool of resources available for integrating information across words, extracting themes, and storing relevant information for later recall.

Of course, it could equally well be the case that the reason why older adults may find it more difficult to comprehend and recall information presented in short lectures is that they are also experiencing cognitive declines in the processes that have to be engaged for comprehension and recall of lecture material or that the executive is not as efficient in marshalling the resources required for this task. How one can assess the relative contributions of these two sources (age-related hearing declines versus cognitive declines) is considered in Section 7.6.

7.3.2 Listening in Noise

Although hearing loss can account for most of the speech-recognition problems experienced by healthy older adults in quiet (see Humes and Dubno, Chapter 8), the elevated thresholds and reduced spectral acuity associated with presbycusis can only account for part of the difficulties that older adults experience in noisy situations. This has led a number of investigators to examine the potential contribution of age-related declines in temporal acuity to the speech-recognition problems of older adults in noisy situations.

Temporal cues relevant to speech processing have been described at three main levels (Greenberg 1996): subsegmental (phonetic), segmental (phonemic), and suprasegmental (syllabic and lexicosyntactic). Subsegmental fine-structure cues include periodicity cues based on the fundamental frequency and harmonic structure of the voice. Some types of segmental information are provided by local gap and duration cues in the envelope that contribute to phoneme identification (e.g., presence of a stop consonant, voice onset time). Suprasegmental cues, such as amplitude fluctuations in the region of 3- to 20-Hz, convey prosodic information related to the rate and rhythm of speech and support both syntactic and lexical processing of the information in the speech signal. For example, Shannon et al. (1995), in their classic study of noise-vocoded speech, showed that speech could still be recognized when most of the segmental and subsegmental information in the speech signal was largely removed by (1) breaking the speech signal into a relatively small number of frequency bands, (2) extracting the amplitude envelope in each band, and (3) using these amplitude envelopes to modulate bands of noise whose bandwidths were identical to those in (1). Shannon et al. were able to show that speech recognition was possible even when as few as two to four bands were used in vocoding. This study clearly demonstrates that listeners can use the envelope characteristics of the speech signal in different spectral regions for word recognition. Recently, Sheldon et al. (2008a) have shown that good-hearing older adults need a larger number of frequency bands in the vocoded speech to perform as well as younger adults in a speech-recognition task. Hence, there is some evidence to indicate that at least some older adults are beginning to experience difficulties at this level of temporal processing.

Over the past few decades, a large number of studies (see Fitzgibbons and Gordon-Salant, Chapter 5) have shown that older adults find it more difficult to detect gaps, to discriminate between different gap durations, or to detect a change in the duration of a sound. Losses in temporal acuity at this level have obvious implications for phoneme recognition. Reduced gap discrimination ability could lead to problems detecting stop consonants and loss of sensitivity to vowel duration removes an allophonic cue to vowel identity. Hence, there is good reason to believe that losses in temporal acuity at a segmental level would make speech recognition more difficult for older than for younger adults.

Finally, a few studies are beginning to suggest that some older adults may be experiencing losses in neural synchrony (e.g., Boettcher et al. 1996; Mills et al. 2006). It has long been known the firing pattern in primary auditory afferents is phase locked to the signal, with the degree of phase locking decreasing as frequency increases (see Schmiedt, Chapter 2). Age-related losses in synchrony would, for example, make it more difficult for an older adult to identify a talker or discriminate between talkers based on their characteristic fundamental frequency and/or to track that talker's voice in a complex auditory scene. Recently, Pichora-Fuller et al. (2007) were able to reduce young adults' performance on the R-SPIN test to that characteristic of older adults by artificially increasing the degree of asynchrony in the speech signal, thereby mimicking a loss of neural synchrony. Hence, losses in neural synchrony, especially in noisy situations, could make it more difficult for older than for younger adults to comprehend and remember speech in difficult or noisy situations.

7.4 Effects of Age-Related Changes in More Central Auditory Processes on the Comprehension of Spoken Language

To fully comprehend the auditory scene, listeners have to locate and perceptually segregate the sound sources in their environment (auditory scene analysis; Bregman 1990) so that they can focus their attention on target sources and ignore or suppress the processing of information from irrelevant sources. This is especially difficult to accomplish in reverberant environments because listeners in such situations typically receive not only the direct wave from each sound source but also myriad reflections off environmental surfaces. To successfully parse the auditory scene in these environments, the auditory system has to be able to recognize when, for example, a waveform arriving at one ear is a filtered and time-delayed version of the same waveform that arrived at the other ear a few milliseconds earlier, so that the information available in both waves can be fused into a single source and distinguished from other sound sources (see Eddins and Hall, Chapter 6).

The ability to successfully parse auditory scenes is influenced by a number of factors. First, the greater the spectral differences among sources, the easier it will be to segregate them. Brungart (2001) has shown that it is easier to segregate one talker from another when the two talkers differ substantially with respect to their fundamental frequencies and other acoustic features of their voices. Second, differences in harmonic structure between two sound sources can facilitate source segregation.

For example, changing the frequency of one of the harmonics of a complex tone leads to its emergence as a second auditory event (Alain et al. 2001, 2006). Third, spatially separating sound sources also will lead to improved source separation (see Freyman et al. 1999). In general, the more two sounds differ with respect to their acoustic properties and the greater their separation in space, the easier it is to perceptually segregate them.

Once the auditory scene has been parsed into its component sound sources, listeners find it easier to focus their attention on the target talker and disregard or suppress information from competing talkers. Otherwise, the intrusion of information from irrelevant talkers might interfere with the target talker's message. Such interference is often referred to as perceptual or informational masking (see Freyman et al. 1999; Schneider et al. 2007).

Several cognitive theories suggest that older adults might be more susceptible to intrusions from irrelevant or distracting stimuli than younger adults because of age-related changes in cognitive functioning (e.g., the inhibitory deficit hypothesis; Hasher and Zacks 1988; Hasher et al. 1999). If older adults are more susceptible to distraction, then even if they are able to segregate sources as well as younger adults, they may still benefit less from cues (such as spatial separation) that release younger listeners from informational masking. Alternatively, signal degradation that results from a deteriorating sensory system may make it more difficult to perceptually segregate the target talker from irrelevant talkers, thereby leading to a greater degree of informational masking in older listeners (also discussed in Humes and Dubno, Chapter 8).

Finally, successful scene analysis could be adversely affected by age-related processing limitations in central auditory processes (Snyder and Alain 2007; Canlon, Illing, and Walton, Chapter 3). For example, an age-related diminution in the ability to fuse left ear and right ear correlated signals could significantly reduce the ability of an older person to parse an auditory scene. This, in turn, would lead to greater interference by competing talkers, i.e., to a greater degree of informational masking. This section focuses first on how age-related changes in more central auditory processes might hamper perceptual segregation of sound sources. Then it explores the effects of age on source segregation and informational masking.

7.4.1 *Effects of Age on Processing Capacity and Executive Control Over Auditory Processes*

Spoken language comprehension requires the integration of information coming from different auditory channels. For example, to segregate and correctly identify two simultaneously spoken vowels that differ in fundamental frequency, the listener groups together all frequencies that are harmonically related. The ability to do this could be limited by age-related declines in the capacity of the auditory system to process and integrate information from several auditory filters. In addition, age-related declines in the ability to control the gain in these auditory channels or to control other aspects of auditory processing could also lead to age-related comprehension problems in situations where there are multiple talkers and/or other sound sources. What is known about how age affects auditory channel capacities and influences how information processing is controlled in these channels is discussed in this section.

7.4.1.1 Age Differences in Channel Capacity

Miller (1956) showed that the ability of listeners to identify stimuli varying along a single dimension is limited by that dimension's bandwidth or channel capacity. Murphy et al. (2006b) measured the channel capacities of younger and older adults by having them identify pure tones that differed only in intensity by pressing a button corresponding to the intensity of the tone (an absolute identification paradigm). If older adults were less able to distinguish intensities because of cochlear degeneration, then their ability to identify the tones would have been poorer than that of younger adults. To avoid possible confounds arising from such age-related sensory declines, the intensity differences between adjacent tonal intensities were always large enough that they were nearly perfectly discriminable to both younger and older adults. When the adjacent tones were perfectly discriminable, there were no age differences in the accuracy with which individuals were able to identify among two to eight pure tones. Hence, with respect to stimulus intensity, it does not appear that there is any diminution in channel capacity with age.

There is some evidence suggesting that the bandwidth and/or the processing strategy differs between younger and older adults when they are asked to identify the temporal duration of pure tones by pressing a button corresponding to the duration. Specifically, McCormack et al. (2002) reported that older adults were less accurate than were younger adults in identifying pure tones that varied only in duration. However, the tonal durations used by these investigators were such that some of the pairs of adjacent stimuli were likely to be below the discrimination thresholds of older adults (Bergeson et al. 2001). Hence, the poorer performance of older adults in identifying tones differing only in duration could have reflected the inability of older adults to distinguish between pairs of adjacent stimuli rather than age-related diminution in the channel capacity for stimulus duration. Further studies are needed to see if there are significant reductions in the bandwidth of this and other kinds of auditory channels.

7.4.1.2 Age-Related Differences in Top-Down Control Over Auditory Gain

A number of studies have documented the existence of a nonlinear cochlear amplifier (see Nobili et al. 1998; Schmiedt, Chapter 2). This mechanism is thought to amplify low-intensity sounds and compress high-intensity sounds (Dallos 1997; Robles and Ruggero 2001). Parker et al. (2002) and Gordon and Schneider (2007) have presented evidence that this nonlinear amplifier is under top-down control and argued that the gain of the amplifier is set to maximize discriminability and to protect the auditory system from overload. Specifically, these investigators argued that the degree of compression imposed on a stimulus was modulated by the participant's expectations concerning how intense the stimulus to be presented might be. In other words, the listener's expectations appeared to control, in a top-down fashion, the gain of the nonlinear amplifier. Hence, loss of top-down control over this nonlinear amplifier could limit the ability of the auditory system to function over a wide range of amplitudes and could lead to declines in discriminability. Because there is a widespread loss of outer hair cells and changes in endocochlear potentials in aging, it is possible that there is a loss in gain control with age. Nevertheless, Murphy et al. (2006b) did not find any age-related changes in top-down control over this system in older adults with good hearing. Hence, it appears that top-down control is preserved in older adults with good hearing.

7.4.1.3 Age-Related Changes in Automatic Versus Controlled Processing

Although age-related decrements in early auditory processing should lead to an impoverished representation of the signal, it is always possible that the perception of the signal can be enhanced and performance improved by exerting compensatory top-down control over how information is processed. For example, attention could be focused on the signal's frequency components (auditory attention bands; Dai et al. 1991) or spatial position (e.g., Mondor et al. 1998; Boehnke and Phillips 1999), thereby enhancing signal quality. Hence, when hearing becomes difficult, either because of poor acoustics or because of hearing loss, one would expect listeners to rely more heavily on controlled processing to compensate for poor signal quality. Accordingly, one might expect older adults to more frequently engage in controlled processing than younger adults because of age-related deficiencies in the automatic (bottom-up) processing of speech in competing noise.

Evidence for support of this notion comes from a study by Alain et al (2004) that compared the mismatched negativity (MMN) wave elicited when young, middle-aged, and older adults listened for a deviant stimulus in an oddball paradigm. The standard stimulus was a tone pip; the deviant stimulus was a gap in an otherwise continuous tone pip. Event-related potentials (ERPs) were collected under two conditions: (1) when the listeners were instructed to ignore the auditory stimulus and perform a concomitant visual serial-choice reaction-time task (passive listening) and (2) when the listeners were instructed to pay attention to the stimulus and to respond as quickly and as accurately as possible by pressing a button whenever they heard a stimulus with a silent gap (active listening). Figure 7.1 shows average MMN waves in the passive listening condition for young, middle-aged, and old adults to a deviant stimulus that could be easily detected under active listening conditions. Specifically, the gap duration of the deviant stimulus presented to an individual in this passive listening condition was the shortest gap duration that

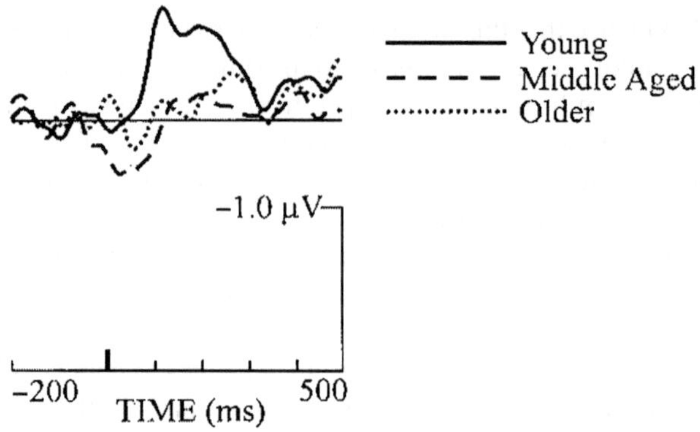

Fig. 7.1 Mismatched negativity waves for young, middle-aged, and older adults in the passive listening condition. The gap size in the deviant stimulus was the shortest gap duration that produced a hit rate within 2% of the asymptotic value reached by the psychometric function relating hit rate to gap detection from the active listening condition. Even though all deviant stimuli were approximately equally detectable in the active listening condition, a significant mismatched negativity (MMN) was only observed for young listeners in the passive condition. (Adapted from Alain et al. 2004 with permission of the author.)

produced a hit rate for that individual in the active listening condition that was within 2% of the asymptotic value of her or his psychometric function, relating hit rate to gap duration. Figure 7.1 shows that a significant MMN in the passive listening condition was observed for younger adults when the deviant stimulus was one that they could easily detect in the active listening condition. In contrast to the findings of a clear MMN for younger adults, in middle-aged and older adults, stimuli that were clearly detectable in active listening conditions failed to produce an MMN in the passive listening condition. The lack of a clear MMN in all but younger adults in this passive listening condition indicates an age-related decrement in automatic processing. However, under active listening conditions, in which listeners presumably engage in controlled processing, the same near-threshold stimuli were readily detected by all age groups. This pattern of results suggests that older adults are able to use top-down processing to compensate when they are actively listening to detect the auditory signal, whereas younger adults apparently do not need to engage such processes. Although such compensation during active listening may overcome age-related problems in passive listening, the increase in effort required to listen in this way may deplete limited cognitive resources in older adults (Pichora-Fuller 2006).

7.4.2 Effects of Age on Source Segregation

The most comprehensive study to date of age-related changes in the ability to segregate two competing speech sources was conducted by Humes et al. (2006).

These investigators pitted two sentences from the coordinate response measure (CRM; Bolia et al. 2000) against each other. The sentences were of the form "ready (call sign), go to (color and number) now" and were played simultaneously. The listener was instructed to attend to the sentence containing the designated call sign and to report the color and number associated with that call sign. The call sign to be attended was presented before or after the two CRM sentences were played. For example, in the before condition, the call sign (e.g., "baron") was presented for three seconds before the two sentences were played (e.g., "ready baron, go to red eight now" vs. "ready charlie, go to green one now"). In the after condition, the call sign was presented for three seconds immediately after the two sentences were played. In both conditions, the correct response for this example is "red eight." Note that the after condition places a greater load on working memory because the listener must retain the information from both sentences until the call sign is presented, whereas in the before condition, listeners can focus their attention on the sentence containing the call sign and either inhibit the processing of the competing sentence or delete its content from working memory. Hence, as seen in Section 7.5.1, according to the prevailing view concerning age-related changes in working memory, age differences should be larger in the after condition than in the before condition.

All sentences were spectrally shaped to adjust for any hearing loss that a participant might be experiencing. This manipulation significantly reduces the probability that any age-related differences in performance on this task are due to age-related cochlear pathologies. The two CRM sentences used in a trial could also be presented to the same ear or to different ears and be spoken by talkers of the same or different genders. Hence, there were four conditions and two age groups. Figure 7.2 depicts the rationalized arcsine transform of the percentage of correct identifications in the before condition. Three features of these data should be noted. First, for both age groups, performance is lowest for same gender sentences presented monaurally; i.e., accuracy is lowest when there are no spatial or gender cues to support source segregation. Second, for both age groups, switching from monaural to dichotic presentations produces a larger improvement in performance for same gender than for different gender talkers. Thus the beneficial effect of spatial separation is larger when there are fewer cues of other sorts to support source segregation. Third, the only significant age difference occurred when the presentation was monaural but the two talkers differed in gender. This suggests that younger adults benefit more than older adults from differences in voice characteristics. However, this age difference in sensitivity to voice characteristics only becomes evident in the absence of other cues to support source segregation.

Figure 7.3 presents the equivalent data in the after condition. As was the case for the before condition, the worst performance is observed when the cues for source segregation are minimal (same gender, monaural presentations) and the effect of spatial separation is larger for same gender than for different gender speakers. Here, however, there are significant age effects in three of the four conditions, suggesting that older adults find it more difficult than do younger adults to retain information in working memory until the call sign cue is presented.

The results of this experiment indicate that both younger and older adults experience a great deal of difficulty in attending to the target sentence when the cues supporting source segregation are minimal (same gender for target and competitor, monaural presentation). When a gender cue is included (but the presentation is monaural), younger adults benefit more from this cue than do older adults. One possibility is that age-related losses in temporal synchrony make it more difficult for older adults to discriminate between two voices based on their fundamental frequency and harmonic structure. An explanation of this sort implies that the failure to be able to segregate the voices based on gender reflects age-related deficiencies in the earlier stages of auditory processing rather than a failure in the later stages of processing involving attention and/or working memory.

Finally, as might be expected from studies of working memory, age differences appear to be greater when the call sign is given after CRM sentence presentation as opposed to before it. Note, however, that age differences in performance in the before case under dichotic listening conditions may have been limited by ceiling effects in the young (see Fig. 7.2). Hence, older adults appear to be able to perform as well as younger adults on speech-recognition tasks, provided that there are

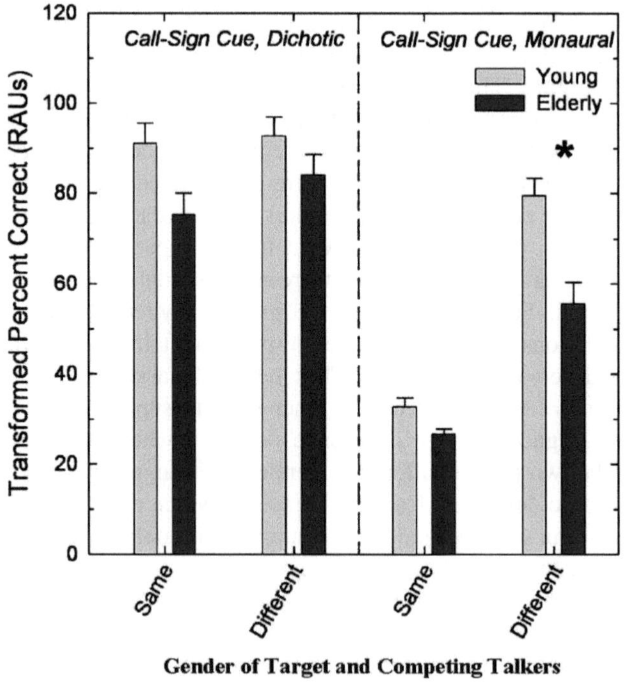

Fig. 7.2 Transformed percentage correct scores (means ± SE) for the coordinate response measure (CRM) target for each of 8 listening conditions when the call sign was presented before the target and competing sentence. RAU, rationalized arcsine unit. *Significant difference between the young and old listeners ($p < 0.01$). (Adapted from Humes et al. 2006 with permission of the author.)

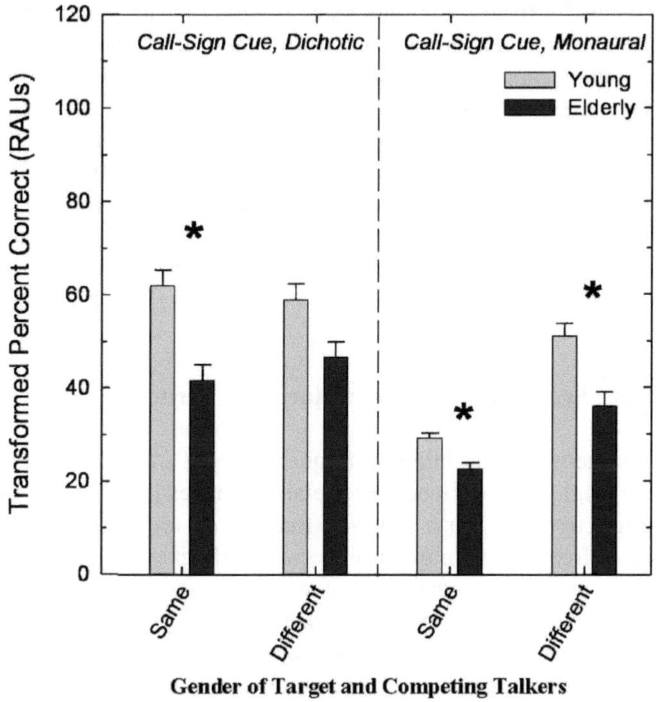

Fig. 7.3 Transformed percentage correct scores (means ± SE) for the CRM target for each of 8 listening conditions when the call sign was presented after the target and competing sentence. *Significant difference between the young and old listeners ($p < 0.01$). (Adapted from Humes et al. 2006 with permission of the author.)

sufficient cues to source segregation and that the working memory requirements of the task do not exceed their working memory capacity.

In a recent study, Singh et al. (2008) explored the effects of location certainty in younger and older listeners in a paradigm developed by Kidd et al. (2005). In one condition of their study, three sentences were presented simultaneously from three different "perceived" locations (perceived location of a sentence in the sound field was manipulated using the "precedence" effect; see Eddins and Hall, Chapter 6). Again, the task was to identify the color and number associated with a particular call sign. The effects of two within-subject factors were explored: (1) temporal position of the call sign (before or after the three sentences were presented) and (2) probability (100%, 80%, 60%, 33%) of the target sentence appearing at the central location (announced before a block of trials). Interestingly, although there was a main effect of age, the effects of location certainty and prior knowledge of the target's call sign were identical for younger and older adults. Because Singh et al. did not compensate for possible cochlear declines, the most likely reason for the poorer performance of older adults was an age-related decline in peripheral auditory processing. The lack of any age interaction suggests that both younger and older adults benefit equally from prior knowledge of target location and target call sign. Hence, although younger adults may outperform older adults when there is competing speech, older adults appear in some cases to be able to utilize information as effectively as younger adults, suggesting that the ways in which listeners handle competing auditory information streams may be relatively unaffected by age.

However, before reaching the conclusion that older adults may be as adept at source segregation as younger adults once age-related changes in cochlear processes have been taken into account, it would be prudent to consider how age might affect other factors that are known to facilitate auditory scene analysis and source segregation.

7.4.2.1 Source Segregation Based on Harmonic Structure

The auditory system tends to interpret frequencies that are harmonically related as belonging to a single source. Hence, age-related changes in the ability to determine whether a single frequency or a group of frequencies are harmonically related to another group of frequencies might be expected to affect scene analysis. Alain et al. (2001) tested the ability of younger and older adults to detect a mistuned harmonic. They found that older adults required a greater degree of mistuning than younger adults to detect a change, with the age difference greater for shorter than for longer sounds. Hence, older adults would presumably require a greater separation in fundamental frequency (and, correspondingly, in harmonic structures) to segregate two voices. Indeed, when two vowels are presented simultaneously, the ability of listeners to correctly identify both vowels increases with the separation between the fundamental frequencies of the two vowels (e.g., Summerfield and Assmann 1991; de Cheveigné 1997; Summers and Leek 1998). A recent study found that older adults benefit less than younger adults from differences in fundamental frequency between two concurrent voices, pointing to age differences in bottom-up auditory processes (Vongpaisal and Pichora-Fuller 2007). Interestingly, an examination of the pattern of brain activity in a different group of participants performing the same task indicated that there was top-down engagement of compensatory cognitive processes in older adults compared with younger adults (Snyder and Alain 2005). Hence, it appears that older adults are less sensitive to differences in the harmonic structure of voices and engage compensatory mechanisms to partially offset age-related peripheral deficits. This would help explain why, in the Humes et al. (2006) study, younger adults profited more than older adults from gender differences between the target and competitor CRM sentences.

7.4.2.2 Source Segregation Based on Attentional Focus

Another possible reason why older adults might not be as efficient as younger adults in using the acoustic cues associated with gender differences to segregate voices or in capitalizing on differences in fundamental frequency when identifying simultaneously presented vowels is that they might not be able to focus their attention on specific-frequency regions as well as younger adults. Studies of auditory attentional focus (e.g., Dai et al. 1991; Hafter and Schlauch 1991) have shown that younger adults are capable of focusing their attention on a specific frequency region if they are expecting a target to have energy in that region. This capacity is usually demonstrated using a probe-frequency paradigm. In this paradigm, the listener is asked to detect a pure tone of a known frequency in a background of bandlimited Gaussian noise. Typically, the intensity of the pure tone (e.g., 1 kHz) is adjusted so that the listener is performing at a targeted degree of accuracy (usually between 85 and 90% correct). The experimenter then occasionally replaces the target tone (e.g., 1 kHz) with a probe tone of another frequency but at the same SPL (for instance, a 980-Hz tone), and then computes how accurate listeners were in detecting the probe tone. Typically, performance on those trials containing a probe decreases with increasing separation between the probe tone's frequency and the target tone's frequency.

Because older adults are often thought to have attention deficits, one might expect their attention bands to be broader than those of younger adults. A possible physiological basis for a broader attentional focus has come from the work of Scharf et al. (1994, 1997), who found that individuals who have had their olivocochlear bundle severed do not demonstrate attentional selectivity. The olivocochlear bundle is believed to be important in allowing for the top-down control of the micromechanical properties of the cochlea and may play a significant role in detecting signals in noise (Pickles 1988). Because the olivocochlear bundle synapses mostly with outer hair cells and because outer hair cells are known to suffer widespread damage with age and/or noise exposure over the lifespan (Willott 1991), one might expect a broadening of attentional focus in older adults. However, Ison et al. (2002) and Murphy et al. (2006b) have shown that the attentional filter has the same bandwidth in younger and older adults. Figure 7.4 plots the percentage of tones correctly detected as a function of the difference in frequency between the probe and target tones for younger and older adults. Clearly, the greater the separation between the probe tone's frequency and the frequency of the target tone, the less likely it is that the probe tone will be detected. Note also that the bandwidth of this "attentional" filter is quite narrow. As Figure 7.4 shows, a probe tone that differs in frequency from the target tone by as little as 100 Hz simply is not detected (the detection rate for a probe tone whose frequency is 100 Hz less than that of the target tone is the same as the probability of reporting a tone when no tone was presented). Hence, both younger and older adults are equally capable of narrowing their attentional focus to a very small region on the basilar membrane (approximately the size of a critical band).

7.4.2.3 Source Segregation Based on Spatial Separation

Perhaps one of the most important cues to source segregation is spatial separation. Consider the case in which there is a single talker in a noisy background, with the

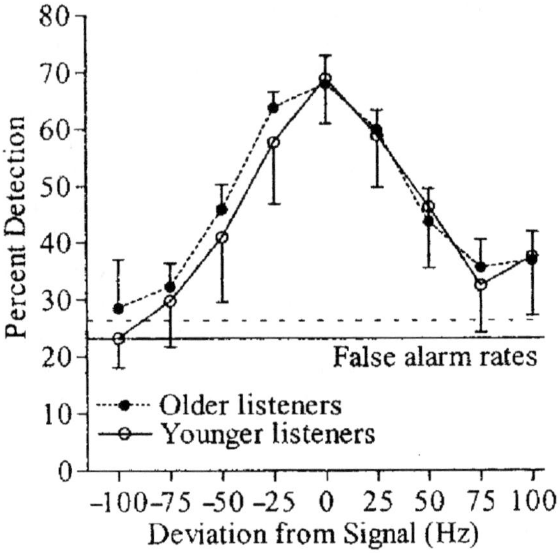

Fig. 7.4 Percent detection (means ± SE) for the signal and for off-frequency probes at the same level that deviated from the signal by ±25 to 100 Hz in younger and older listeners. Also shown are the false alarm rates (responding signal present when no signal was presented). (Adapted from Ison et al. 2002 with permission of the author.)

location of both the talker and the noise source being to the listener's left. When both talker and noise are colocated, the SNR in each ear will be approximately the same. However, if the talker moves to the right of the listener, with the noise remaining on the left, there will be a significant increase in the SNR ratio in the right ear, which could readily facilitate source segregation. In addition, the change in target position could lead to interaural timing differences that might unmask the talker's voice. Hence, spatial separation should be a powerful cue to source segregation.

Because a number of studies have indicated that older adults are less sensitive to binaural cues than are younger adults (see Eddins and Hall, Chapter 6), one might expect that older adults would be less able to use such cues to achieve the same degree of source segregation as younger adults. Hence, age-related changes in binaural processing could lead to poorer source segregation. This issue is addressed in Section 7.4.3 on informational masking.

7.4.2.4 Source Segregation Based on Prior Knowledge

Listening to a conversation in a noisy environment is much easier when the listener is familiar with the topic of conversation. The most likely reason for this is that a priori knowledge of the topic creates expectations about the semantic content and linguistic structure of the unfolding speech signal that facilitates processing.

In particular, if listeners miss some of the words or phrases because the listening situation is difficult, they may be able to recover the lost information from the context provided by the parts of the conversation they have heard and knowledge stored in long-term memory. However, it also is possible that such knowledge helps the listener to focus attention on the relevant voice. For example, if the topic of the conversation of interest is the impending marriage between Emily and Tom and the listener perceives the following sentence fragment "the bridesmaids will," it is quite likely that this voice is a relevant part of the conversation. Hence, it would make sense for listeners to focus their attention on this auditory stream.

Knowledge and expectations regarding aspects of the unfolding speech signal can also be used to advantage by listeners. Such benefit is illustrated in a study by Freyman et al. (2004) showing that the final word of a masked phrase was recognized better if listeners had already heard the initial part of a phrase than if they were hearing the phrase for the first time. Hence, partial knowledge of the sentence improved recognition of the sentence-final word. It is possible that listeners derived some benefit from priming because they acquired knowledge of the acoustical properties of the initial portion of the utterance; however, the same reduction in informational masking was obtained when the participant read the prime instead of listening to it. Clearly, knowledge of the words alone is sufficient to lead to an improvement in performance, independent of the modality in which that knowledge is acquired. Another possibility is that knowing part of the sentence helps the listener to identify and focus in on the target stream. Hence, it is reasonable to hypothesize that a listener in a complex acoustic environment is capable of using knowledge about the nature of conversation to identify and focus in on the talkers participating in a discussion. It would be very interesting to determine whether older adults are as good as are younger adults in using this kind of prior knowledge to achieve source segregation. To our knowledge, there are no published studies in this area.

7.4.2.5 Source Segregation Based on Other Aspects of the Acoustic Scene

Bregman (1990) has argued that the auditory system will capitalize on virtually any features of the auditory scene that will aid in source segregation. One additional cue for source segregation that should be mentioned is "auditory image size." Freyman et al. (1999) demonstrated the usefulness of this cue in a study in which they compared performance when both masker and target were presented over the same frontal loudspeaker to performance when the target was presented over the frontal loudspeaker but the masker was presented over both the frontal loudspeaker and 4 ms later over a loudspeaker located to the listener's right. Note that in the latter condition, the masker was perceived to be frontally located because of the precedence effect (see Section 7.4.3). The target sentences were repeated with much greater accuracy in the latter condition. Even though the location of all images in both conditions remained in the frontal position, the image of the target in the latter condition was more compact than that of the masker, whereas the masker and target had the same degree of compactness when both were presented only over the frontal loudspeaker. This comparison suggests that differences in the compactness of target and masker will improve speech recognition, presumably because it enables a listener to more accurately parse the auditory scene into two different sound sources. How age might alter the effectiveness of this cue to source segregation is not currently known.

In general, there are likely to be a number of acoustic features that could be used to achieve source segregation. How source segregation enhances spoken language comprehension, i.e., how it helps to unmask the speech signal, is considered next.

7.4.3 Effects of Age on Informational Masking

Before discussing the effects of age on informational masking, it is important to distinguish informational masking from energetic masking. When the SNR is low in a spectral region, the energy in competing sound sources can simply overwhelm (mask) the energy in the signal, making it difficult for the listener to extract the target signal from the noise background. This kind of masking is often referred to as "energetic" or "peripheral" masking, and it has been studied extensively (see, e.g., Plomp and Mimpen 1979). In contrast, "informational" or "perceptual" masking occurs when competing signals and background noises interfere with speech recognition at more central auditory- and/or cognitive-processing levels. For example, consider the case in which someone is attempting to attend to one person who is talking when there are two other people who are also talking. To understand what is being said in this situation, the listener must either 1) focus attention on one stream and suppress the other streams of information or 2) attempt to simultaneously process more than one stream at a time. If it becomes difficult for the listener to inhibit the processing of irrelevant information or to simultaneously process more than two information streams, the listener may require a higher SNR for speech recognition and comprehension than would be required if the maskers were acoustically matched nonspeech maskers. Clearly, any factor that facilitates source segregation should make it easier to either (1) focus attention on one source and ignore or suppress the processing of information from the other sources or (2) simultaneously process more than one source at a time. Hence, source segregation should facilitate the release from informational masking more than it would facilitate release from energetic masking.

Many competing sound sources are likely to give rise to some combination of both energetic and informational masking. A common approach to determining what portion of total masking is due to informational versus energetic masking is to compare performance for speech and nonspeech maskers in conditions manipulating auditory or cognitive factors. For instance, spatial separation of the target and masker is one factor that reduces speech-on-speech masking. This release from masking could be due to a reduction in peripheral (energetic) masking and/or a reduction in the amount of interference produced at more central (cognitive) levels. To determine how much of the release from masking is due to the binaural advantage in overcoming acoustical masking versus improved attentional focus in overcoming distraction, the release in speech-on-speech masking seen in conditions of spatial separation may be compared with release when the masker is an unmodulated speech-spectrum noise. It is assumed that there will be equivalent energetic masking by the speech and spectrally matched speech-spectrum maskers. Because speech-spectrum noise is unlikely to initiate any competing phonetic, semantic, or linguistic processing, it should not interfere with speech processing at these more central levels. Therefore, if the release from masking due to spatial separation is larger when the masker is speech than when the masker is speech-spectrum noise, one can infer that the manipulation is effective in reducing interference at more central levels, i.e., in reducing informational masking. Presumably spatial separation is beneficial in reducing energetic masking for both the speech and speech-spectrum noise maskers because interaural cues enable the listener to better segregate the target and masker. In addition, presumably spatial separation is beneficial in reducing informational masking because it enables the listener to better isolate the target speech signal from competing speech in the auditory scene.

An important question is the degree to which age-related changes in either peripheral or central processes reduce the effectiveness of source segregation (however achieved) in providing relief from informational masking. Given that age-related losses in peripheral auditory functioning are likely to reduce the effectiveness of cues to source segregation, one might expect to find less of a release from informational masking in older than in younger adults.

An alternative explanation for possible age differences in release from informational masking could be that older adults are less able than younger adults to benefit from source segregation once it has been achieved because of declines in cognitive capacity. To directly test cognitive theories of age-related declines in spoken language comprehension and bypass age-related differences in sensory processing, Li et al. (2004) used the paradigm developed by Freyman et al. (1999) in which younger and older adults were asked to repeat meaningless target sentences (e.g., "A rose could paint the fish.") presented in either a noise background (energetic masker) or a background in which two other people were also speaking nonsense sentences (informational masker). The target and masker were perceived as coming from the same spatial location or from different spatial locations. Rather than actually changing the physical location of two loudspeakers to achieve the perception that the target and masker were spatially separated, the paradigm capitalizes on the precedence phenomenon to change the perceived locations. In the precedence paradigm, all stimuli were presented over each of two loudspeakers. If a signal is presented simultaneously from both loudspeakers, it is perceived to be located centrally; however, if the signal presented from one loudspeaker leads the same signal presented from a second loudspeaker, the listener perceives that there is only a single source located at the position of the loudspeaker from which the leading sound was presented (see Zurek 1987 for a review). In the experiment of Li et al.

(2004), the target sentence was presented over both loudspeakers, with the right speaker leading the left by 3 ms so that the target sentence was always perceived as coming from the right. The masker was presented either in the same fashion as the target or with the lag between the presentation of the masker from one loudspeaker relative to the other changed so that the masker was perceived as originating from a different location. Because changing the perceived location in this way does not change the acoustic stimulation at either ear in any significant way (see the Appendix in Li et al.), the amount of energetic masking should not change in any significant way with perceived spatial separation. Nevertheless, changing the perceived location of the masker should provide a release from informational masking if it facilitates segregation of the target and masker.

Li et al. (2004) found that both younger and older adults benefited equally from spatial separation when spatial separation was induced using the precedence effect. Interestingly, the only age difference in their experiment was that older adults required a 2.8-dB higher SNR to perform equivalently to younger adults in all conditions. Hence, once age-related differences in peripheral auditory processing are controlled for, the two age groups appear to benefit by the same amount from spatial separation of target and masker.

The fact that older adults in this experiment needed a higher SNR to perform equivalently to younger adults suggests that it should be possible to compensate for a number of age-related deficits in processing speech by improving the SNR for older adults either by improvements in the acoustic environment or by the appropriate use of noise-reduction algorithms in assistive listening technologies.

7.4.4 Tentative Conclusions Concerning Age-Related Changes in More Central Auditory Processes

The available studies of how age-related changes in more central auditory processes might affect spoken language comprehension suggest that older adults with clinically normal audiometric thresholds throughout most of the speech range may be as good as younger adults at source segregation, scene analysis, and release from informational masking once the effects of subclinical age-related deficits in lower-level processing are taken into account. Moreover, there do not appear to be any significant declines in these good-hearing older adults with respect to the bandwidths of auditory channels, ability to focus attention, and/or top-down control over auditory processing. In contrast, a number of findings do point to age-related declines in the automatic processing of near-threshold stimuli. Importantly, it seems that older adults may make more extensive use of controlled top-down processing to compensate during listening. Thus it appears that many central auditory processes are preserved in aging but that they may play a more extensive compensatory role because of age-related declines in lower-level (peripheral or brainstem) auditory processing.

7.5 Effects of Age-Related Changes in Cognitive Processes on Comprehension of Spoken Language

Although there is a fairly large body of research aimed at investigating the extent to which age-related changes in cognitive mechanisms account for age-related declines in language processing (for reviews, see Light 1990; Kemper 1992; Stine 1995; Johnson 2003), it is important to keep in mind that most of the research has focused on the comprehension and recall of written discourse rather than spoken discourse (although see, e.g., Tun et al. 1991; Wingfield and Stine 1992; Titone et al. 2000). As will be seen, there is evidence to suggest that declines in working memory capacity, inhibitory control, and processing speed play a role in the effects of aging on language processing. However, there are vigorous debates as to which of the three plays the primary role (for a review, see Van der Linden et al. 1999).

7.5.1 *Working Memory*

In most contemporary models of language comprehension, working memory represents "the critical bottleneck in which signals are decoded, concepts are activated, linguistic constituents are parsed, thematic roles are assigned and coherence among text-based ideas is sought" (Stine et al. 1995, p. 1). Consequently, it is not surprising that age-related declines in language comprehension are frequently attributed to age-related declines in working memory capacity. The working memory deficit hypothesis is well supported in the literature. Older adults perform more poorly than their younger counterparts on tasks that assess the combined processing and storage capacity of working memory (see Bopp and Verhaeghen 2005 for a meta analysis), and these age-related working memory span differences account for a significant proportion of the age-related variance on written and spoken language comprehension tasks (e.g., Van der Linden et al. 1999; Brébion 2003; DeDe et al. 2004).

7.5.2 *Inhibitory Control*

Some researchers prefer to attribute age-related declines in language comprehension to age-related declines in the ability to inhibit the processing of irrelevant stimuli. According to the inhibition-deficit hypothesis, aging is associated with reduced inhibitory mechanisms for suppressing the activation of goal-irrelevant information (Hasher and Zacks 1988), allowing interfering information to intrude into working memory or preventing no longer relevant information from being purged from working memory. The irrelevant information squanders working memory resources and disrupts the processing of goal-relevant information (Hasher et al. 1999), thereby impairing the reader's or listener's ability to construct a coherent representation of the discourse. Support for the inhibition-deficit hypothesis comes from several studies that have shown that measures of inhibition efficiency (e.g., interference on the Stroop color-word task) appear to mediate age-related differences in written language comprehension (Zacks and Hasher 1994; Kwong See and Ryan 1995; Van der Linden et al. 1999). However, not all the reported data are consistent with the inhibition-deficit hypothesis and even the consistent data are compatible with alternative interpretations (see Burke 1997 for a critical review).

7.5.3 Processing Speed

Finally, in all models of language comprehension, the processes involved in the construction of a complete and coherent discourse representation are assumed to be time-consuming. Consequently, it is not surprising that age-related slowing is frequently viewed as the primary contributor to age-related declines in language comprehension. The processing-speed hypothesis is well supported in the literature for both written and spoken language comprehension (e.g., Cohen 1979; Wingfield, et al. 1985; Tun et al. 1992; Stine and Hindman 1994; Stine et al. 1995). Studies of reading comprehension have shown that age differences in text memory are much larger when reading is experimenter paced rather than self paced (Verhaeghen et al. 1993; Stine-Morrow et al. 2001; Johnson 2003). Fine-grained analyses of self-paced reading times suggest that older adults need to allocate more processing time to new information (Stine et al. 1995) and propositionally dense sentences (Stine and Hindman 1994) than do their younger counterparts.

The contribution of speed of processing to spoken language comprehension has been studied by comparing the performance of younger and older adults when speech is artificially speeded (Wingfield et al. 1985; Gordon-Salant and Fitzgibbons 1993, 1997, 1999; Wingfield 1996), and the typical finding has been that comprehension declines more rapidly for older adults than for younger adults as speech rate increases. Although such a finding is consistent with a slowing hypothesis, there is another possible reason for why older adults find it more difficult to handle rapid rates of speech. Speeding speech, in addition to increasing the rate of flow of information, also tends to degrade and/or distort consonant phonemes in the speech signal (Gordon-Salant and Fitzgibbons 1999; Wingfield et al. 1999). Therefore, it is possible that the reason why older adults are more affected by speeding is that the auditory systems of older adults are less able to handle these distortions than are the auditory systems of younger adults. Indeed, recent studies have found that if speech is speeded in a way that minimizes the adverse effects of speed-induced acoustic distortions on the auditory systems of older adults, increasing the rate of speech has the same effect on speech recognition (Schneider et al. 2005) and spoken language comprehension (Gordon et al. 2009) in younger and older adults. These results support the view that auditory decline rather than cognitive slowing may be responsible for older adults' poorer performance in speeded-speech conditions.

7.5.4 Tentative Conclusions Concerning Cognitive Mediators of Age-Related Changes in Spoken Language Comprehension

Although there is a significant body of evidence to suggest that deficits in working memory capacity, inhibition control, and processing speed could all contribute to age-related differences in spoken language comprehension performance, there is conflicting evidence concerning the relative contributions of the three factors. Most researchers acknowledge that these three indices of processing efficiency are interdependent, but the debate continues as to which of the three plays the primary role in accounting for age-related declines in comprehension. For example, Kwong See and Ryan (1995) have argued that the influence of working memory is secondary to the influences of inhibition control and processing speed. On the other hand, Van der Linden et al. (1999) argue that age-related reductions in processing speed and resistance to interference have an indirect influence on comprehension that is mediated by reductions in working memory capacity. Unraveling the relative and independent contributions of these cognitive mechanisms remains a tricky enterprise.

7.6 Auditory-Cognitive Interactions

Equally tricky is the task of investigating the complex interactions between the aging auditory and cognitive systems and how these interactions contribute to the speech-understanding difficulties of older listeners (e.g., van Rooij and Plomp 1992; Humes 1996; Schneider et al. 2000; Murphy et al. 2006a; George et al. 2007). In this section, correlational and experimental approaches to investigating these complex auditory-cognitive interactions are described.

Several studies have used a correlational approach to investigate the relative contributions of auditory and cognitive factors to speech-understanding difficulties in older adults (e.g., van Rooij and Plomp 1992; Humes 1996). To assess auditory and cognitive competence, younger and older listeners were administered tests of basic auditory abilities (e.g., pure-tone sensitivity, frequency, and duration discrimination) and basic cognitive function (e.g., digit span, Wechsler Adult Intelligence Scale-revised). Scores on these auditory and cognitive tests were then correlated with performance on a number of tests of speech recognition in which listeners were required to detect, discriminate, or identify nonsense syllables, phonemes, spondees, isolated words, words presented in sentences, or whole sentences in quiet and noise. The best single predictor of word and sentence recognition across the studies was the listener's pure-tone threshold function. Most of the cognitive measures correlated poorly with speech recognition. Results such as these led Humes (1996) to conclude that auditory declines are primarily responsible for age-related declines in speech-understanding performance. As provocative as these findings are, there is always the concern about making causal inferences from correlational designs. Moreover, it is possible that these particular correlational studies underestimated the contribution of cognitive factors because (1) the particular choice of cognitive measures may not have been the most appropriate and (2) it is unlikely that simple speech detection and discrimination tests fully engage the linguistic and cognitive processes that operate in everyday listening situations.

Some of the limitations found in correlational studies can be redressed by taking an experimental approach that controls for age-related hearing differences and uses more natural listening tasks. For example, Schneider et al. (2000) approximated more naturalistic listening conditions in the laboratory by having participants listen to complex single-talker discourse in quiet or in a background of conversational noise (12-talker babble), conditions that would be similar to attending a 10- to 15-minute lecture with an audience that is either very attentive or a lot less so. The methodology involved presenting the monologues and noise under identical physical conditions to the younger and older listeners, which is the typical approach in cognitive aging research, or adjusting the listening conditions to compensate for the poorer hearing abilities of the older listeners. When the younger and older adults listened to the monologues under identical stimulus conditions (passages were presented at the same sound pressure level to all participants and the noise, when present, was identical for all participants), the older adults provided fewer correct responses to questions about the discourse than did the younger adults. One might be inclined to attribute the negative age difference in this study to declines in cognitive mechanisms such as working memory capacity, inhibition control, or processing speed. However, the notion that age differences were due primarily to cognitive factors was challenged by the results of a second experiment that adjusted the listening situation to make it equally difficult for both young and old adults to identify individual words. In conditions in which it was equally difficult for young and old to recognize individual words, age-related differences in comprehension and recall of the monologues were largely eliminated. The latter finding suggested that the speech-understanding difficulties of older adults may be largely a consequence of age-related auditory declines rather than age-related cognitive-linguistic declines. Presumably, perceptual declines in older adults result in inadequate or error-prone representations of external events. These inadequacies and errors at the perceptual level then cascade upward and lead to errors in comprehension (see also McCoy et al. 2005).

Of course, natural listening situations do not simply involve listening to a single talker in a noisy background. Murphy et al. (2006a) investigated potential interactions between perceptual and cognitive demands on central resources by asking younger and older adults to comprehend and remember details from two-person conversations when there were varying degrees of spatial separation between the two talkers. In this study, younger and older adults listened to 2-person plays against a background of 12-speaker babble. The voices of the two actors were presented either over the same central loudspeaker (colocation condition) or over separate loudspeakers located 45° to the left and the right of the listener (spatial separation condition). In addition, in both conditions, 12-talker babble could be presented over the central loudspeaker only. The SNR in this situation was individually adjusted so that all individuals, both young and old, were equally able to recognize individual words presented over the left, right, or central loudspeakers when these words were unsupported by context. Thus, in both conditions, younger and older adults were tested in conditions in which they performed equivalently well with respect to word recognition.

Figure 7.5 plots the percentage of detailed information correctly recalled as a function of noise level for both younger and older adults when the two voices were spatially separate versus colocated. When spatial position cues are absent (Talker 1, Talker 2, and babble played over the central loudspeaker), younger and older adults performed equivalently, suggesting that younger and older adults are equally adept at executing the cognitive processes that are required for comprehension, memory, and recall in this task. However, older adults were not as good as younger adults in the same task when the two voices were spatially separate from each other and from the source of the babble.

One possible explanation of this result is that perceived spatial segregation in older adults is not as robust and stable as it is in younger adults. Alternatively, because adding spatial separation to the auditory scene could increase the cognitive load (by requiring listeners to switch attention between spatial positions), older listeners might find it more difficult than younger listeners to handle the increased cognitive demands because of resource limitations. In general, because complex listening tasks (such as comprehending a conversation in a noisy environment with competing talkers) requires the smooth integration of a number of perceptual and cognitive components, one is more likely to observe age-related deficits in spoken language comprehension in such situations.

In the beginning of this section, it was suggested that the study of auditory-cognitive interactions using more ecologically valid stimuli is a tricky business. So far it appears that when the auditory scene is rather simple (all sound sources originating from the same location) and once one controls for age differences in hearing (by making individual words equally difficult for younger and older adults to recognize), age differences in word recognition and in comprehension of spoken discourse tend to disappear. However, when the auditory scene becomes more complex (spatial separation between talkers and masking noise), age differences emerged, presumably because the task of integrating information coming from two different spatial locations while suppressing the processing of information from a third location placed additional demands on working memory and attentional resources. Hence, to the extent that processing resources are more limited in older than in younger adults, one would expect to find that age differences increase as the complexity of the auditory scene increases.

The interpretation above implies that increasing the complexity of the auditory scene either indirectly or directly draws on some of the resources that are involved in spoken language comprehension. A recent study (Heinrich et al. 2008) on the effects of noise on memory suggests that this is indeed the case. Previous studies (Rabbitt 1968, 1991; Murphy et al. 2000) have shown that a continuous background noise or babble (12 people talking simultaneously) affects memory even when the words are clearly audible. Heinrich et al. (2008), in a series of studies (using young adults) designed to determine why background babble interfered with memory consolidation, concluded that listeners were attending to the auditory stream to facilitate the extraction of the signal from the background babble, thereby diverting attentional resources from the task of memory consolidation. Thus the results of this study support the notion that the diversion of attentional resources to the task of parsing the auditory scene leaves fewer resources available for more higher-level processing of the information in the signal.

In conclusion, these studies indicate a rather complex pattern of interdependency between peripheral and central processing of speech. Age-related changes in peripheral processing degrade the speech signal and impede auditory scene analysis. When the listening situation is simple (all sound sources emanating from the same location) and the processing demands light (e.g., speech recognition, processing a monologue), it is possible to show that once these age-related peripheral auditory declines are taken into account, age differences in performance become minimal or disappear altogether. Some evidence has also been presented that central attentional resources may be required, in some instances, either to parse the auditory scene or to make effective use of the information in the parsed scene. This draw on resources, in turn, may deplete the pool of resources available for the processing of language.

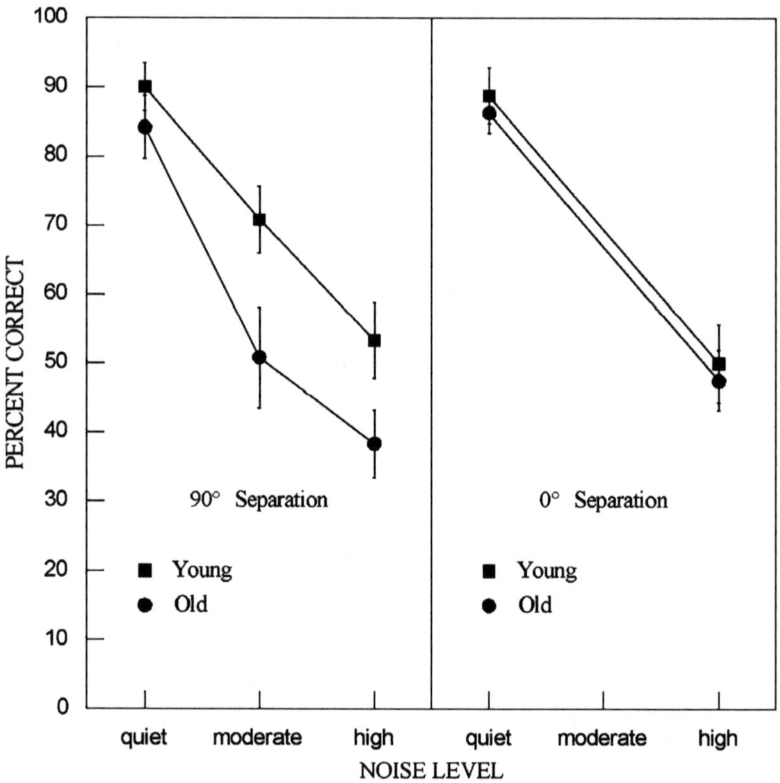

Fig. 7.5 Percentage of detailed information correctly recalled for younger and older adult listeners in quiet and in noise for two conditions of spatial separation between the voices. (Adapted from Murphy et al. 2006a with permission of the author.)

Hence, if older adults have a smaller pool of available resources (e.g., a reduced working memory capacity), one would expect negative age differences to emerge as auditory scenes become complex. Alternatively, one might expect such age differences to emerge when the cognitive demands placed on the individual are increased (e.g., listening to a lecture while responding to e-mail). In all cases, it appears that, in the presence of age-related declines in lower-level auditory processing, there is, correspondingly, a greater need to deploy central processing resources to compensate for these lower-level auditory processing deficiencies. This, in turn, leaves fewer resources available for the processing of spoken language. Conversely, if there are age-related declines in higher-level processes, this could have an adverse effect on auditory scene analysis and signal extraction. The limited number of studies currently available suggest that higher-level, more central auditory processes (attention bands, channel capacity, etc.) may be spared in healthy older adults. Moreover, age-related declines in the cognitive-level processes deemed essential for language comprehension do not appear to have any substantial effect on performance when the auditory scene is simple and the task requirements are not excessively demanding. Importantly, age differences begin to emerge as the auditory scene becomes more complex and/or the comprehension task becomes more demanding.

7.7 Auditory-Cognitive Interactions in Different Populations

The argument has been advanced that age differences in speech recognition and spoken language comprehension in a population of healthy and cognitively-intact individuals are largely due to age-related changes in auditory processes. The available evidence also suggests that age differences are more likely to emerge as the auditory scene becomes more complex and/or the comprehension task becomes more demanding because listeners in such situations will have to depend more and more on the top-down deployment of attentional resources to compensate for adverse listening situations and/or subclinical declines in auditory processing. Moreover, because hearing does not decline at a uniform rate in all individuals, one would also expect larger variations in performance in older than in younger individuals, with some older individuals performing as well as their younger counterparts. Finally, one might expect to find the extent of the decline in both speech recognition and comprehension and the nature and/or kind of top-down compensatory mechanisms that will be engaged when listening becomes difficult to differ depending on the specific nature of the auditory or cognitive decline (Pichora-Fuller 2007). In general, regardless of age, any listener will shift from automatic processing of incoming speech information to more effortful or controlled processing when the listening conditions or task demands become sufficiently challenging. The key questions are when does this breaking point occur (see Rönnberg et al. (2008) for a possible model of this process) and what are the consequences of switching from automatic to controlled processing in older populations with hearing and/or cognitive losses.

Previous studies have shown that older adults with good audiograms require a SNR that is ~3 dB higher than that needed by their younger counterparts to perform equivalently on tests of speech recognition and comprehension. Presumably, the additional 3 dB are needed to overcome the effects of subclinical age-related auditory declines. Older adults with relatively good hearing, however, are not representative of the general population because their linguistic (e.g., vocabulary) and cognitive (e.g., working memory spans) abilities as well as their general health status are likely to be much better than the average or median of older adults in the general population. Such high-performing older adults may excel in using compensatory processing to an extent that is beyond the capabilities of other older adults in the general population. Hence, speech recognition and comprehension are likely to be more drastically affected in older adults with hearing losses or with declining cognitive capacities. The remainder of the chapter considers how auditory-processing deficits associated with presbycusis may interact with cognition during spoken language comprehension in populations of older adults who have clinically significant levels of hearing loss and/or who may not be as cognitively competent as those older adults typically studied in the laboratory. Before doing so, however, it is important to note that competence at the two levels of processing (sensory and cognitive) are strongly linked in older adults.

In a seminal study, Lindenberger and Baltes (1994) reported a very strong link between sensory and cognitive functioning in a large-scale correlational study of adults aged 70 to 103 years (the Berlin Aging study). Specifically, basic measures of hearing sensitivity and visual acuity accounted for 93.1% of the age-related variance in cognitive functioning (for a review, see Schneider and Pichora-Fuller 2000). This strong linkage, irrespective of reasons for its existence, is quite likely to affect the nature of the interaction between the auditory and cognitive systems in any individual with either auditory or cognitive problems. The Berlin group proposed four hypotheses concerning possible explanations for the powerful intersystem connections between perception and cognition in aging: (1) the declines are symptomatic of widespread neural degeneration (common cause hypothesis); (2) cognitive declines result in perceptual declines (cognitive load on perception hypothesis); (3) perceptual declines result in long-term cognitive declines (deprivation hypothesis); or (4) impoverished perceptual input results in compromised cognitive performance (information degradation hypothesis). A number of studies have been conducted to evaluate the contributions of each of these explanations to the linkage between perception and cognition in aging (for a review, see Gallacher 2005) with many of the laboratory studies described earlier, providing evidence in support of the information degradation hypothesis. Nevertheless, when dealing with an aging individual, it could be that the nature of the interaction between hearing and cognition in speech processing has been altered because of widespread neural degeneration, because cognitive declines have led to inefficient sensory processing (e.g., lack of top-down control over perceptual processes), because long-term sensory deprivation has led to cognitive deterioration, or because information degradation has compromised cognitive performance. Whatever the reason, it is likely that the pattern of auditory-cognitive interaction will differ depending on the specific subpopulation of older adults being studied.

In addition, "presbycusis" is a catchall term that refers to hearing loss in an older adult with no known specific cause (e.g., disease, trauma). Recently, researchers have begun to distinguish among a number of different subtypes of presbycusis (see Schmiedt, Chapter 2). Therefore, it is likely that the interplay between auditory and cognitive factors will vary with the particular subtype. Better diagnosis of various subtypes of presbycusis should reduce the apparent hetereogeneity seen in older groups tested in the lab and in the clinic. Furthermore, if the nature of the auditory-cognitive interactions in spoken language comprehension can be determined for each of the different subtypes, rehabilitation strategies could be developed that are specifically tailored to the abilities and potentials of older adults. In the meantime, without being able to differentiate patterns that might be associated with different subtypes of presbycusis, one should proceed with caution in considering the literature regarding the interactions between auditory and cognitive processing in older adults with clinically significant hearing impairment, including research regarding the experiences of older adults using hearing aids or other assistive technology.

7.7.1 Interaction of Auditory and Cognitive Factors in Older Listeners With Hearing Loss

In studies where simple signals and tasks are used to measure speech understanding, audiometric thresholds explain more of the variance than is explained by cognitive variables (see Humes and Dubno, Chapter 8). However, as the listening situation and/or the task becomes more difficult, cognitive factors appear to account for more of the variance in performance of older adults (for a review, see Humes 2007). For example, in the study concerning source segregation described in detail in Section 7.4.2 (Humes et al. 2006), in which amplification was used to compensate for hearing loss, individual differences within the older group with hearing loss were associated with a measure of working memory (digit span) when the call sign was presented after the CRM sentences were played (a condition that placed a high load on working memory) but not when the call sign was presented before the sentences were played (low load on working memory). By way of contrast, performance on the CRM tasks did not correlate significantly with high-frequency hearing loss in either the low- or high- working memory load conditions. The latter result is not too surprising because amplification was used to compensate for high-frequency hearing losses. The former result, however, supports the notion that cognitive factors, such as working memory load, play an increasingly important role as the demands of the task increase.

7.7.2 Interventions

Over the last half century, those with sensorineural hearing loss, especially presbycusics, were considered to present special problems as candidates for audiological rehabilitation (Davis and Silverman 1970, pp. 321-322). When treating older adults, clinicians must frequently reckon with poor health and the gradual failure of other faculties, particularly vision, which is quite helpful as a supplement to hearing. Not only do older people often become very dependent on others, but they may also find themselves unable to "keep up with the times." In addition, many of them live alone or with children who "have their own lives to lead." All these factors may lead to tensions, fears, and anxiety, which may discourage the use of new technologies (Czaja et al. 2006). All of these factors might lead to less successful interventions and hearing aid outcomes in older than in younger adults.

Tremendous technological advances have resulted in present-day hearing aids being more effective than past technologies in overcoming the peripheral auditory deficits associated with sensorineural hearing loss. However, overcoming the central auditory-processing and cognitive-processing problems that affect older listeners is an ongoing topic of research concerning hearing aid design (Edwards 2007). Furthermore, in addition to needed improvements in hearing aid technology, older adults continue to require other rehabilitation approaches that are tailored to their specific needs and ecologies (e.g., Kiessling et al. 2003; Kricos 2006; Willott and Schacht, Chapter 10). Some of the issues that can modulate the effectiveness of interventions in older populations with hearing losses are consideredin Section 7.7.2.

7.7.2.1 Hearing Aids

A typical finding is that there may be a decade or more delay between when a person becomes aware of hearing problems and when the first hearing aid is obtained (e.g., Hétu 1996). The stigma and negative stereotypes associated with wearing a hearing aid and being old, along with other psychological and social factors, seem to conspire to promote the denial of hearing loss, prolong the period before rehabilitation is sought, and foster the rejection and discontinuation of hearing aid use (Kochkin 1993; Garstecki and Erler 1998). Consequently, the average first-time hearing aid user has already reached retirement age (see Cruickshanks, Zhan, and Zhong, Chapter 9).

Not surprisingly, the rate of hearing aid use is even lower in those with dementia than it is in those whose with normal cognitive function (Cohen-Mansfield and Infeld 2006). Indeed, those with no prior hearing aid experience are often considered to be unlikely candidates for a first-time hearing aid fitting. Those with dementia who learned to use a hearing aid before suffering cognitive loss need increasing support from others to maintain hearing aid use as their general health and cognitive abilities continue to deteriorate (Hoek et al. 1997; Jennings 2005). Therefore, it seems that early audiological intervention is worthwhile not only because of the immediate benefits to the older person with hearing loss but also because early intervention may alleviate some of the negative consequences of further age-related declines in auditory and cognitive abilities. The importance of early intervention for age-related hearing loss has led to the development of new rehabilitative strategies that, in addition to providing a hearing aid, also include a health promotion component and new training techniques that are tailored to fit the needs of the individual (Chisolm et al. 2003; Gates and Mills 2005; Kricos 2006).

A basic hope today is that audiological intervention and rehabilitation in older adults will not only improve auditory function and spoken language comprehension but also increase competence in performing activities of daily living (e.g., Brennan et al. 2005) and result in improvements in an individual's overall quality of life (e.g., Stark and Hickson 2004; Chia et al. 2007; Chisolm et al. 2007). Indeed, on the basis of a large-scale longitudinal study of a comprehensive set of outcome measures administered to older adults who had been fit with hearing aids, Humes (2003) concluded that improvements could be demonstrated in three distinct categories: speech perception, hearing aid use, and satisfaction. There is also some evidence that sensory-focused interventions, such as hearing aid fitting (or cataract surgery to correct vision), result in significant posttreatment improvements on cognitive measures administered in the same modality as the sensory intervention (Mulrow et al. 1990; Van Boxtel et al. 2000; Valentijn et al. 2005). However, improvements after hearing aid fitting have not been seen on cognitive tests administered using visual stimuli (Tesch-Romer 1997; van Hooren et al. 2005). It is worth noting that within-modality improvements provide further evidence for the information degradation hypothesis.

Whether the strong correlation between auditory and cognitive functioning is explained in terms of the common-cause hypothesis, the information-degradation hypothesis, or a combination of these hypotheses, another emerging assumption motivating rehabilitative approaches is that compensation for sensory loss may slow or attenuate the symptoms of cognitive decline (Wahl and Heyl 2003). In turn, both perceptual and cognitive declines have been linked to emotional and psychosocial problems and even longevity (e.g., Appolonio et al. 1996; Cacciatore et al. 1999). Thus it seems that audiological rehabilitation may have positive consequences for cognitive and social function in older communicators (e.g., Mulrow et al. 1992).

Another perspective on the connection between hearing aid use and cognition that has gained recent attention among rehabilitative audiologists is that cognitive function may influence the suitability of candidates for particular types of hearing aid processing. In recent landmark studies, it has been shown that an individual's cognitive capacity is significantly related to the degree to which he or she benefits from more demanding types of signal-processing algorithms used in hearing aids when more demanding speech and background signals are used to test performance (Gatehouse et al. 2003, 2006; Lunner 2003; Lunner and Sundewall-Thorén 2007). Importantly, traditional measures of hearing such as pure-tone average account for most of the variance in speech understanding when the signal-processing and background signals are less challenging, but cognitive measures account for most of the variance in performance when the listening situation is more challenging. The connection between cognitive ability and candidacy for different types of hearing aids has compelled audiologists to begin to consider which cognitive measures could be incorporated into audiological practice and how such measures could be used to either guide hearing aid fitting or to evaluate outcomes of rehabilitation (see Pichora-Fuller and Singh 2006).

7.7.2.2 The need for Comprehensive Rehabilitation

Given the inevitable failure of technology to solve all communication problems experienced by an older adult with hearing loss, a more comprehensive approach to rehabilitation is required, especially for older adults who have central auditory and cognitive declines in addition to cochlear hearing loss. Moreover, communication difficulties due to hearing loss (cochlear or central) are often exacerbated by the attitudes and behavior of communication partners and by the context within which communication takes place (Pichora-Fuller and Robertson 1997). Therefore, comprehensive rehabilitation would necessarily include training and counseling components that would improve the communicative behavior of both the impaired listener and his or her communicative partners and the ecology or situations in which they communicate (Pichora-Fuller and Schow 2007). Specifically, intervention may be required to enable the person with hearing loss to cope with their reactions to communication difficulties, to develop self-efficacy, and to use contextual information and conversational strategies to circumvent or compensate for the poor quality of perceptual input. Intervention may also be required to enable the person's communication partners to support them in a variety of ways including producing more easily understood speech and language. In addition, it may be important to optimize environmental factors such as reducing ambient noise and distraction or improving lighting. The effectiveness of these forms of intervention has been convincingly documented (e.g., Hickson and Worrall 2003; Kramer et al. 2005; Boothroyd 2007).

In general, the value of a more comprehensive approach to audiological rehabilitation has received renewed attention as the importance of the interactions between auditory and cognitive factors has become better understood (see Willott and Schacht, Chapter 10).

7.7.3 Interaction of Auditory and Cognitive Factors in Older Listeners with Cognitive Loss

7.7.3.1 Prevalence

As with hearing loss, reports on the prevalence of cognitive impairment vary depending on many research parameters. For Alzheimer's disease, prevalence estimates range from 1% at age 65 years to 75% at age 90 years (Hy and Keller 2000). In a sample of over 1,800 adults living in the community who participated in the Canadian Study of Health and Aging (Ebly et al. 1994), the prevalence of dementia increased with age, with ~15% being affected in the age group from 75 to 84 years, 23% in the group aged 85 to 89 years, 40% in those 90 to 94 years, and 58% in those 95 years and older. Probable or possible Alzheimer's disease accounted for 75% of the cases of dementia, and vascular etiology alone accounted for another 13%. Of course, these figures would be much higher in the segment of the population who are living in residential care.

7.7.3.2 Correlations Between Sensory and Cognitive Impairment in Population Studies

Hearing impairment is associated with cognitive function even when the sample is deemed to be clinically normal on cognitive screening tests such as the Mini-Mental State Examination (MMSE) (Golding et al. 2006). Hence, there is an obvious practical need for taking sensory function into account when assessing cognitive function (e.g., Uhlmann et al. 1989b). Hearing impairment may even contribute or accelerate clinically significant cognitive decline and it has been suggested that sensory intervention could slow cognitive decline (Peters et al. 1988; Wahl and Heyl 2003). In a large-scale study, Uhlmann et al. (1989a) examined the relationship between hearing loss and cognitive decline in 100 people with and 100 people without Alzheimer's type dementia. (The two groups were matched with respect to age, sex, and education.) These investigators found that cognitive function was correlated with hearing loss in both the demented and nondemented groups, with the prevalence of hearing loss being greater in the demented group. These results led the investigators to identify hearing loss (especially in the moderate-to-severe range) as a risk factor for Alzheimer's type dementia. In a large-scale, multicenter longitudinal study of older women who were participating in research concerning osteoporotic fractures, combined vision and hearing loss was associated with the greatest odds for cognitive decline on the MMSE and functional decline in five everyday activities over a period of four years (Lin et al. 2004). In a smaller longitudinal study of patients with dementia of various etiologies, cognitive decline in Alzheimer's patients with impaired hearing over a period of about one year was more rapid than in Alzheimer's patients with relatively good hearing (Peters et al.1988).

It is noteworthy that in the studies investigating the relationships between cognitive impairment and sensory impairment, the criterion for vision impairment is typically based on measures of corrected vision; however, the criterion for hearing impairment is usually based on unaided audiometric thresholds. This may be a consequence of a relatively small percentage of participants who are hearing aid users and the reluctance on the part of the researchers to obtain performance measures from them. In addition, only a handful of studies have included tests of central auditory processing. Those that have included such tests have indicated that people with probable Alzheimer's disease performed worse on the tests of central auditory processing even though they were matched to a nondemented control group with respect to age, gender, and pure-tone average (Strouse et al. 1995). In a recent large-scale longitudinal study, speech tests, which were employed to measure central auditory dysfunction, proved to be predictive of the likelihood of developing Alzheimer's disease even after the contribution of audiometric sensitivity had been taken into account (Gates et al. 2002). Further work is clearly needed to identify the nature of the linkage between hearing losses of various types and cognitive declines and to determine the effectiveness of audiological interventions in stemming the tide of cognitive decline in the population.

7.7.3.3 Intervention

The literature suggests that providing comprehensive audiological rehabilitation may be especially important for older adults with cognitive impairment. As cognitive impairment progresses, more responsibility for communication will shift from the listener with hearing loss to the communication partners, and it will become even more important to prevent communication problems by optimizing communication contexts rather than simply relying on technology to overcome auditory problems. Importantly, hearing aid use should not be precluded by dementia. Indeed, reduced rate of decline in MMSE scores over a six-month period has been documented after intervention with hearing aids (Allen et al. 2003). Use of hearing aids has also been related to reduced caregiver-identified problem behaviors in patients with Alzheimer's disease living in the community (Palmer et al. 1999).

7.8 Summary and Recommendations for Future Research

7.8.1 Summary

Spoken language comprehension requires the smooth and rapid functioning of an integrated system of perceptual and cognitive processes. Studies of high-functioning younger and older adults with good hearing have shown that the pattern of interdependency between lower-level and higher-level processing of speech can be quite complex. It also appears that any factor (such as background noise, competing speech) that degrades the speech signal will place a greater processing burden on the higher-level components involved in speech processing and most likely will engage working memory and other attentional processes as the listener attempts to process impoverished or distorted speech signals. Conversely, any cognitive-level demands (such as trying to compose an e-mail while listening over the telephone) will also lead to poorer spoken language comprehension because it divides attention between two tasks. The importance of these interactions has been demonstrated in highly selected experimental samples of younger and older adults. However, relatively little is known about the nature of these sensory-cognitive interactions in the general population where there is a much broader range of both auditory and cognitive abilities. Because of the strong linkages that exist between auditory and cognitive functioning, one might expect to find more complex and varied patterns of sensory-cognitive interactions in the general population than those found in high-functioning, good-hearing older adults.

7.8.2 Recommendations for Future Research

First, systematic laboratory research is needed to more fully understand the nature of the complex interactions that occur between the auditory and cognitive processes involved in spoken language comprehension. This could be done in two parts: (1) continuing to explore interactions in high-functioning older adults and (2) extend these laboratory studies to include specific impaired subpopulations having different degrees and types of both auditory and cognitive impairments. Special attention should be paid to individuals with different subtypes of presbycusis, different subtypes of cognitive impairment, or combinations of more precisely subtyped auditory and cognitive impairments.

Second, more research is needed on how to translate experimental findings into clinical practice and models of health service delivery. Experimental research has advanced our knowledge of interactions between auditory and cognitive factors and developed experimental tools to measure various aspects of auditory and cognitive processing. Research is needed on how to develop cost-effective, clinically feasible versions of these experimental techniques. Assuming that this new generation of auditory and cognitive measures has been developed, they could be used to devise better and more individually tailored interventions.

Acknowledgments This work was supported by grants from the Natural Sciences and Engineering Research Council of Canada (RGPIN 2690, RGPIN 138472, RGPIN 9952) and Canadian Institutes of Health Research (MGC 42665, MT 15359).

References

Alain C, McDonald KL, Ostroff JM, Schneider B (2001) Age-related changes in detecting a mistuned haromonic. J Acoust Soc Am 109:2211–2216.

Alain C, McDonald KL, Ostroff JM, Schneider BA (2004) Aging: a switch from automatic to controlled processing of sounds? Psychol Aging 19:125–133.

Alain C, Dyson BJ, Snyder JS (2006) Aging and the perceptual organization of sounds: a change of scene? In: Conn MP (ed) Handbook of Models for Human Aging. Amsterdam: Elsevier Academic Press, pp. 759–769.

Allen NH, Burns A, Newton V, Hickson F, Ramsden R, Rogers J, Butler S, Thistlewaite G, Morris J (2003) The effects of improving hearing in dementia. Age Ageing 32:189–193.

Appolonio I, Carabellese C, Frattola L, Trabucchi M (1996) Effects of sensory aids on the quality of life and mortality of elderly people: a multivariate analysis. Age Ageing 25:89–96.

Baddeley AD (1986) Working Memory. New York: Oxford University Press.

Baddeley AD (1993) Working memory or working attention? In: Baddeley AD, Weiskrantz L (eds) Attention: Selection, Awareness and Control. New York: Oxford University Press, pp. 157–170.

Bergeson TR, Schneider BA, Hamstra SJ (2001) Duration discrimination in younger and older adults. Can Acoust 29:3–9.

Boehnke SE, Phillips DP (1999) Azimuthal tuning of human perceptual channels for sound location. J Acoust Soc Am 106:1948–1955.

Boettcher FA, Mills JH, Swerdloff JL, Holley BL (1996) Auditory evoked potentials in aged gerbils: responses elicited by noises separated by a silent gap. Hear Res 102:167–178.

Bolia RS, Nelson WT, Ericson MA, Simpson BD (2000) A speech corpus for multitalker communications research. J Acoust Soc Am 107:1065–1066.
Boothroyd A (2007) Adult aural rehabilitation: what is it and does it work? Trends Amplif 11:63–71.
Bopp KL, Verhaeghen P (2005) Aging and verbal memory span: a meta-analysis. J Gerontol B Psychol Sci Soc Sci 60:P223-P233.
Brébion G (2003) Working memory, language comprehension, and aging: four experiments to understand a deficit. Exp Aging Res 29: 269–301
Bregman AS (1990) Auditory Scene Analysis: The Perceptual Organization of Sound. Cambridge, MA: MIT Press.
Brennan M, Horowitz A, Su Y-P (2005) Dual sensory loss and its impact on everyday competence. Gerontologist 45:337–346.
Brungart DS (2001) Informational and energetic masking effects in the perception of two simultaneous talkers. J Acoust Soc Am 109:1101–1109.
Burke DM (1997) Language, aging, and inhibitory deficits: evaluation of a theory. J Gerontol B Psychol Sci Soc Sci 52:P254-P264.
Cacciatore F, Napoli C, Abete P, Marciano E, Triassi M, Rengo F (1999) Quality of life determinants and hearing function in an elderly population: Osservatorio Geriatrico Campano Study Group. Gerontology 45:323–328.
Cerella J (1990) Aging and information-processing rate. In: Birren JE, Schaie KW (eds) Handbook of the Psychology of Aging, 3rd Ed. San Diego, CA: Academic Press, pp. 201–221.
Chia E-M, Wang JJ, Rochtchina R, Cumming RR, Newall P, Mitchell P (2007) Hearing impairment and health-related quality of life: The Blue Mountain Hearing Study. Ear Hear 28:187–195.
Chisolm TH, Willott JF, Lister JJ (2003) The aging auditory system: anatomic and physiologic changes and implications for rehabilitation. Int J Audiol 42:S3-S10.
Chisolm TH, Johnson CE, Danhauer JL, Portz LJP, Abrams H, Lesner S, McCarthy PA, Newman CW (2007) A systematic review of health-related quality of life and hearing aids: final report of the American Academy of Audiology Task Force on the Health-Related Quality of Life Benefits of Amplification on Adults. J Am Acad Audiol 18:151–183.
Cohen G (1979) Language comprehension in old age. Cognit Psychol 11:412–429.
Cohen-Mansfield J, Infeld DL (2006) Hearing aids for nursing home residents: current policy and future needs. Health Policy 79:49–56.
Craik FIM, Byrd M (1982) Aging and cognitive deficits: the role of attentional resources. In: Craik FIM, Trehub S (eds) Aging and Cognitive Processes. New York: Plenum, pp. 191–211.
Craik FIM, Lockhart RS (1972) Levels of processing: a framework for memory research. J Verb Learn Verb Behav 11:671–684.
Czaja SJ, Charness N, Fisk AD, Hertzog C, Nair SN, Rogers WA, Sharit J (2006) Factors predicting the use of technology: findings from the Center for Research and Education on Aging and Technology Enhancement (CREATE). Psychol Aging 21:333–352.
Dai H, Scharf B, Buus S (1991) Effective attenuation of signals in noise under focused attention. J Acoust Soc Am 89:2837–2842.
Dallos P (1997) Outer hair cells: the inside story. Ann Otol Rhinol Laryngol 106:16–22.
Daneman M, Carpenter PA (1980) Individual differences in working memory and reading. J Verb Learn Verb Behav 19:450–466.
Daneman M, Merikle PM (1996) Working memory and comprehension: a meta-analysis. Psychon Bull Rev 3:422–433.
Davis H, Silverman R (1970) Hearing and Deafness. New York: Holt, Rinehart and Winston.
de Cheveigné A (1997) Concurrent vowel identification. III. A neural model of harmonic interference cancellation. J Acoust Soc Am 101:2857–2865.
DeDe G, Caplan D, Kemtes K, Waters G (2004) The relationship between age, verbal working memory, and language comprehension. Psychol Aging 19:601–616.
Dubno JR, Ahlstrom JB, Horwitz AR (2000) Use of context by young and aged adults in normal hearing. J Acoust Soc Am 107:538–546.
Edwards B (2007) The future of hearing aid technology. Trends Amplif 11:31–45.

Elby EM, Parhad IM, Hogan DB, Fung TS (1994) Prevalence and types of dementia in the very old: results from the Canadian Study of Health and Aging. Neurology 44:1593–1600.
Freyman RL, Helfer KS, McCall DD, Clifton RK (1999) The role of perceived spatial separation in the unmasking of speech. J Acoust Soc Am 106:3578–3588.
Freyman RL, Balakrishnan U, Helfer KS (2004) Effect of number of masking talkers and auditory priming on informational masking in speech recognition. J Acoust Soc Am 115:2246–2256.
Gallacher J (2005) Hearing, cognitive impairment and aging: a critical review. Rev Clin Gerontol 14:1–11.
Garstecki DC, Erler SF (1998) Hearing and aging. Topics Geriatr Rehabil 14:1–17.
Gatehouse S, Naylor G, Elberling C (2003) Benefits from hearing aids in relation to the interaction between the user and the environment. Int J Audiol 42, Suppl 1:S77-S85.
Gatehouse S, Naylor G, Elberling C (2006) Linear and nonlinear hearing aid fittings - 2. Patterns of candidature. Int J Audiol 45:153–171.
Gates GA, Mills JH (2005) Presbycusis. Lancet 366:1111–1120.
Gates GA, Beiser A, Rees TS, Agostino RB, Wolf PA (2002) Central auditory dysfunction may precede the onset of clinical dementia in people with probable Alzheimer's disease. J Am Geriatr Soc 50:482–488.
George ELJ, Zekveld AA, Kramer SE, Goverts ST, Festen JM, Houtgast T (2007) Auditory and nonauditory factors affecting speech reception in noise by older listeners. J Acoust Soc Am 121:2362–2375.
Golding M, Taylor A, Cupples L, Mitchell P (2006) Odds of demonstrating auditory processing abnormality in the average older adult: the Blue Mountains hearing study. Ear Hear 27:129–138.
Gordon MS, Schneider BA (2007) Gain control in the auditory system: absolute identification of intensity within and across two ears. Percept Psychophys 69:232–240.
Gordon MS, Daneman M, Schneider BA (2009). Comprehension of speeded discourse by younger and older listeners. Exp Aging Res 35:277–296.
Gordon-Salant S, Fitzgibbons PJ (1993) Temporal factors and speech recognition performance in young and elderly listeners. J Speech Hear Res 36:1276–1285.
Gordon-Salant S, Fitzgibbons PJ (1997) Selected cognitive factors and speech recognition performance. J Speech Lang Hear Res 40:423–431.
Gordon-Salant S, Fitzgibbons PJ (1999) Profile of auditory temporal processing in older adults. J Speech Lang Hear Res 44:709–719.
Greenberg S (1996) Auditory processing of speech. In: Lass NJ (ed) Principles of Experimental Phonetics. St. Louis, MO: Mosby, pp. 362–407.
Hafter ER, Schlauch RS (1991) Cognitive factors and selection of auditory listening bands. In: Dancer AL, Henderson D, Salvi RJ, Hammernik RP (eds) Noise-Induced Hearing Loss. Philadelphia, PA: B. C. Decker, pp. 303–310.
Hasher L, Zacks RT (1988) Working memory, comprehension, and aging: a review and a new view. In: Bower GH (ed) The Psychology of Learning and Motivation, Vol. 22. San Diego, CA: Academic Press, pp. 193–225.
Hasher L, Zacks RT, May CP (1999) Inhibitory control, circadian arousal, and age. In: Gopher D, Koriat A (eds) Attention and Performance XVII: Cognitive Regulation of Performance: Interaction of Theory and Application. Cambridge, MA: MIT Press, pp. 653–675.
Heinrich A, Schneider BA, Craik FIM (2008) Investigating the influence of continuous babble on auditory short-term memory performance. Q J Exp Psychol 5:735–751.
Hétu R (1996) The stigma attached to hearing impairment. Scand Audiol Suppl 43:12–24.
Hickson L, Worrall L (2003) Beyond hearing aid fitting: improving communication for older adults. Int J Audiol 42:S84-S91.
Hoek D, Pichora-Fuller MK, Paccioretti D, MacDonald MA, Shyng G (1997) Community outreach to hard-of-hearing seniors. J Speech Lang Pathol Audiol 21:199–208.
Humes LE (1996) Speech understanding in the elderly. J Am Acad Audiol 7:161–167.
Humes LE (2003) Modeling and predicting hearing aid outcome. Trends Amplif 7:41–75.
Humes LE (2007) The contributions of audibility and cognitive factors to the benefit provided by amplified speech to older adults. J Am Acad Audiol 18:590–603.

Humes LE, Lee JH, Coughlin MP (2006) Auditory measures of selective and divided attention in young and older adults using single-talker competition. J Acoust Soc Am 120:2926–2937.
Hy LX, Keller DM (2000) Prevalence of AD among whites: a summary by levels of severity. Neurology 55:198–204.
Ison JR, Virag TM, Allen PD, Hammond GR (2002) The attention filter for tones in noise has the same shape and effective bandwidth in the elderly as it has in younger listeners. J Acoust Soc Am 112:238–246.
Jennings MB (2005) Audiologic rehabilitation needs of older adults with hearing loss: views on assistive technology uptake and appropriate support services. J Speech Lang Pathol Audiol 29:112–124.
Johnson RE (2003) Aging and the remembering of text. Dev Rev 23:261–346.
Kemper S (1992) Language and aging. In: Craik FIM, Salthouse TA (eds) The Handbook of Aging and Cognition. Hillsdale, NJ: Erlbaum, pp. 213–270.
Kidd G Jr, Arbogast TL, Mason CR, Gallun FJ (2005) The advantage of knowing where to listen. J Acoust Soc Am 118:3804–3815.
Kiessling J, Pichora-Fuller MK, Gatehouse S, Stephens D, Arlinger S, Chisolm T, Davis AC, Erber NP, Hickson L, Holmes A, Rosenhall U, von Wedel H(2003). Candidature for and delivery of audiological services: special needs of older people. Int J Audiol volume 42, issue 6, supplement 2:S92-S101.
Kochkin S (1993) MarkeTrak III: why 20 million in the US don't use hearing aids for their hearing loss. Hear J 46:28–31.
Kramer SE, Allessie GHM, Dondorp AW, Zekveld AA, Kapteyn TS (2005) A home education program for older adults with hearing impairment and their significant others: a randomized trial evaluating short- and long-term effects. Int J Audiol 44:255–264.
Kricos PB (2006) Audiologic management of older adults with hearing loss and compromised cognitive/psychoacoustic auditory processing capabilities. Trends Amplif 10:1–28.
Kwong See ST, Ryan E (1995) Cognitive mediation of adult age differences in language performance. Psychol Aging 10:458–468.
Li L, Daneman M, Qi J, Schneider BA (2004) Does the information content of an irrelevant source differentially affect spoken word recognition in younger and older adults? J Exp Psychol Hum Percept Perform 30:1077–1091.
Light L (1990) Interactions between memory and language in old age. In: Birren JE, Schaie KW (eds) Handbook of the Psychology of Aging, 3rd Ed. San Diego, CA: Academic Press, pp. 275–290.
Lin MY, Gutierrez PR, Stone KL, Yaffe K, Ensrud KE, Fink HA, Sarkisian CA, Coleman AL, Mangione CM (2004) Vision impairment and combined vision and hearing impairment predict cognitive and functional decline in older women. J Am Geriatr Soc 52:1996–2002.
Lindenberger U, Baltes PB (1994) Sensory functioning and intelligence in old age: a sensory connection. Psychol Aging 9:339–355.
Lunner T (2003) Cognitive function in relation to hearing aid use. Int J Audiol 42:S49–S58.
Lunner T, Sundewall-Thorén E (2007) Interactions between cognition, compression, and listening conditions: effects on speech-in-noise performance in a two-channel hearing aid. J Am Acad Audiol 18:604–617.
Mayr U, Spieler DH, Kliegl R (2001) Aging and Executive Control. New York: Routledge.
McCormack T, Brown GDA, Maylor EA, Richardson LBN, Darby RJ (2002) Effects of aging on absolute identification of duration. Psychol Aging 17:363–378.
McCoy SL, Tun PA, Cox I, Colangelo M, Stewart RA, Wingfield A (2005) Hearing loss and perceptual effort: downstream effects on older adults' memory for speech. Q J Exp Psychol 58:22–33.
McDowd JM, Shaw RJ (2000) Attention and aging: a functional perspective. In: Craik FIM, Salthouse TA (eds) The Handbook of Aging and Cognition, 2nd Ed. Mahwah, NJ: Erlbaum, pp. 221–292.
Miller GA (1956). The magical number seven, plus or minus two: some limits on our capacity for processing information. Psychol Rev 63: 81–97.
Mills JH, Schmiedt RA, Schulte BA, Dubno JR (2006) Age-related hearing loss: a loss of voltage, not hair cells. Semin Hear 27:228–236.

Miyake A, Shah P (1999) Models of Working Memory: Mechanisms of Active Maintenance and Executive Control. New York: Cambridge University Press.

Mondor TA, Zatorre RJ, Terrio NA (1998) Constraints on the selection of auditory information. J Exp Psychol Hum Percept Perform 24:66–79.

Mulrow CD, Aguilar C, Endicott JE, Tuley MR, Velez R, Charlip WS, Rhodes MC, Hill JA, DeNino LA (1990) Quality-of-life changes and hearing impairment. Ann Intern Med 113:188–194.

Mulrow CD, Tuley MR, Aguilar C (1992) Sustained benefits of hearing aids. J Speech Hear Res 35:1402–1405.

Murphy DR, Craik FIM, Li KZH, Schneider BA (2000) Comparing the effects of aging and background noise on short-term memory performance. Psychol Aging 15:323–334.

Murphy DR, Daneman M, Schneider BA (2006a) Why do older adults have difficulty following conversations? Psychol Aging 21:49–61.

Murphy DR, Schneider BA, Speranza F, Moraglia G (2006b) A comparison of higher-order auditory processes in younger and older adults. Psychol Aging 21:763–773.

Nobili R, Mammano F, Ashmore J (1998) How well do we understand the cochlea? Trends Neurosci 21:159–167.

O'Mahoney CFO, Stephens SDG, Cadge BA (1996) Who prompts patients to consult about hearing loss? Br J Audiol 30:153–158.

Palmer CV, Adams SW, Bourgeois M, Durrant J, Ross M (1999) Reduction in caregiver-identified problem behaviors in patients with Alzheimer disease post-hearing-aid fitting. J Speech Lang Hear Res 42:312–328.

Parker S, Murphy DR, Schneider B (2002) Top-down gain control in the auditory system: evidence from identification and discrimination experiments. Percept Psychophys 64:598–615.

Peters CA, Potter JF, Scholer SG (1988) Hearing impairment as a predictor of cognitive decline in dementia. J Am Geriatr Soc 36:981–986.

Pichora-Fuller MK (2006) Perceptual effort and apparent cognitive decline: implications for audiologic rehabilitation. Semin Hear 27:284–293.

Pichora-Fuller MK (2007) Audition and cognition: what audiologists need to know about listening. In: Palmer C, Seewald R (eds) Hearing Care for Adults. Stäfa, Switzerland: Phonak, pp. 71–85.

Pichora-Fuller MK (2008) Use of supportive context by younger and older adult listeners: balancing bottom-up and top-down information processing. Int J Audiol volume 47, issue 1, supplement 2: S72-S82.

Pichora-Fuller MK, Robertson L (1997) Planning and evaluation of a hearing rehabilitation program in a home-for-the-aged: use of hearing aids and assistive listening devices. J Speech Lang Pathol Audiol 21:174–186.

Pichora-Fuller MK, Schow R (2007) Audiologic rehabilitation for adults: assessment and management. In: Schow RL, Nerbonne MA (eds) Introduction to Audiologic Rehabilitation, 5th Ed. Boston: Allyn Bacon, pp. 367–434.

Pichora-Fuller MK, Singh G (2006) Effects of age on auditory and cognitive processing: Implications for hearing aid fitting and audiological rehabilitation. Trends Amplif 10:29–59.

Pichora-Fuller MK, Schneider BA, Daneman M (1995) How young and old adults listen to and remember speech in noise. J Acoust Soc Am 97:593–608.

Pichora-Fuller MK, Schneider BA, MacDonald E, Pass HE, Brown S (2007) Temporal jitter disrupts speech intelligibility. Hear Res 223:114–121.

Pickles JO (1988) An Introduction to the Physiology of Hearing, 2nd Ed. London: Academic Press.

Plomp R, Mimpen AM (1979) Speech-reception threshold for sentences as a function of age and noise level. J Acoust Soc Am 66:1333–1342.

Rabbitt PMA (1968) Channel-capacity, intelligibility and immediate memory. Q J Exp Psychol 20:241–248.

Rabbitt PMA (1991) Mild hearing loss can cause apparent memory failures which increase with age and reduce with IQ. Acta Otolaryngol 476:167–176.

Robles L, Ruggero MA (2001) Mechanics of the mammalian cochlea. Physiol Rev 81:1305–1352.

Rönnberg J, Rudner M, Foo C, Lunner T (2008). Cognition counts: A working memory system for ease of language understanding. Int J Audiol., volume 47, issue 1, supplement 2: S99-S105.

Ryan EB, Giles H, Baroclucci G, Henwood K (1986) Psycholinguistic and social psychological components of communication by and with the elderly. Lang Commun 6:1–24.

Salthouse TA (1996) The processing-speed theory of adult age differences in cognition. Psychol Rev 103:403–428.

Scharf B, Magnun J, Collett L, Ulmer E, Chays A (1994) On the role of the olivocochlear bundle in hearing: a case study. Hear Res 75:11–26.

Scharf B, Magnun J, Chays A (1997) On the role of the olivocochlear bundle in hearing: 16 case studies. Hear Res 103:101–122.

Schneider BA, Pichora-Fuller MK (2000) Implications of perceptual deterioration for cognitive aging research. In: Craik FIM, Salthouse TA (eds) The Handbook of Aging and Cognition, 2nd Ed. Mahwah, NJ: Erlbaum, pp. 155–219.

Schneider BA, Daneman M, Murphy DR, Kwong See S (2000) Listening to discourse in distracting settings: the effects of aging. Psychol Aging 15:110–125.

Schneider BA, Daneman M, Murphy DR (2005) Speech comprehension difficulties in older adults: cognitive slowing or age-related changes in hearing? Psychol Aging 20:261–271.

Schneider BA, Li L, Daneman M (2007) How competing speech interferes with speech comprehension in everyday listening situations. J Am Acad Audiol 18:578–591.

Shallice T, Burgess P (1993) Supervisory control of action and thought selection. In: Baddeley A, Weiskrantz L (eds) Attention: Selection, Awareness and Control. New York: Oxford University Press, pp. 171–187.

Shannon RV, Zeng F, Kamath V, Wygonski J (1995) Speech recognition with primarily temporal cues. Science 270:303–304.

Sheldon S, Pichora-Fuller MK, Schneider BA (2008a) Effect of age, presentation method, and training on identification of noise-vocoded words. J Acoust Soc Am 123:476–488.

Sheldon S, Pichora-Fuller MK, Schneider BA (2008b) Priming and sentence context support listening to noise-vocoded speech by younger and older adults. J Acoust Soc Am 123:489–499.

Singh G, Pichora-Fuller MK, Schneider BA (2008).The effect of age on auditory spatial attention in conditions of real and simulated spatial separation by younger and older adults. J Acoust Soc Am 124: 1294–1305.

Snyder JS, Alain C (2005) Age-related changes in neural activity associated with concurrent vowel segregation. Cogn Brain Res 24:492–499.

Snyder JS, Alain C (2007) Toward a neurophysiological theory of auditory stream segregation. Psychol Bull 133:780–799.

Stark P, Hickson L (2004) Outcomes of hearing aid fitting for older people with hearing impairment and their significant others. Int J Audiol 43:390–398.

Stine EAL (1995) Aging and the distribution of resources in working memory. In: Allen PA, Bashore TA (eds) Age Differences in Word and Language Processing. Amsterdam: Elsevier, pp. 171–187.

Stine EAL, Hindman J (1994) Age differences in reading time allocation for propositionally dense sentences. Aging Cogn 1:2–16.

Stine EAL, Cheung H, Henderson D (1995) Adult age differences in the on-line processing of new concepts in discourse. Aging Cogn 2:1–18.

Stine-Morrow EAL, Milinder LA, Pullara O, Herman B (2001) Patterns of resource allocation are reliable among younger and older readers. Psychol Aging 16:69–84.

Strouse AL, Hall JW III, Burger MC (1995) Central auditory processing in Alzheimer's disease. Ear Hear 16:230–238.

Summerfield Q, Assmann PF (1991) Perception of concurrent vowels: effects of harmonic misalignment and pitch-period asynchrony. J Acoust Soc Am 89:1364–1377.

Summers V, Leek MR (1998) F_0 processing and the separation of competing speech signals by listeners with normal hearing and with hearing loss. J Speech Lang Hear Res 41:1294–1306.

Tesch-Romer C (1997) Psychological effects of hearing aid use in older adults. J Gerontol B Psychol Sci Soc Sci 52:P127-P138.

Titone D, Prentice KJ, Wingfield A (2000) Resource allocation during spoken discourse processing: effects of age and passage difficulty as revealed by self-paced listening. Mem Cogn 28:1029–1040.
Tun PA, Wingfield A, Stine EAL (1991) Speech processing capacity in young and older adults: a dual-task study. Psychol Aging 6:3–9.
Tun PA, Wingfield A, Stine EAL, Mecsas C (1992) Rapid speech and divided attention: processing rate versus processing resources as an explanation of age effects. Psychol Aging 7:546–550.
Uhlmann RF, Larson EB, Rees TS, Koepsell TD, Duckert LG (1989a) Relationship of hearing impairment to dementia and cognitive dysfunction in older adults. J Am Med Assoc 261: 1916–1919.
Uhlmann RF, Teri L, Rees TS, Mozlowski KJ, Larson EB (1989b) Impact of mild to moderate hearing loss on mental status testing. J Am Geriatr Soc 37:223–228.
Valentijn SAM, Van Boxtel MPJ, Van Hooren SAH, Bosma H, Beckers HJM, Ponds RWHM, Jolles J (2005) Change in sensory functioning predicts change in cognitive functioning: results from a 6-year follow-up in the Maastricht aging study. J Am Geriatr Soc 53:374–380.
Van Boxtel MPJ, van Beijsterveldt T, Houx PJ, Anteunis LJC, Metsemakers JFM, Jolles J (2000) Mild hearing impairment can reduce verbal memory performance in a healthy older adult population. J Clin Exp Neuropsychol 22:147–154.
Van der Linden M, Brédart S, Beerten A (1994) Age-related differences in updating working memory. Br J Psychol 85:145–152.
Van der Linden M, Hupet M, Feyereisen P, Schelstraete M, Bestgen M, Bruyer G L, Abdessadek EA, Seron X (1999) Cognitive mediators of age-related differences in language comprehension and verbal processing. Aging Neuropsychol Cogn 6:32–55.
Van Hooren SAH, Anteunis LJC, Valentijn SAM, Bosma H, Ponds RWHM, Jolles J, van Boxtel MPJ (2005) Does cognitive function in older adults with hearing impairment improve by hearing aid use? Int J Audiol 44, issue 5,:265–271.
van Rooij JCGM, Plomp R (1992). Auditive and cognitive factors in speech perception by elderly listeners: III. Additional data and final discussion. J Acoust Soc Am 91:1028–1033.
Verhaeghen P, Salthouse TA (1997). Meta-analysis of age-cognition relations in adulthood: estimates of linear and non-linear age effects and structural models. Psychol Bull 122:231–249.
Verhaeghen P, Marcoen A, Goossens L (1993) Facts and fiction about memory aging: a quantitative integration of research findings. J Gerontol B Psychol Sci Soc Sci 48:157–171.
Verhaeghen P, Steitz DW, Sliwinski MJ, Cerella J (2003) Aging and dual-task performance: a meta-analysis. Psychol Aging 18:443–460.
Verhaeghen P, Cerella J, Bopp KL, Basak C (2005) Aging and varieties of cognitive control: a review of meta-analyses on resistance to interference, coordination, and task switching, and an experimental exploration of age-sensitivity in the newly identified process of focus switching. In: Engle RW, Sedek G, Von Hecker U, McIntosh DN (eds) Cognitive limitations in aging and psychopathology. New York: Cambridge University Press, pp. 160–189.
Vongpaisal T, Pichora-Fuller MK (2007) Effect of age on F_0 difference limen and concurrent vowel identification. J Speech Lang Hear Res 50:1139–1156.
Wahl H-W, Heyl V (2003) Connections between vision, hearing, and cognitive function in old age. Generations 27:39–45.
West MJ (1996) An application of prefrontal cortex function theory to cognitive aging. Psychol Bull 120:272–292.
Willott JF (1991) Aging and the Auditory System: Anatomy, Physiology, and Psychophysics. San Diego, CA: Singular Publishing Group.
Wingfield A (1996) Cognitive factors in auditory performance: context, speed of processing, and constraints of memory. J Am Acad Audiol 7:175–182.
Wingfield A, Stine EAL (1992) Age differences in perceptual processing and memory for spoken language. In: Simon JD, West RL (eds) Everyday Memory and Aging: Current Research and Methodology. New York: Springer-Verlag, pp. 101–123.
Wingfield A, Tun PA (2007) Cognitive supports and cognitive constraints on comprehension of spoken language. J Am Acad Audiol 18:548–558.

Wingfield A, Poon LW, Lombardi L, Lowe D (1985) Speed of processing in normal aging: effects of speech rate, linguistic structure, and processing time. J Gerontol 40:579–585.
Wingfield A, Tun PA, Koh CK, Rosen MJ (1999) Regaining lost time: adult aging and the effect of restoration on recall of time-compressed speech. Psychol Aging 14:380–389.
Zacks RT, Hasher L (1994) Directed ignoring. Inhibitory regulation of working memory. In: Dagenbach D, Carr TH (eds) Inhibitory Processes in Attention, Memory, and Language. San Diego, CA: Academic Press, pp. 241–264.
Zurek PM (1987) The precedence effect. In: Yost WA, Gourevitch G (eds) Directional Hearing. New York: Springer-Verlag, pp. 85–105.

Chapter 8
Factors Affecting Speech Understanding in Older Adults[1]

Larry E. Humes and Judy R. Dubno

8.1 Introduction

This chapter reviews various factors that affect the speech-understanding abilities of older adults. Before proceeding to the identification of several such factors, however, it is important to clearly define what is meant by "speech understanding." This term is used to refer to either the open-set recognition or the closed-set identification of nonsense syllables, words, or sentences by human listeners. Many years ago, Miller et al. (1951) demonstrated that the distinction between open-set recognition and closed-set identification blurs as the set size for closed-set identification increases. When words were used as the speech material, Miller et al. (1951) demonstrated that the closed-set speech-identification performance of young normal-hearing listeners progressively approached that of open-set speech recognition as the set size doubled in successive steps from 2 to 256 words. Clopper et al. (2006) have also demonstrated that lexical factors (e.g., word frequency and acoustic-phonetic similarity) impacting word identification and word recognition are very similar when the set size is reasonably large for the closed-set identification task and the alternatives in the response are reasonably confusable with the stimulus item. Thus the processes of closed-set speech identification and open-set speech recognition are considered to be very similar and both are referred to here as measures of "speech understanding."

L.E. Humes (✉)
Department of Speech and Hearing Sciences, Indiana University, 200 South Jordan Avenue, Bloomington IN, 47405-7002,
e-mail: humes@indiana.edu

J.R. Dubno
Department of Otolaryngology-Head and Neck Surgery, Medical University of South Carolina, 135 Rutledge Avenue, MSC 550, Charleston, SC 29525–5500
e-mail: dubnojr@musc.edu

[1]This chapter is dedicated to Donald D. Dirks, a wonderful mentor who contributed to the development of the research careers of both authors and helped to shape the ideas contained therein.

S. Gordon-Salant et al. (eds.), *The Aging Auditory System*, Springer Handbook of Auditory Research 34,
DOI 10.1007/978-1-4419-0993-0_8, © Springer Science+Business Media, LLC 2010

Additional insight into what is meant by "speech understanding" in this chapter can be gained by considering some of the things that it is not. For example, considerable speech-perception research has been conducted over the past half century in both young and old adults in which the discrimination of minimally contrasting speech sounds, often consonant-vowel or vowel-consonant syllables, has been measured. To illustrate the differences between speech recognition, speech identification, and speech-sound discrimination as used here, consider the following three tasks using the same consonant-vowel syllable, /ba/, as the speech stimulus. In open-set speech recognition, the listener is asked to say or write down the syllable that was heard after the /ba/ stimulus was presented. The range of alternative responses is restricted only by the experimenter's instructions or the listener's imagination. In closed-set speech identification, the same /ba/ stimulus is presented, but the listener is given a list of possible answers, such as /ba/, /pa/, /ta/ and /da/, only one of which corresponds to the stimulus that was presented. Finally, in minimal contrast speech-sound discrimination, some acoustic characteristic of the /ba/ stimulus is altered systematically, with the altered and one or more unaltered /ba/ stimuli presented in random sequence. The listener simply indicates whether the stimuli presented in the sequence were the same or different or may be asked to select the "different" stimulus from those presented in the sequence. In the discrimination task, the listener is never asked to indicate the sound(s) heard, only whether they were the same or different. As a result, this task is considered to be conceptually distinct from the processes of speech recognition and speech identification, and we do not consider this form of speech-sound discrimination to be a part of "speech understanding" as used here.

Finally, although definitions of the terms "understanding" and "comprehension" may be similar, speech comprehension is not used interchangeably with speech understanding in this review. Most often, comprehension is assessed with phrases or sentences and involves the deciphering of the talker's intended meaning behind the spoken message. For example, consider the following spoken message: "What day of the week follows Thursday?" An open-set recognition task making use of this question might ask the listener to repeat the entire sentence and a speech-recognition score could be determined by counting the number of words correctly repeated (or the entire sentence could be scored as correct or incorrect). A closed-set identification task making use of this same speech stimulus might have seven columns of words (one for each word in the sentence), with four alternative words in each column, and the listener would be required to select the correct word from each of the seven columns. Or, for sentence-based scoring, four alternative sentences could be displayed orthographically from which the listener must select the one matching the sentence heard. For a comprehension task, either an open-set or closed-set version can be created. In the open-set version, the listener simply answers the question and a correct answer ("Friday," in this example) implies correct comprehension. Likewise, four days of the week could be listed as response alternatives and the listener asked to select the correct response to the question ("Friday") from among the alternatives. Comprehension is a higher-level process than either speech recognition or speech identification.

Whereas it is unlikely that accurate comprehension can occur without accurate recognition or identification (e.g., imagine that "Tuesday" was perceived by the listener instead of "Thursday" in this example), accurate recognition or identification does not guarantee accurate comprehension.

In summary, the term "speech understanding" is used here to refer to the task of either open-set speech recognition or closed-set speech identification. It does not include tasks of minimal-contrast speech-sound discrimination or speech comprehension. (The latter topic is treated by Schneider, Pichora-Fuller, and Daneman, Chapter 7).

8.2 Peripheral, Central-Auditory, and Cognitive Factors

As noted in earlier chapters, as humans age, several changes may take place in their auditory systems that could have a negative impact on speech understanding. At the periphery, although age-related changes may occur in the outer and middle ears, these are largely inconsequential for everyday speech communication by older adults. On the other hand, age-related hearing loss (presbycusis) occurs in ~30% of persons over the age of 60 (see Cruickshanks, Zhan, and Zhong, Chapter 9), which has serious consequences for speech understanding for most individuals. For example, functional deficits that accompany age-related hearing loss (elevated thresholds, reduced cochlear compression, broader tuning) can result in reduced speech understanding. These are related to anatomic and physiological changes in the auditory periphery, primarily affecting the cochlea and auditory nerve of older adults (reviewed in detail by Schmiedt, Chapter 2).

As noted by Canlon, Illing, and Walton, (Chapter 3), age-related changes can also occur at higher centers of the auditory portion of the central nervous system. These modality-specific deficits can, in turn, have a negative impact on speech understanding in older listeners. Moreover, as noted by Schneider, Pichora-Fuller, and Daneman (Chapter 7), there may be amodal changes in the cortex of older listeners that impact cognitive functions, such as speed of processing, working memory, and attention. These changes alone may result in reduced speech understanding in older listeners. However, age-related changes to the auditory periphery also degrade the speech signal delivered to the central nervous system for cognitive and linguistic processing so that the speech understanding problems of older listeners may represent the combined effects of peripheral, central-auditory, and cognitive factors.

In fact, prior reviews of the speech-understanding problems of older adults have framed their reviews around a similar "site of lesion" framework, with peripheral (primarily cochlear), central-auditory, and cognitive "sites" hypothesized as the primary contributors (Committee on Hearing, Bioacoustics, and Biomechanics [CHABA] 1988; Humes 1996). Although there is probably little disagreement among researchers as to the extent of speech-understanding difficulties experienced by older adults, the challenge has been in identifying the nature of the underlying cause(s) of these difficulties.

Of the peripheral, central-auditory, and cognitive explanations of age-related speech-understanding declines, the peripheral explanation in its simplest form (elevated hearing thresholds) is probably the most straightforward. In addition, given that the prevalence of hearing loss in older adults is greater than either central-auditory deficits (Cooper and Gates 1991) or mild cognitive impairment (Lopez et al. 2003; Portet et al. 2006), peripheral hearing loss is the most likely explanatory factor of the speech-understanding problems observed in older adults. If the peripheral explanation fails to account for the measured speech-understanding performance of older adults, then alternative and more complex mechanisms can be considered further.

8.2.1 Peripheral Factors

When evaluating the extent to which hearing loss of cochlear origin accounts for speech-understanding problems in older adults, several approaches have been pursued. Some of the more commonly pursued approaches are reviewed here.

8.2.1.1 Articulation Index Framework

One of the most powerful approaches used to explain the average speech-understanding performance of human listeners is the Articulation Index (AI) framework championed by Harvey Fletcher and his colleagues at Bell Labs in the 1940s and 1950s (French and Steinberg 1947; Fletcher and Galt 1950; Fletcher 1953). There have been many variations of the AI framework developed in the intervening 50 to 60 years. Although too numerous to review in detail here, notable variations include the first American National Standard of the AI (ANSI 1969); the Speech Transmission Index, which used the modulation transfer function to quantify the signal-to-noise ratio (SNR; Steeneken and Houtgast 1980; Houtgast and Steeneken 1985); and the most recent version of the American National Standard (ANSI 1997), in which the index was modified and renamed the Speech Intelligibility Index. Because there is more in common among these various indices than differences, the general expression "AI framework" will be used throughout this review to refer to any one of these indices.

Regardless of the specific version of the AI framework, the key conceptual components are very similar across versions. Central to this framework, for example, is the long-term average spectrum for speech, illustrated by the line labeled "rms" in Figure 8.1. This idealized spectrum represents the root-mean-square (rms) amplitude at each frequency measured using a 125-ms rectangular window and averaged over many seconds of running speech for a large number of male and female talkers. If this framework is to be used to estimate or predict speech-understanding performance of groups of individuals listening to specific speech stimuli, the long-term average spectrum of the actual speech materials spoken

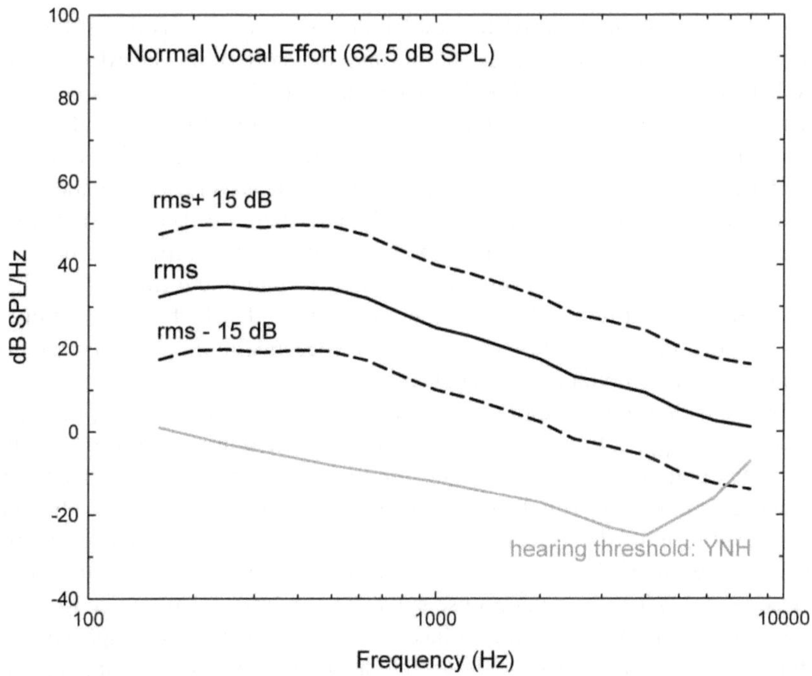

Fig. 8.1 Root-mean-square (rms) long-term average speech spectrum for normal vocal effort from ANSI (1997) and the 30-dB range of speech amplitudes, 15 dB above and below the rms spectrum, which contribute to speech understanding according to the Articulation Index framework. For comparison, average hearing thresholds for young normal-hearing (YNH) adults along the same coordinates are shown.

by the talker(s) used in the measurements should be substituted for the generic version shown in Figure 8.1. For comparison purposes, the average hearing thresholds for young normal-hearing (YNH) adults have also been provided in Figure 8.1

A second critical concept in the AI framework is the identification of the range of speech sound levels above and below the rms long-term average spectrum that contribute to intelligibility. One way in which this range has been established has been strictly from acoustical measurements of the distribution of speech levels measured with the 125-ms window during running speech. The acoustical range is then described in terms of the 10th and 90th, 5th and 95th, or 1st and 99th percentiles from these speech-level distributions. Depending on the definition of speech peaks and speech minima used, the ranges derived have varied typically between 30 and 36 dB, with the speech peaks represented in an idealized fashion at either 12 or 15 dB above the long-term average spectrum (French and Steinberg 1947; Fletcher and Galt 1950; Pavlovic 1984). In the current American National Standard (ANSI 1997), a 30-dB range from 15 dB to −15 dB re the long-term average speech spectrum has been adopted, and this is illustrated in Figure 8.1.

The basic concept behind the AI framework and its successors is that if the entire 30-dB range of speech intensities is audible and undistorted (i.e., heard and useable by the listener) from ~100 to 8,000 Hz, then speech understanding will be optimal (typically >95% for nonsense syllables). As the area represented by this 30-dB range from 100 to 8,000 Hz, referred to here as the "speech area," is reduced or distorted (meaning that portions of the speech are not heard or useable), then the AI decreases, which predicts a decrease in speech understanding. There are many ways in which the speech area can be diminished, including filtering, masking noise, and hearing loss. The power of this approach is that these multiple forms of degradation can be combined and reduced to a single metric, the AI, which is proportional to speech understanding regardless of the specific speech stimuli or the specific form of degradation or combination of degradations.

Before considering two other concepts central to the AI framework, perhaps an example of its application would be useful here. Imagine spectrally shaping a steady-state random noise so that its rms long-term average spectrum matches that of the speech stimulus. Given the matching of the long-term spectra, the adjustment of the overall level of the speech or noise to produce a specific SNR results in an identical change in SNR across the entire spectrum (i.e., the SNR is the same at each frequency). When using nonsense syllables as stimuli to minimize the contributions of higher-level processing, it has generally been found that speech-understanding performance increases above 0% at a SNR of −15 dB and increases linearly with increases in SNR until reaching an asymptote at 15 dB (Fletcher and Galt 1950; Pavlovic 1984; Kamm et al. 1985). SNRs of 10, 0 and −10 dB are illustrated in Figure 8.2, along with the AI values for these conditions (ANSI 1997). In this special case of identical long-term spectra between the noise and the speech stimuli (and no other factors limiting the speech area), the AI can be easily calculated from the SNR alone as follows: AI = (SNR [in dB] + 15 dB)/30 dB. In this case, the AI is the proportion of the speech area in Figure 8.2 that remains visible (audible) in the presence of the competing background noise.

Another central concept in this framework is acknowledging that actual speech-understanding performance for a given acoustical condition (such as any of the three conditions illustrated in Fig. 8.2) will vary with the type of speech materials used to measure speech understanding. This is illustrated by the transfer functions relating percent correct performance by YNH adults to the AI in Figure 8.3. The functions in the top panel were derived for nonsense syllables, using either an open-set speech-recognition task (Fletcher and Galt 1950) or a closed-set speech-identification task having seven to nine response alternatives per test item (Kamm et al. 1985). The two functions for nonsense syllables in the top panel are very similar and both show monotonic increases in speech understanding with increases in the AI. The functions in Figure 8.3, bottom, were derived from those developed by Dirks et al. (1986) for the Speech Perception in Noise (SPIN) test (Kalikow et al. 1977; Bilger et al., 1984). This test requires open-set recognition of the final word in the sentence. The words preceding the final word provide either very little semantic context (predictability-low [PL] sentences) or considerable semantic context (predictability-high [PH] sentences). The dotted vertical lines in Figure 8.3

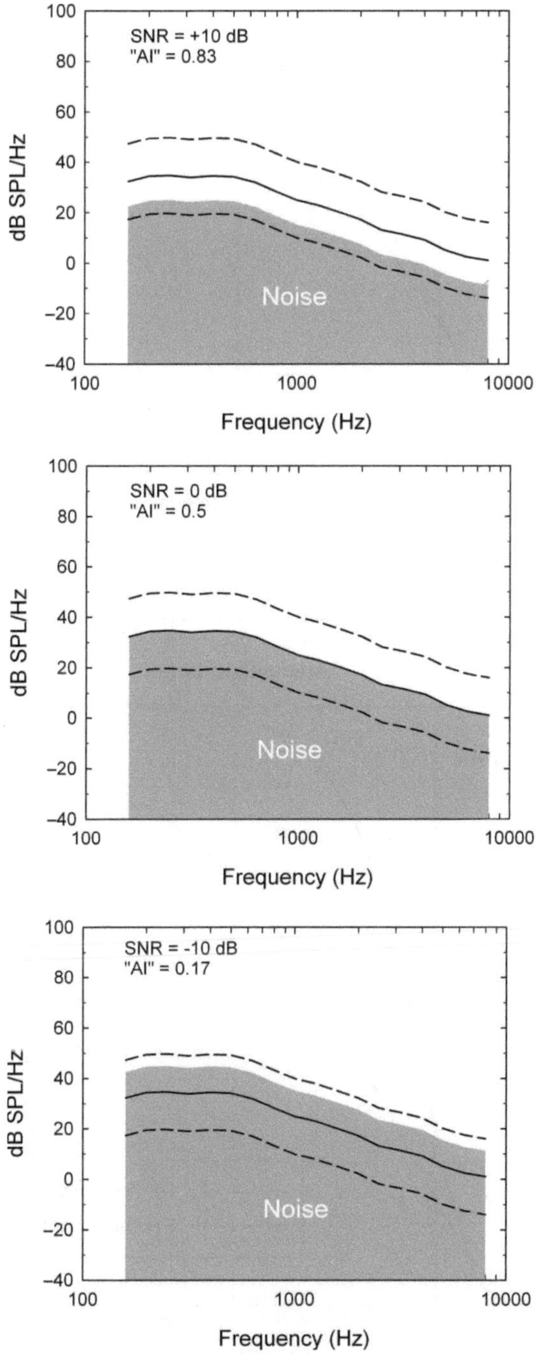

Fig. 8.2 Audibility of the 30-dB speech area across frequency as the signal-to-noise ratio (SNR) is varied from 10 dB (top) to 0 dB (middle) to −10 dB (bottom). The competing noise is steady state and is assumed to have a spectrum identical to that of the speech stimulus. Its amplitude is represented by the shaded grey area. The values calculated for the Articulation Index (AI) framework from ANSI (1997) are also shown.

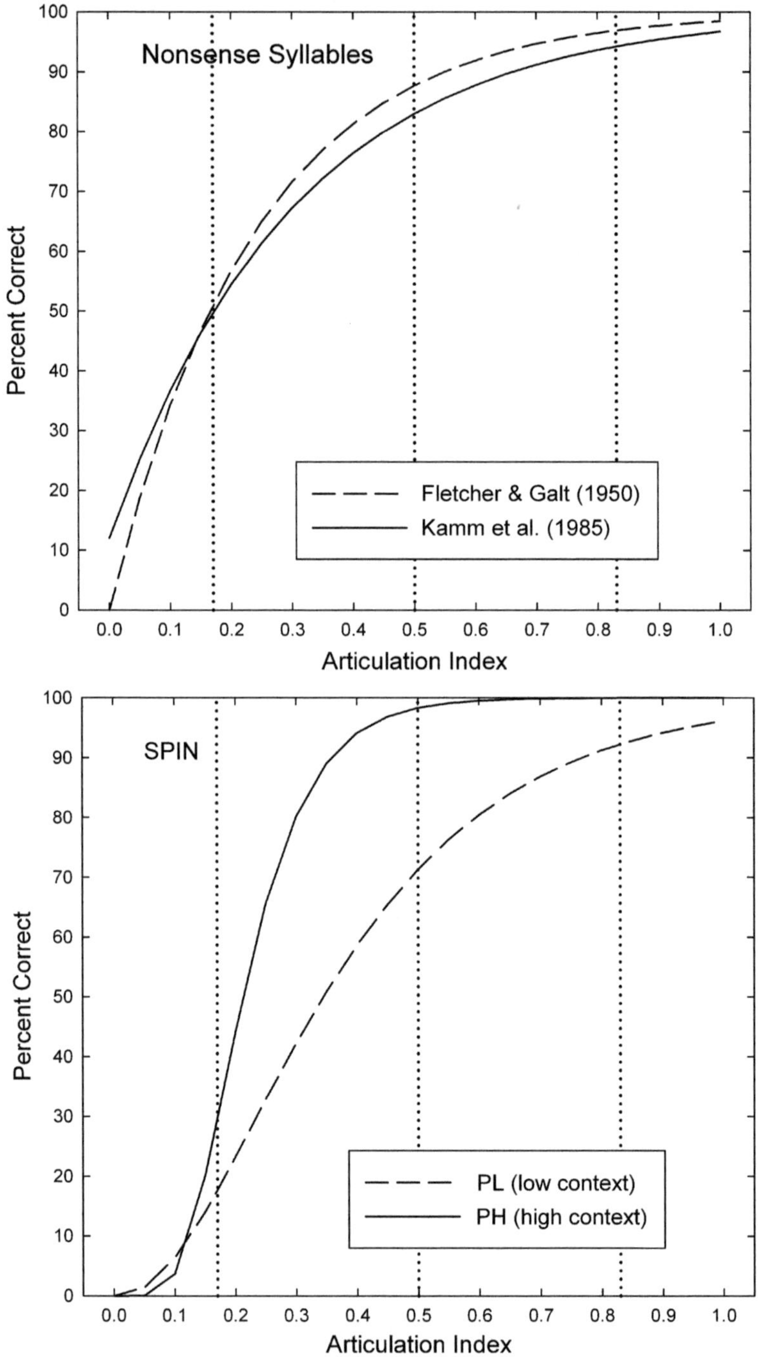

Fig. 8.3 Transfer functions relating the percent correct speech-understanding score to the AI for nonsense syllables (top) and the predictability-low (PL) and predictability-high (PH) sentences of the Speech Perception in Noise (SPIN) test (bottom; Kalikow et al. 1977; Dirks et al. 1986). Dotted vertical lines are the AI values corresponding to those in Figure 8.2 for SNRs of –10, 0, and 10 dB.

correspond to the AI values of 0.17, 0.5, and 0.83 associated with the SNRs of −10, 0, and 10 dB displayed previously in Figure 8.2. The intersections of these dotted vertical lines with the various transfer functions in Figure 8.3 indicate the predicted speech-understanding scores for these conditions and speech materials. Thus when desiring to map AI values to scores for specific sets of speech materials, the transfer function relating the AI to percent correct speech understanding for those materials must be known. In all cases, however, there is a monotonic increase in speech-understanding performance, with increases in AI (except for AI values > ~0.6 where the SPIN-PH function asymptotes at 100%).

The final central concept of the AI framework reviewed here is referred to as the band-importance function. As noted, in addition to the noise masking illustrated in Figure 8.2, another way to reduce the audible speech area is via low-pass or high-pass filtering. Early in the development of this framework, based largely on extensive studies of low-pass and high-pass filtered speech, it was recognized that certain frequency regions contributed more to speech understanding than others (French and Steinberg 1947; Fletcher and Galt 1950). For example, for the open-set recognition of nonsense syllables, acoustic cues in the frequency region from 1,000 to 3,000 Hz contribute much more to speech understanding than do higher- or lower-frequency regions. This is managed within the AI framework by dividing the horizontal (frequency) dimension of the speech area in Figure 8.1 into a specific number of bands and then weighting each band according to its empirically measured importance. The weights assigned to each band sum to 1.0 and are used as multipliers of the SNR measured in each of the bands.

One approach to partitioning the frequency axis would be to simply adjust the upper and lower frequencies of each band so that each band contributed equally to speech understanding; for example, with 20 bands from 100 to 8,000 Hz, each would contribute 1/20 or 0.05 AI (French and Steinberg 1947). If a low-pass filter rendered the 10 highest-frequency bands inaudible while the SNR was 15 dB in the remaining lower-frequency bands, there would be 10 bands with full contribution (0.05) and 10 with no contribution, for a total AI value of 0.5 [(10 × 0.05) + (10 × 0.0)]. Note that the same would hold true for a complementary high-pass filter that rendered the lowest 10 bands inaudible while keeping the SNR optimal (15 dB) in the 10 higher-frequency bands. As demonstrated by the 0-dB SNR in Figure 8.2, middle, this condition also yielded an AI of 0.5. Thus given a single transfer function relating the value of the AI to speech-understanding performance in percent correct (e.g., Fig. 8.3), all three of these conditions (low-pass filtering, high-pass filtering, and 0-dB SNR) resulted in an AI value of 0.5 and would be expected to yield the same speech-understanding score. This again is the power behind the AI framework. Different acoustical factors that impact speech understanding in various ways can be reduced to a single value that is predictive of speech-understanding performance regardless of the way in which the speech area has been modified. It is important to recognize, however, as did the developers of the AI framework, that the predictions are made for average speech-understanding scores for a group of listeners (YNH listeners when initially developed). The index was not designed to predict performance for an individual, although it has been used in this manner, and it does not suggest that specific errors resulting from a given form of distortion will be the same for all listeners, even though their speech-understanding scores might be identical.

How can the sensorineural hearing loss associated with aging be included in the AI framework? This is illustrated in Figure 8.4, top, where the median hearing thresholds for males 60, 70, and 80 years of age (ISO 2000) have been added to the information shown previously in Figure 8.1. In this situation, with conversational level (62.5-dB SPL) speech presented in quiet, the loss of high-frequency hearing with advancing age progressively reduces the audibility of the speech area and decreases speech understanding. The bottom panel illustrates a situation with speech at the same level and the level of a speech-shaped steady-state background noise set to achieve a 0-dB SNR. AI values were calculated from ANSI (1997) for each of the three age groups and both of the listening conditions shown in Figure 8.4, and appear in Table 8.1. The SPIN transfer functions in Figure 8.3, bottom, were then used to estimate speech-understanding scores for low- and high-context sentences from the calculated AI values and these also are provided in Table 8.1. With regard to the low-context SPIN-PL materials, note that as age and corresponding high-frequency hearing loss increase, the AI decreases and so do the predicted speech-understanding scores. This is true for both quiet and noise and is consistent with the progressively diminishing portions of the speech area available to the listener, as illustrated in Figure 8.4. Note that the pattern is somewhat different, however, for the high-context sentences (SPIN-PH). In this case, even though the value of the AI for the 80 year olds in quiet is about half that of the young adults in quiet, the estimated speech-understanding performance is 99-100% for all listener groups. This is because of the broad asymptote at 100% performance in the transfer function for the SPIN-PH materials (Fig. 8.3, bottom). When noise is added, however, a considerable decline in performance is observed for the oldest age group. If one considers the high-context sentences to be most like everyday communication, then the latter pattern of predictions for the 80 year olds is consistent with their most frequent clinical complaint: they can hear speech but can't understand it, especially with noise in the background.

There have been numerous attempts to make use of the AI framework in research on speech understanding in older adults. In general, for quiet or steady-state background-noise listening conditions, including multitalker babble (≥8 talkers), group results have been fairly well captured by the predictions of this framework (e.g., Pavlovic 1984; Kamm et al. 1985; Dirks et al. 1986; Pavlovic et al. 1986; Humes 2002). Even here, however, large individual differences have been observed (Dirks et al. 1986; Pavlovic et al. 1986; Humes 2002). In a study of 171 older adults listening to speech (nonsense syllables and sentences) with and without clinically fit hearing aids worn in quiet and background noise (multitalker babble), Humes (2002) found that 53.2% of the variance in speech-understanding performance of the older adults could be explained by the AI values calculated for these same listeners. An additional 11.4% of the total variance could be accounted for by measures of central-auditory or cognitive function. For clinical measures of speech understanding, including those used by Humes (2002), test-retest correla-

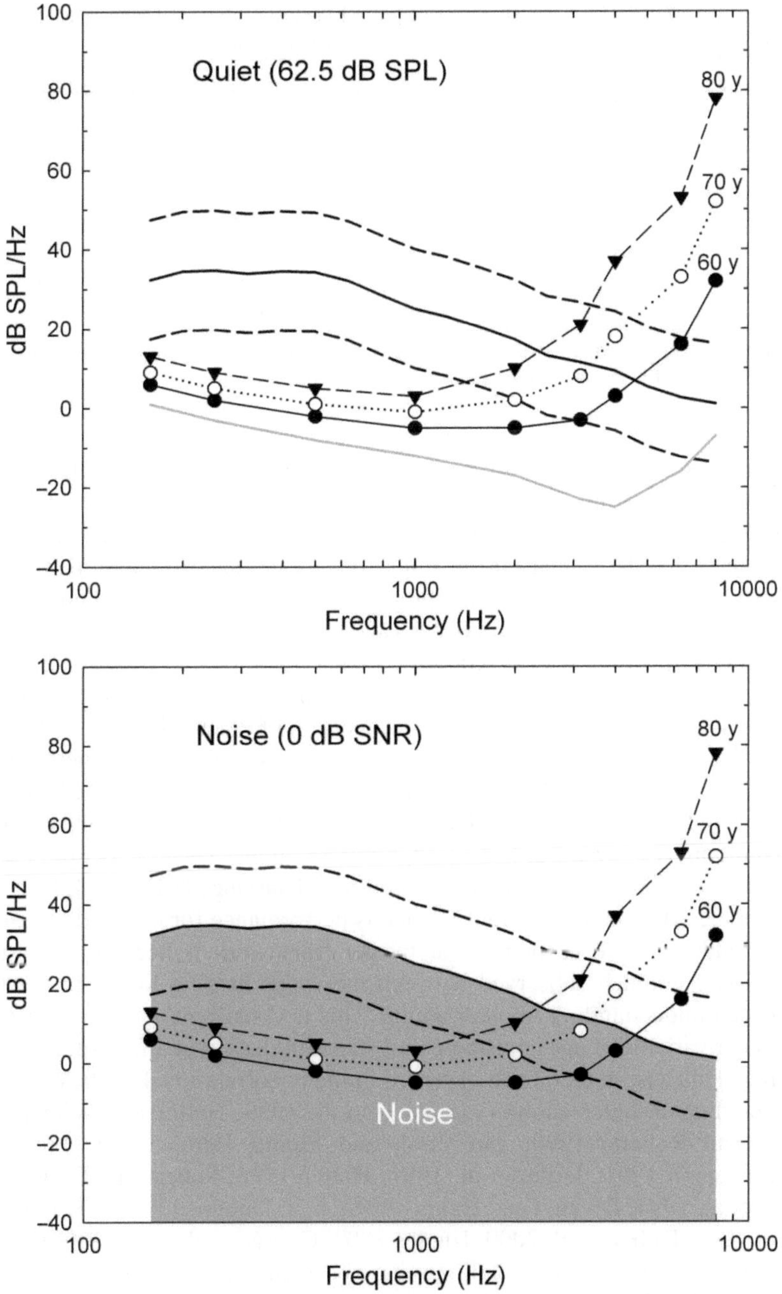

Fig. 8.4 Depiction of the speech area in quiet (top) and in a background of speech-shaped noise (bottom). Also shown are median hearing thresholds of 60-, 70- and 80-year-old men from ISO (2000).

Table 8.1 Percent correct speech-understanding scores for young normal-hearing adults and three age groups of older adults with high-frequency hearing loss as predicted from transfer functions in Figure 8.3. Conversational speech level (62.5-dB SPL) is assumed.

Condition	Group	AI	% SPIN-PL	% SPIN-PH
Quiet	YNH	1.00	96	100
	60 year olds	0.84	93	100
	70 year olds	0.66	85	100
	80 year olds	0.53	74	99
0-dB SNR	YNH	0.47	68	98
	60 year olds	0.43	63	96
	70 year olds	0.36	53	90
	80 year olds	0.28	39	75

YNH, young normal hearing; SNR, signal-to-noise ratio; SPIN, Speech Perception in Noise; PL, predictability low; PH, predictability high. AI, Articulation Index

tions for the scores are typically between 0.8 and 0.9. This suggests that the total systematic variance (nonerror variance) that one could hope to account for would vary between 64 and 81%. Accounting for 53.2% of the total variance in this context suggests that the primary factor impacting speech-understanding performance in these 171 older adults was the audibility of the speech area. The AI concept only takes into consideration speech acoustics and the hearing loss of the listener. Other peripheral dysfunction, such as reduced cochlear nonlinearities (Dubno et al. 2007; Horwitz et al. 2007) or higher-level deficits such as central-auditory or cognitive difficulties, are not incorporated into this conceptual framework. Thus to the extent that the AI framework provides accurate descriptions of the speech-understanding performance of older adults in these listening conditions, age-related threshold elevation and resulting reductions in speech audibility underlie their speech-understanding performance. (As noted, however, higher-level factors did account for significant, albeit small, amounts of additional variance in Humes [2002].)

When the AI framework largely explains performance for older adults and the only listener variable accounted for by the framework is hearing loss, strong correlations should be observed between measured hearing loss for pure tones and speech understanding in older adults. This is consistent with the notion that the listener variable of age contributes substantially less, if at all, to speech understanding. This has, in fact, been observed in many correlational studies of speech understanding in older adults (van Rooij et al. 1989; Helfer and Wilber 1990; Humes and Roberts 1990; van Rooij and Plomp 1990, 1992; Humes and Christopherson 1991; Jerger et al. 1991; Helfer 1992; Souza and Turner 1994; Humes et al. 1994; Divenyi and Haupt 1997a, b, c; Dubno et al. 1997; Jerger and Chmiel 1997; Dubno et al. 2000; Humes 2002; George et al. 2006, 2007). In such studies, the primary listening conditions have been quiet or steady-state background noise (occasionally including multitalker babble). Hearing loss was the primary factor identified as a contributor to the speech-understanding performance of the older adults in each of these studies and typically accounted for 50-90% of the total variance.

Given the primacy of hearing loss as an explanatory factor for the speech-understanding performance of older adults and the fact that hearing loss is the only listener-related variable entered into typical AI frameworks, why bother with a more complex AI framework? There are several reasons to do so. First, there are special listening situations that might dictate the entry of other listener-related variables to obtain a more complete description of speech-understanding performance in noise. For example, Dubno and Ahlstrom (1995a, b) have measured the speech-understanding performance of older adults in the presence of intense low-pass noise. Rather than relying on acoustic measures of SNR in each frequency region, the AI framework provides a means to incorporate expected upward spread of masking from intense low-pass noise and its impact on the speech area when calculating AI values. This has been a feature of this predictive framework from the outset. However, Dubno and Ahlstrom (1995a, b) found that AI predictions were more accurate when the actual masked thresholds of the listener in the presence of the low-pass noise were used rather than relying solely on expected masked thresholds and hearing thresholds. This means that the higher-than-normal masked thresholds of older listeners with hearing loss further reduced their audible speech area, which accounted for their poorer-than-normal speech understanding. Note that these results are consistent with the earlier conclusion that elevated thresholds (in this case, masked thresholds) and limited audibility of the speech area are the primary factors underlying the speech-understanding performance of older adults.

Another example of the utility of the AI framework relates to the dependence of speech-understanding performance on speech presentation level. From the outset, Fletcher and colleagues built into the framework a decrease in the AI and, therefore, speech understanding, at high presentation levels. This has been confirmed more recently by Studebaker et al. (1999), Dubno et al. (2005a, b), and Summers and Cord (2007). For a fixed acoustical SNR, as the speech presentation level exceeds ~70-dB SPL, speech-understanding performance decreases in YNH listeners (Dubno et al. 2005a,b) as well as in older listeners with mildly impaired hearing (Dubno et al. 2006). With SNR held constant, declines in performance with increases in level were largely accounted for by higher-than-normal masked thresholds and concurrent reductions in speech audibility, as explained earlier. In some studies conducted previously, a constant SPL might be used that is either the same for all listeners, young and old, normal hearing, and hearing impaired or varied with hearing status (lower levels for YNH listeners and higher levels for those with hearing loss). In the latter case, level differences across groups may result, and these level differences may impact the measured speech-understanding performance. In other studies, a constant sensation level might be employed such that individual listeners within a group receive different presentation levels. In such cases, it is critical to model the corresponding changes in speech presentation level and the parallel changes in the AI.

Aside from these specific cases for which the complete AI framework might be needed, the central concepts of the framework are critical to obtaining a clear picture of the impact of various listening conditions on the audibility of the speech area (e.g., Humes 1991; Dubno and Schaefer 1992; Dubno and Dirks 1993; Lee and

Humes 1993; Dubno and Ahlstrom 1995a, b; Humes 2002, 2007, 2008). To illustrate this, consider the following common misconception in research on speech understanding in older adults. It is commonly assumed that increasing the presentation level of the speech stimulus without changing its spectrum can result in optimal audibility of the speech area. This misconception is illustrated in Figure 8.5, which is basically identical to Figure 8.2 except that the speech presentation level has been increased to 90-dB SPL. Clearly, although speech energy at this level is well above the low- and mid-frequency hearing thresholds, the median high-frequency thresholds for 70- and 80-year-old men render portions of the speech area inaudible, even at this high presentation level of 90-dB SPL. Furthermore, the reader should keep in mind that median thresholds are shown in Figure 8.5. That is, for 50% of the older adults with hearing thresholds higher than those shown for each age group, still less of the speech area will be audible in the higher frequencies. It should also be kept in mind that this illustration is for undistorted speech in quiet. Recall, moreover, that use of high presentation levels alone can result in decreased speech-understanding performance, even in YNH adults (e.g., Studebaker et al. 1999). It is

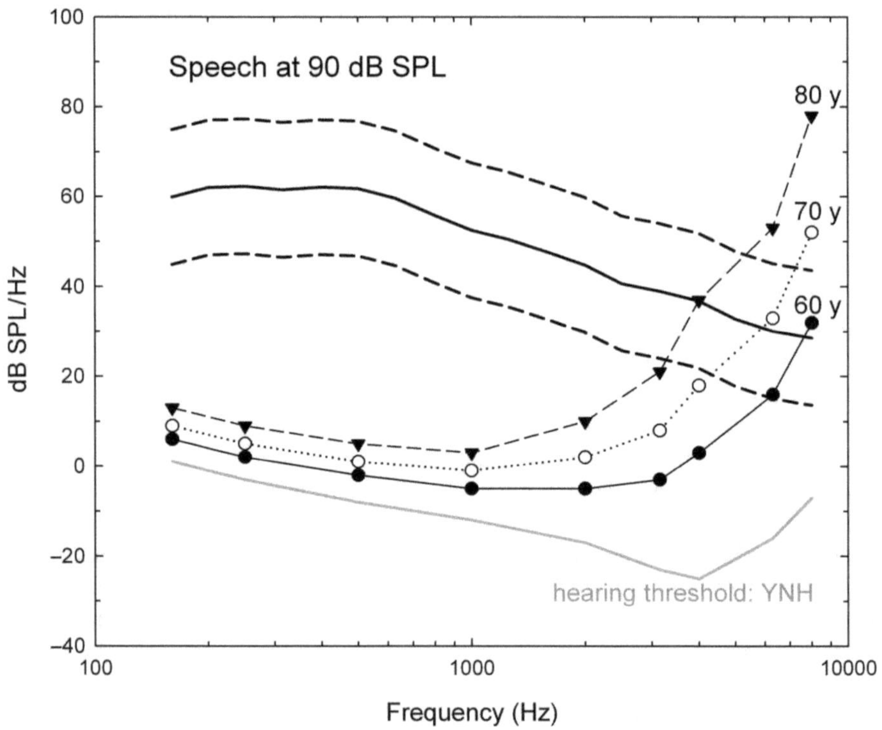

Fig. 8.5 Median hearing thresholds for 60-, 70- and 80-year-old men superimposed on the speech area when the level of speech has been increased from normal vocal effort (62.5-dB SPL) to a level of 90-dB SPL. Note that even at this high presentation level the high-frequency portions of the speech stimulus are not fully audible.

apparent that use of a high presentation level alone is insufficient to ensure the full audibility of the speech area for listeners with high-frequency hearing loss.

In summary, the AI framework is capable of making quantitative predictions of the speech-understanding performance of older listeners, which are reasonably accurate for quiet listening conditions and conditions including steady-state background noise. To the extent that the AI framework provides an adequate description of the data, elevated thresholds and a reduction in speech audibility (the simplest of the peripheral factors) explain the speech-understanding difficulties of older adults.

Although the AI framework has proven to be an extremely useful tool in understanding the speech-understanding problems of older adults, it is not without limitations. For example, it has primarily been developed from work on speech in quiet at different presentation levels, filtered speech, and speech in steady-state background noise. It has also been applied to reverberation, either alone or in combination with some of the factors noted in the preceding statement, but with less validation of this application. It has not been developed for application to many other forms of temporally distorted speech, such as interrupted or time-compressed speech, or for temporally fluctuating background noise, although there have been attempts to extend the framework to at least some of these cases (e.g., Dubno et al. 2002b; Rhebergen and Versfeld 2005; Rhebergen et al. 2006). Moreover, the AI framework has not been applied and validated on a widespread basis to sound-field listening situations in which binaural processing is involved, although, again, there have been attempts to extend it to some of these situations (e.g., Levitt and Rabiner 1967; Zurek 1993; Ahlstrom et al. 2009). Finally, the framework makes identical predictions for competing background stimuli as long as their rms long-term average spectra are equivalent, but performance can vary widely in these same conditions. For example, AI predictions would be the same for speech with a single competing talker in the background whether that competing speech was played forward or in reverse, yet listeners find the latter masker to be much less distracting and have higher speech-understanding scores as a result (e.g., Dirks and Bower 1969; Festen and Plomp 1990; Humes et al. 2006).

8.2.1.2 Plomp's Speech-Recognition Threshold Model

The focus in this review thus far has been on listening conditions that are acoustically fixed in terms of speech level and SNR, with the percent correct speech-understanding performance determined for nonsense syllables, words, or sentences. An alternate approach is to adaptively vary the SNR (usually, the noise level is held constant and the speech level varied) to achieve a criterion performance level, such as 50% correct (Bode and Carhart 1974; Plomp and Mimpen 1979a; Dirks et al. 1982). Plomp (1978, 1986) specifically made use of sentence materials and a spectrally matched competing steady-state noise and developed a simple, but elegant, model of speech-recognition thresholds (SRTs) in quiet and noise. Briefly, the SRT (50% performance criterion) for sentences is adaptively measured in quiet and then in increasing levels of noise in a group of YNH listeners. This provides a reference function that is illustrated by the solid line in Figure 8.6. Next, comparable measurements are made in older adults with varying degrees of hearing loss. Hypothetical functions for two older hearing-impaired listeners (HI1 and HI2) are shown as dashed lines in Figure 8.6. Note that both have somewhat similar amounts of hearing loss for speech (elevated SRTs) in quiet and in low noise levels (far left) but that as noise level increases, HI1's SRT is equal to that of the YNH adults, whereas HI2's SRT runs parallel to but higher than the normal-hearing function. The elevated parallel function reflects the observation that at high noise levels, HI2 needs a better-than-normal SNR to achieve 50% correct. Plomp (1978, 1986) argued that this reflected additional "distortion" above and beyond the loss of audibility, most likely associated with other aspects of cochlear pathology such as decreased frequency resolution. This SRT model was subsequently supported by a large amount of data on the SRT in quiet and in noise in older adults obtained by Plomp and Mimpen (1979b). The need for a better-than-normal SNR by some older adults for unaided SRT in noise has been clearly established in these studies. However, the need to invoke some form of "distortion," other than that associated with reduced speech audibility due to hearing loss in the high frequencies, is much less clear.

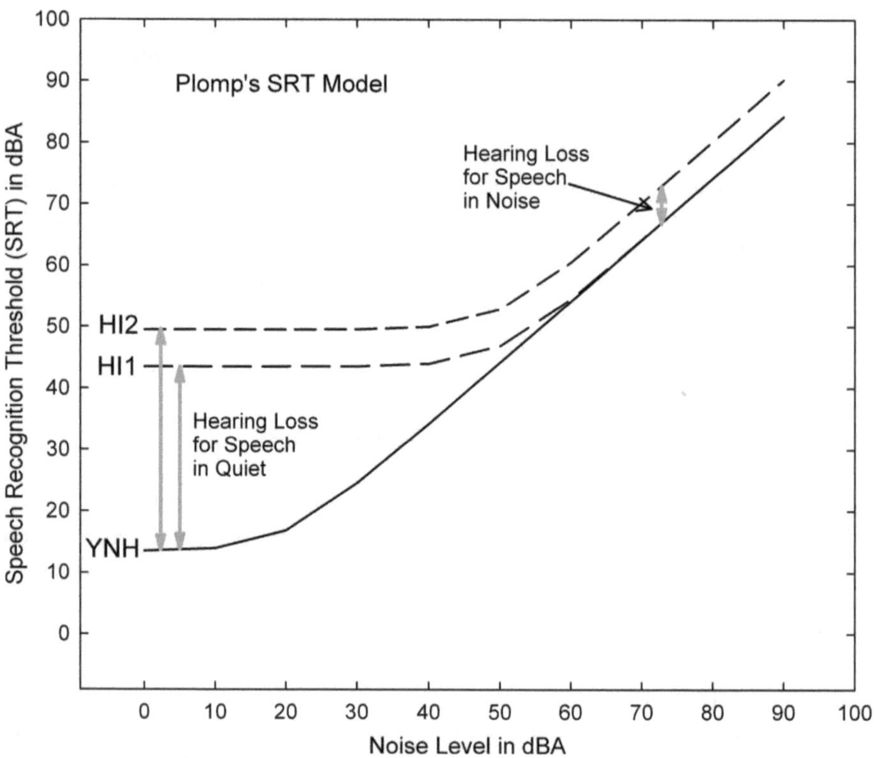

Fig. 8.6 Plomp's speech reception threshold (SRT) model for YNH and two hearing-impaired (HI) listeners. Hearing loss for speech in quiet and in noise is indicated by the gray arrows.

The AI framework described in Section 2.1.1 offers an alternate interpretation of the higher-than-normal SRT in noise observed in some older adults. Van Tasell and Yanz (1987) and Lee and Humes (1993), for instance, argued that, given the high-frequency sloping nature of the hearing loss in older adults, the "distortion" could instead be a manifestation of the loss of high-frequency portions of the speech area. The argument presented by Lee and Humes (1993) is illustrated schematically in Figure 8.7. For the SRT in noise measurements, YNH listeners typically reach SRT (50% correct) for sentences at a –6-dB SNR. This is shown in Figure 8.7, top, in which the rms level of the noise (shaded region) exceeds the rms level of the speech by 6 dB. This corresponds to an AI value of 0.3. When the speech and noise remain unchanged and are presented to a typical 80-year-old man, his high-frequency sensorineural hearing loss makes portions of the speech spectrum inaudible, as shown in Figure 8.7, middle. This reduction in audible bandwidth lowers the AI to a value of 0.2 and performance can be expected to be below 50% (SRT). As a consequence, the speech level must be raised or the noise level reduced to improve the SNR and reach an AI value of 0.3 for 50% correct performance (or SRT). Using the AI (ANSI 1997) and reducing the noise level to improve the SNR (Fig. 8.7, bottom) resulted in a SNR of –1 dB for the 80-year-old man to achieve 50% correct performance (SRT). Thus the AI framework indicates that to achieve an AI value equal to that of the YNH adults, the 80-year-old man will require a 5-dB better SNR. This can be accounted for entirely, however, by the reduced audibility of the higher frequencies due to the peripheral hearing loss. There is no need to invoke other forms of "distortion" to account for these findings. In fact, Plomp (1986) presented data that support this trend for YNH listeners undergoing low-pass filtering of the speech and noise, with cutoff frequency progressively decreasing in octave steps from 8,000 Hz to 1,000 Hz. That is, as the low-pass filter cut away progressively more of the speech area, the speech level needed to reach SRT (i.e., 50% correct recognition in noise) increased proportionately. The amount of increase in SNR, moreover, is in line with that observed frequently in older adults with high-frequency hearing loss. Identical patterns of results have been observed for YNH listeners listening to low-pass filtered speech using versions of Plomp's test developed using American English speech materials, such as the Hearing in Noise Test (HINT; Nilsson et al. 1994) and the Quick Speech in Noise (Quick-SIN) test (Killion et al. 2004). This audibility-based explanation of "distortion" is also consistent with the findings of Smoorenburg (1992) from 400 ears with noise-induced hearing loss, a pattern of hearing loss primarily impacting the higher frequencies similar to presbycusis in which hearing thresholds at 2,000 and 4,000 Hz were found to be predictive of SRT in noise. Moreover, Van Tasell and Yanz (1987) demonstrated in young adults with moderate-to-severe high-frequency sensorineural hearing loss that if the audibility of the speech area was restored through amplification, the need for a better-than-normal SNR disappeared in most listeners. Thus the "distortion" and "SNR loss" reported in older adults with high-frequency hearing loss may, in fact, simply be manifestations of the restricted bandwidth or loss of the high-frequency portion of the speech area experienced by older listeners.

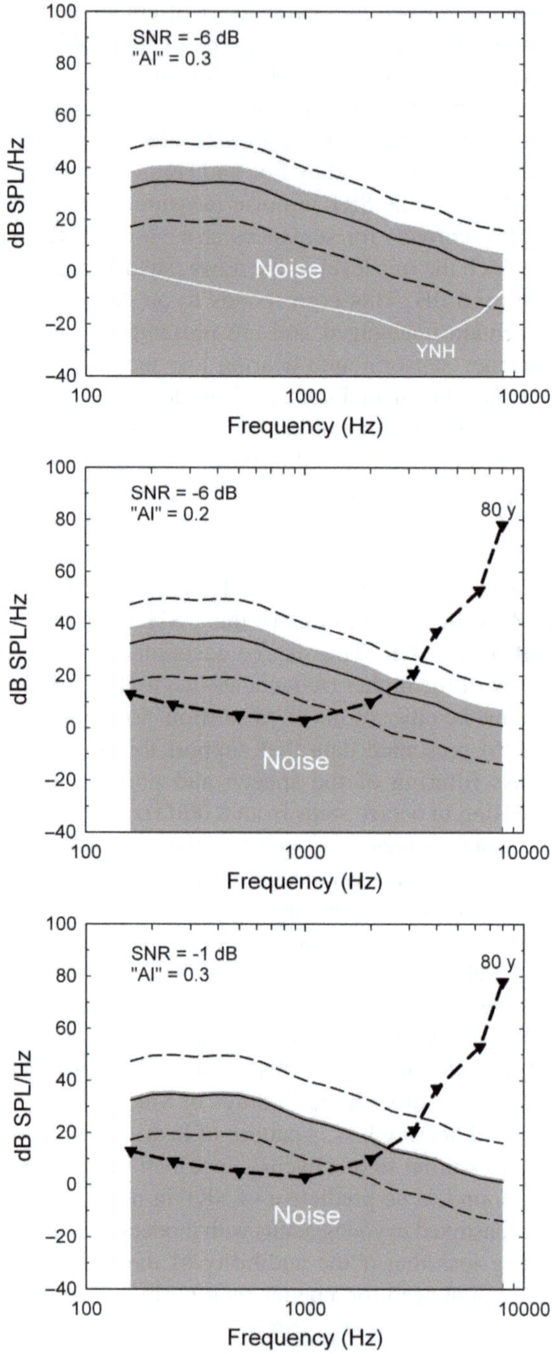

Fig. 8.7 Schematic illustration of the need for a better-than-normal SNR at SRT for listeners with high-frequency hearing loss. Top: based on SRT for YNH listeners in Figure 8.6, for a speech

Of course, the elevated SRT in noise could be a manifestation of some other form of distortion (peripheral or central in origin), but it is not possible to isolate these two possibilities from the basic SRT in noise paradigm alone. Use of the AI framework, together with the SRT in noise measurements, could help disentangle bandwidth restriction from true "distortion." Another way in which this could be accomplished is to make use of low-pass-filtered speech and noise (with complementary high-pass masking noise) in both YNH and older hearing-impaired adults with the cutoff frequency set to ~1,500 Hz. In this case, the high frequencies are inaudible to normal-hearing and hearing-impaired listeners alike and a SNR that is higher than that required for broadband speech should be required for the younger listeners with normal hearing. If older hearing-impaired adults still require a better-than normal SNR under the same test conditions, then it should not be due to the restricted bandwidth that was limiting performance in the broadband listening condition but some other "distortion" process. In fact, Horwitz et al. (2002) did just this experiment in six young adults and six older adults with high-frequency sensorineural hearing loss. For broadband (unfiltered) speech in noise, the older group needed about a 12-dB better SNR than the younger adults. When low-pass-filtered speech and noise were used, however, the older adults only needed a SNR that was ~3-6 dB better than that of the younger group. The interpretation of these results, within the framework described above, is that about half or more of the initial "distortion" measured in the broadband conditions was attributable to bandwidth restriction (high-frequency hearing loss) but that the balance of the deficit could not be accounted for in this way. On the other hand, Dubno and Ahlstrom (1995a, b) found a close correspondence between the performance of YNH and older hearing-impaired listeners for low-pass speech and noise, whereas Dubno et al. (2006) found ~40-50% of older adults with very slight hearing loss in the high frequencies had poorer performance for low-pass speech and noise than YNH adults. These conflicting group data suggest that there may be considerable individual variation in the factors underlying the performance of older adults. This, in turn, suggests that the use of low-pass-filtered speech in noise might prove helpful in determining whether elevated SNRs in older adults with impaired hearing result from restricted bandwidths or from true "distortion." Differentiating these causes is important because bandwidth restrictions could be more easily addressed via well-fit amplification than other forms of distortion.

Fig. 8.7 (continued) level of 62.5 dB SPL (normal vocal effort), a noise level of 68.5 dB SPL would correspond to SRT (50% correct). This corresponds to a SNR of –6 dB and yields an AI value of 0.3. Middle: median hearing thresholds for an 80-year-old man are added to the same –6-dB SNR shown in top panel. Because the high-frequency hearing loss has rendered some of the speech (and noise) inaudible, the AI for this older adult has been reduced to a value of 0.2. Given that an AI of 0.3 was established as corresponding to 50% correct or SRT in top panel, the older adult would perform worse than 50% correct under these conditions. Bottom: keeping the speech level the same as in top and middle panels, the noise level was decreased 5 dB (to an SNR of –1 dB) to make the AI equal to that of the top panel (0.3) and restore performance to SRT (50% correct). Thus the 80-year-old adult needed a 5-dB better-than-normal speech level to reach SRT.

Another approach to disentangling these two explanations for elevated SRTs in noise in older adults is to make sure that the experimental conditions are selected so that the background noise rather than the elevated quiet thresholds is limiting audibility at the higher frequencies in broadband test conditions. This can be accomplished either by the right combination of high noise levels and milder amounts of hearing loss (Lee and Humes 1993) or through spectral shaping of the speech and noise (George et al. 2006, 2007).

8.2.1.3 Simulated Hearing Loss

Another approach to evaluating the role played by peripheral hearing loss in the speech-understanding problems of older adults is more empirical in nature and attempts to simulate the primary features of cochlear-based sensorineural hearing loss in YNH listeners. In most of these studies, noise masking was introduced into the YNH adult's ear to produce masked thresholds that matched the quiet thresholds of an older adult with impaired hearing. According to the AI framework, these two cases would yield equivalent audibility of the speech area and equivalent performance would be predicted. This has been the case in many such studies measuring speech understanding in quiet and steady-state background noise (Fabry and Van Tasell 1986; Humes and Roberts 1990; Humes and Christopherson 1991; Humes et al. 1991; Dubno and Ahlstrom 1995a, b; Dubno and Schaefer 1995) as well as in several studies using this approach to study the recognition of reverberant speech (Humes and Roberts 1990; Humes and Christopherson 1991; Halling and Humes 2000).

When attempting to simulate the performance of older adults in quiet, however, one of the differences between the older adults with actual hearing loss and the young adults with noise-simulated hearing loss is that only the latter group is listening in noise. The noise, audible only to the younger group with simulated hearing loss, could tax cognitive processes, such as working memory (Rabbit 1968; Pichora-Fuller et al. 1995; Surprenant 2007), and impair performance even though audibility was equivalent for both the younger and older listeners. Dubno and colleagues (Dubno and Dirks 1993; Dubno and Ahlstrom 1995a, b; Dubno and Schaefer 1995; Dubno et al. 2006) addressed this by conducting a series of studies of the speech-understanding performance of YNH and older hearing-impaired listeners in which a spectrally shaped noise was used to elevate the hearing thresholds of both groups. In particular, a spectrally shaped noise was used elevated the hearing thresholds of the (older) hearing-impaired listeners slightly (3-5 dB) and then elevated the pure-tone thresholds of the YNH listeners to those same target levels. This produces equivalent audibility and conditions involving listening in noise for both groups. The results obtained by Dubno and colleagues using this approach have generally revealed good agreement between the performance of young adults with simulated hearing loss and older adults with impaired hearing, with some notable exceptions (Dubno et al. 2006).

8.2.1.4 Factorial Combinations of Age and Hearing Status

Probably the experimental paradigm that has yielded results most at odds with the simple peripheral explanation, especially for temporally degraded speech and complex listening conditions, has been the independent-group design with factorial combinations of age (typically, young and old adults) and hearing status (typically, normal and impaired). This approach provides the means to examine not only the main effects of age and hearing loss but also their interaction. Although the details vary across studies, the general pattern that emerges is that hearing loss primarily determines performance in quiet or in steady-state background noise but that age or the interaction of age with hearing loss impacts performance in conditions involving temporally distorted speech (i.e., time compression, reverberation) or competing speech stimuli (Dubno et al. 1984; Gordon-Salant and Fitzgibbons 1993, 1995, 1999, 2001; Versfeld and Dreschler 2002). It should be noted that the effects of age or the combined effects of age and hearing loss on auditory temporal processing (e.g., Schneider et al. 1994; Strouse et al. 1998; Snell and Frisina 2000) or the understanding of temporally degraded speech (especially time-compressed speech; e.g., Wingfield et al. 1985, 1999) have been observed using other paradigms as well.

Figure 8.8 illustrates some typical data for each of the four groups common to such factorial designs for quiet (top) and 50% time compression (bottom). Note that the pattern of findings across the four groups differs in the two panels. Figure 8.8, top, illustrates the pattern of mean data observed for a main effect of hearing loss and no effect of age (and no interaction between these two variables). Both hearing-impaired groups perform worse than the two normal-hearing groups on speech-understanding measures in quiet. This pattern has been typically observed in such factorial independent-group studies for speech understanding in quiet, in steady-state noise, and, occasionally, in reverberation. Figure 8.8, bottom, shows the mean speech-understanding scores from the same study (Gordon-Salant and Fitzgibbons 1993) for 50% time compression. A similar pattern of results was obtained in this same study for reverberant speech and interrupted speech (with several values of time compression, reverberation, and interruption examined). The pattern of mean data in Figure 8.8, bottom, clearly reveals that both age and hearing loss have a negative impact on speech-understanding performance. Interestingly, even in this study by Gordon-Salant and Fitzgibbons (1993), correlational analyses revealed strong associations between high-frequency hearing loss and the speech-understanding scores for time-compressed speech and mildly reverberant speech (reverberation time of 0.2 s).

It can be very challenging to find sufficient numbers of older adults with hearing sensitivity that is precisely matched to that of younger adults. One way to address this limitation is using the factorial design in conjunction with the AI framework to account for effects of any differences in thresholds between groups of younger and older subjects with normal or impaired hearing (as in Dubno et al. 1984).

In the factorial-design studies cited above, for example, there were differences in high-frequency hearing thresholds between the YNH adults and the elderly

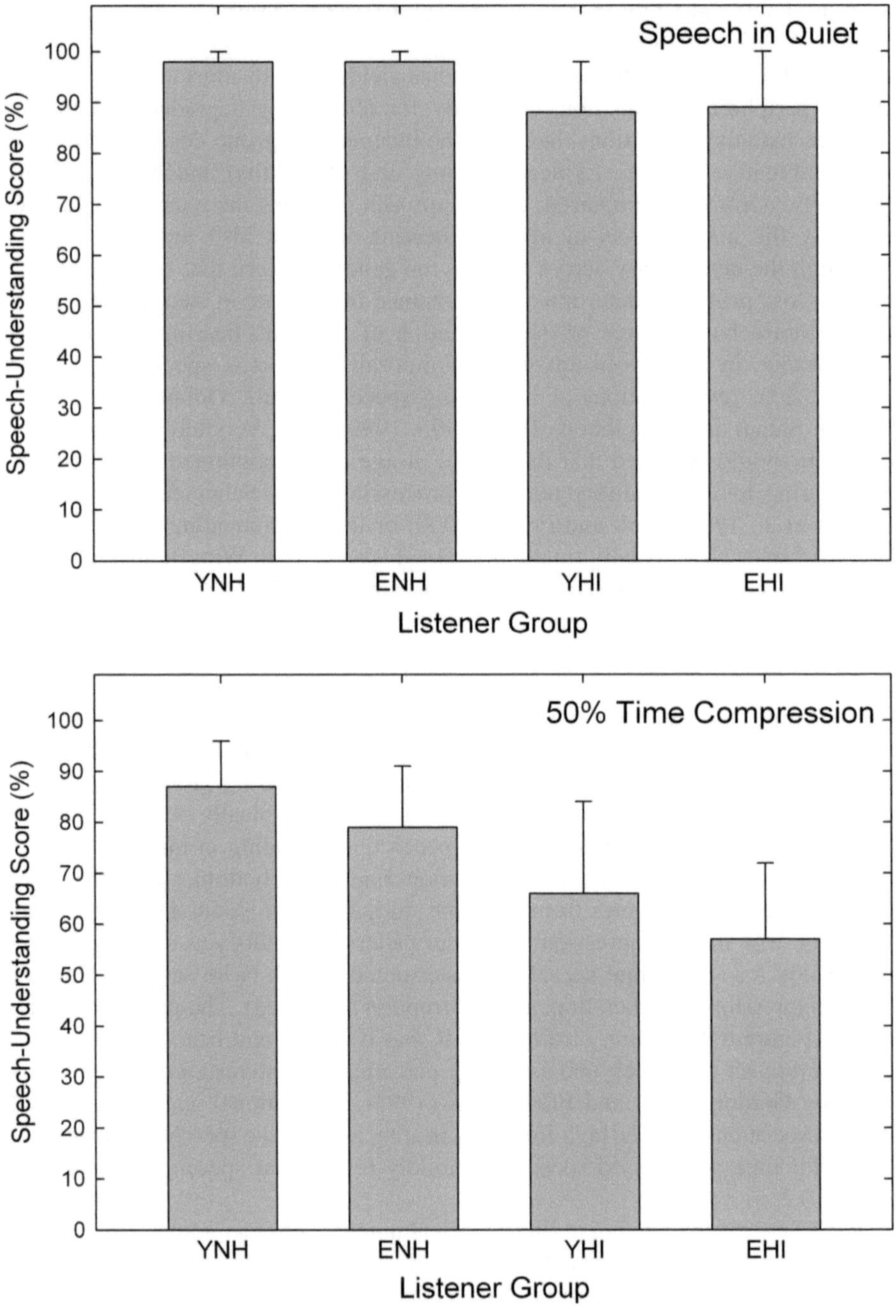

Fig. 8.8 Speech-understanding scores from four listener groups for speech in quiet and 50% time-compressed speech. Values are means ± SD. YNH, young normal hearing; ENH, elderly normal hearing; YHI, young hearing impaired; EHI, elderly hearing impaired. Data are from Gordon-Salant and Fitzgibbons (1993).

normal-hearing adults, especially at the higher frequencies reported. Speech stimuli, however, were presented at stimulus levels that were sufficient to ensure full audibility of the speech stimuli for the normal-hearing groups through at least 4,000 Hz. This assumes that, consistent with the AI framework and the simple peripheral explanation, reduced audibility of the higher frequencies is the only consequence of slight elevations in high-frequency hearing thresholds in the older adults. If, however, elevated thresholds were considered "markers" of underlying cochlear pathology, then use of high presentation levels will not compensate for this difference between the young and old "normal-hearing" groups (Humes, 2007). That is, the young group has no apparent cochlear pathology based on hearing thresholds <5- to 10-dB hearing loss through 8,000 Hz, whereas the older "normal-hearing" group has high-frequency hearing thresholds that often are elevated to at least 15- to 20-dB hearing loss at 4,000 Hz and still higher at 6,000 and 8,000 Hz. Thus although the pattern of findings from factorial independent-group designs for temporally distorted speech are not consistent with the simplest peripheral explanation, given significant differences in high-frequency hearing thresholds between the young and old normal-hearing groups, it is not possible to rule out that second-order peripheral factors were contributing to the age-group differences.

8.2.1.5 Longitudinal Studies

Nearly all of the studies of the speech-understanding difficulties of older adults conducted to date have used cross-sectional designs and multivariate approaches. As noted in this review, these laboratory studies indicate that hearing loss or audibility is the primary factor contributing to individual differences in speech recognition of older adults, with age and cognitive function accounting for only small portions of the variance. These results are consistent with those of large-scale cross-sectional studies reporting that age-related differences in speech recognition were accounted for by grouping subjects according to audiometric configuration or degree of hearing loss (e.g., Gates et al. 1990; Wiley et al. 1998).

Different conclusions were reached by Dubno et al. (2008b), however, in a longitudinal study of speech understanding using isolated monosyllabic words in quiet ($n = 256$) and key words in high- and low-context sentences in babble ($n = 85$). Repeated measures were obtained yearly (words) or every 2-3 years (sentences) over a period of (on average) 7 years (words) to 10 years (sentences). To control for concurrent changes in pure-tone thresholds and speech levels over time for each listener, speech-understanding scores were compared with scores predicted using the AI framework. Recognition of key words in sentences in babble, both low and high context, did not decline significantly with age. However, recognition of words in quiet declined significantly faster with age than predicted by declines in speech audibility. As shown in Figure 8.9, the mean observed scores at younger ages were better than predicted. whereas the mean observed scores at older ages were worse than predicted. That is, as subjects aged, their observed scores deviated increasingly

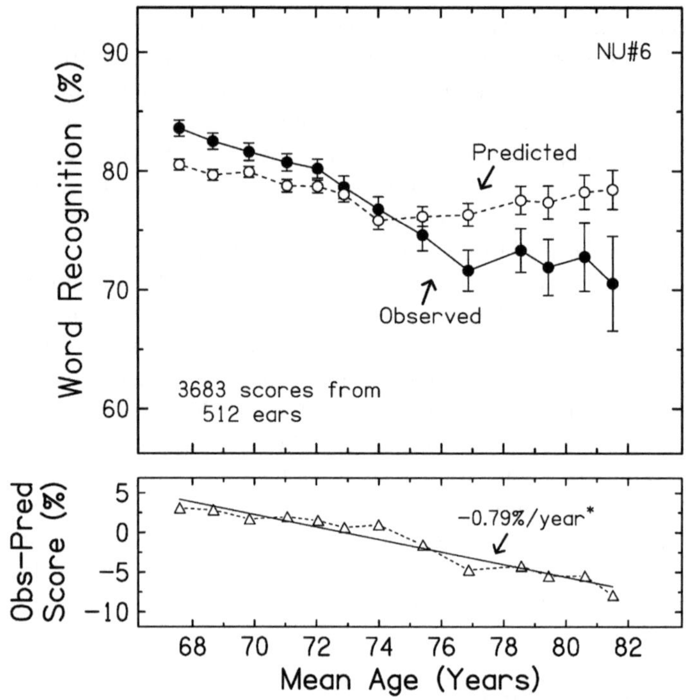

Fig. 8.9 Top: observed word-recognition scores and scores predicted by the AI for laboratory visits 1-13 at which NU#6 scores were obtained plotted as a function of mean age during those visits. Values are means ± SE in percent correct. Bottom: mean differences between observed (obs) and predicted (pred) scores (triangles) for the same laboratory visits plotted as a function of mean age during those visits. A linear regression function fit to obs-pred differences is shown (solid line) along with the slope (%/year). *Significant difference. (Reprinted with permission from Dubno et al. 2008. Copyright 2008, Acoustical Society of America).

from their predicted scores but only for words in quiet. The rate of decline did not accelerate with age but increased with the degree of hearing loss. This suggested that with more severe injury to the auditory system, impairments to auditory function other than reduced audibility resulted in faster declines in word recognition as subjects aged.

These results were consistent with the small (n = 29) longitudinal study of Divenyi et al. (2005), which found that speech-recognition measures in older subjects declined with increasing age more rapidly than their pure-tone thresholds. In the Divenyi et al. (2005) study, however, all 29 older adults selected for follow-up evaluation were among those who had “above-average results” on the large test battery during the initial assessment. Thus it is possible that the decline over time could have been exaggerated by statistical regression toward the mean. In Dubno et al. (2008), on the other hand, all the assessments of speech understanding were unaided, but 31% of the subjects reported owning and wearing hearing aids (only 55% were considered to be hearing-aid candidates). Declines in unaided performance have been observed more frequently among older hearing-aid wearers, even though aided performance remained intact, at least for longitudinal studies over a 2- to 3-year period (Humes et al. 2002; Humes and Wilson 2003). In addition, even though the declines in speech understanding over time observed by Dubno et al. (2008) were more rapid than concomitant declines in pure-tone hearing thresholds, the correlations between high-frequency hearing sensitivity and speech-understanding scores were still –0.60 and –0.64 for the initial and final measurements, respectively (accounting for 36-41% of the variance). More longitudinal studies of speech understanding in older adults are needed before it will be possible to determine if the trends that emerge from this research design are compatible with those from the much more common cross-sectional designs.

8.2.1.6 Limitations of the Peripheral Explanation

To recap, for speech in quiet or speech in steady-state background noise, with the speech presented at levels ranging from 50- to 90-dB SPL, the evidence clearly supports the simplest form of the peripheral explanation (i.e., audibility) as the primary factor underlying the speech-understanding problems of older adults. Even for some forms of temporally distorted speech, such as time compression and reverberation, high-frequency hearing loss is often the primary factor accounting for individual differences in speech-understanding performance among older adults (Helfer and Wilber 1990; Helfer 1992; Gordon-Salant and Fitzgibbons 1993, 1997; Halling and Humes 2000; Humes 2005, 2008). This has also been the case when undistorted speech is presented in temporally interrupted noise (Festen and Plomp 1990; Dubno et al. 2002b, 2003; Versfeld and Dreschler 2002; George et al. 2006, 2007; Jin and Nelson 2006).

In general, however, the relationship between measured speech understanding and high-frequency hearing loss is weaker for temporally distorted speech than for undistorted speech in quiet or steady-state noise. Thus in these conditions involving temporally distorted speech or temporally varying background maskers, high-frequency hearing loss alone accounts for significant, but smaller, amounts of variance, leaving more variance that potentially could be explained by other factors. As noted, the strongest evidence for this probably has come from the studies using independent-group designs with factorial combinations of age and hearing status. In this case, however, given significant differences in high-frequency hearing sensitivity between the younger and older adults with normal hearing, it is possible that second-order effects associated with even mild cochlear pathology (Halling and Humes 2000; Dubno et al. 2006) as well as central or cognitive factors could play a role.

There is other evidence, however, that also supports the special difficulties that older adults have with temporally distorted speech or temporally varying backgrounds. Dubno et al. (2002b, 2003), for example, used a simple modification of the AI to predict speech identification in interrupted noise and ensured that audibility differences among younger and older adults during the "off" portions of the noise were minimized. Although the benefit of masker modulation was predicted to be larger for older than younger listeners, relative to steady-state noise, scores improved more in interrupted noise for younger than for older listeners, particularly at a higher noise level. In this case, factors other than audibility were limiting the performance of the older adults, which may relate to recovery from forward maskers. Similarly, George et al. (2006, 2007) demonstrated that the speech-understanding performance of older adults listening to spectrally shaped speech and noise could not be explained by threshold elevation for fluctuating noise. In addition, they found that other factors such as temporal resolution and cognitive processing could account for significant amounts of individual variations in performance.

Another listening condition for which high-frequency hearing thresholds account for much smaller amounts of variance is the use of dichotic presentation of speech, although, again, hearing sensitivity still accounts for the largest amount of variance among the factors examined (Jerger et al. 1989; Jerger et al. 1991; Humes et al., 1996; Divenyi and Haupt 1997b; Hallgren et al. 2001). For example, in a study involving 200 older adults and 5 measures of speech understanding by Jerger et al. (1991), 3 of the measures included speech in a steady-state background (noise or multitalker babble). Correlations between speech-understanding scores on these three measures and high-frequency hearing loss ranged from –0.73 to –0.78 (accounting for 54-61% of the variance). With monaural presentation of sentences in a background of competing speech (discourse), the correlation with hearing loss decreased slightly to –0.65, accounting for 42% of the variance. Finally, when the latter task was administered dichotically, the correlations with hearing loss dropped further, to –0.55, accounting for 30% of the variance. Interestingly, for these latter two measures of speech understanding, age or a measure of cognitive function emerged as an additional explanatory variable in subsequent regression analyses, whereas hearing loss was the only predictor identified for the three measures of speech understanding in steady-state noise.

8.2.2 Beyond Peripheral Factors

In general, across various studies, pure-tone hearing thresholds account for significant, but decreased, amounts of variance in speech-understanding performance for temporally distorted speech, temporally interrupted background noise, and dichotic competing speech. Therefore, for these conditions, factors beyond simple audibility loss may need to be considered when attempting to explain the speech-understanding problems of older listeners. For example, age-related changes in the auditory periphery (e.g., elevated thresholds, broadened tuning, reduced compression) may reduce the ability to benefit from temporal or spectral "dips" that occur for a single-talker or other fluctuating maskers, beyond simple audibility effects. For each of these listening conditions, higher-level deficits in central-auditory or cognitive processing have also been suggested as possible explanatory factors. For example, older adults have been found to perform more poorly than younger adults on various measures of auditory temporal processing (Schneider et al. 1994; Fitzgibbons and Gordon-Salant 1995, 1998, 2004, 2006; Strouse et al. 1998; Gordon-Salant and Fitzgibbons 1999). Older adults have also demonstrated a general trend toward "cognitive slowing" with advancing age (Salthouse 1985, 1991, 2000; Wingfield and Tun 2001). Either a modality-specific auditory temporal-processing deficit related to auditory peripheral or central changes or general cognitive slowing could explain the difficulties of older adults with temporally degraded speech or temporally interrupted noise.

They could, for that matter, also contribute to the difficulties observed for dichotic processing of speech, but other factors could also be operating here. For example, there are specific auditory pathways and areas activated in the central nervous system when processing dichotic speech and age-related deficits in these areas or pathways could lead to speech-understanding problems (Jerger et al. 1989, 1991; Martin and Jerger 2005; Roup et al. 2006). In addition, there is evidence in support of age-related declines in attention, a cognitive mechanism, which could contribute to the decline in dichotic speech-understanding performance (McDowd and Shaw 2000; Rogers 2000). Furthermore, one could argue that it is not these age-related declines in higher-level auditory or cognitive processing alone but the combination of a degraded sensory input due to the peripheral hearing loss and these higher-level processing deficits that greatly increases the speech-understanding difficulties of older adults (Humes et al. 1993; Pichora-Fuller et al. 1995; Wingfield 1996; Schneider and Pichora-Fuller 2000; Pichora-Fuller 2003; Humes and Floyd 2005; Pichora-Fuller and Singh 2006). In this case, the hypothesis is that there exists a finite amount of information-processing resources and if older adults with high-frequency hearing loss have to divert some of these resources to repair what is otherwise a nearly automatic resource-free process of encoding the sensory input, then fewer resources will be available for subsequent higher-level processing.

One of the challenges in identifying the nature of additional factors contributing to the speech-understanding problems of older adults beyond elevated thresholds is the ability to distinguish between "central-auditory" and "cognitive" factors. Many tests designed to assess "central auditory" function, for example, are, for the most part, considered to be "auditory" because they make use of sound as the sensory stimulus. As one might imagine, this ambiguity can lead to considerable diagnostic overlap between those older adults considered to have "central-auditory" versus "cognitive" processing problems. For example, Jerger et al. (1989), in a study of 130 older adults, identified half (65) of the older adults as having "central-auditory processing disorder," but within that group, 54% (35) also had "abnormal cognitive status."

It has been argued previously (Humes et al. 1992; McFarland and Cacace 1995; Cacace and McFarland 1998, 2005; Humes 2008) that one way of distinguishing between central-auditory and cognitive processing deficits would be to examine the modality specificity of the deficit. A strong correlation between performance on a comparable task performed auditorily and visually argues in favor of a cognitive explanation rather than an explanation unique to the auditory system. Likewise, the lack of correlation between performance in each modality on similar or identical tasks is evidence of modality independence. The "common cause" hypothesis was put forth by Lindenberger and Baltes (1994) and Baltes and Lindenberger (1997) to interpret the relatively high correlations observed among sensory and cognitive factors in older adults. These authors proposed that the link between sensory and cognitive measures increased with age because both functions reflect the same anatomic and physiological changes in the aging brain.

Most often, speech has been the stimulus used to assess central-auditory function in older adults. As noted, the inaudibility of the higher-frequency portions of the speech stimulus can have a negative impact on speech-understanding performance and the presence of such a hearing loss in a large portion of the older population (especially the older clinical population) can make it difficult to interpret the results of speech-based measures of central-auditory function (Humes et al., 1996; Humes 2008). To address this, several recent studies have attempted to eliminate or minimize the contributions of the inaudibility of the higher-frequency regions of the speech stimulus by amplifying or spectrally shaping the speech stimulus (Humes 2002, 2007; George et al. 2006, 2007; Humes et al. 2006, 2007; Zekveld et al. 2007). Moreover, three of these studies also developed visual analogs of the auditory speech-understanding measures to examine the issue of modality specificity (George et al. 2007; Humes et al. 2007; Zekveld et al. 2007) in older adults. For the latter three studies, the trends that emerged were as follows. First, for the recognition of sentences in either a babble background or modulated noise, there was ~25-30% common variance between the auditory and visual versions of the tasks in older adults. Second, even though speech was made audible in these studies, the degree of high-frequency hearing loss was still the best predictor of speech understanding in all conditions, accounting for ~40-50% of the variance. This suggests that the hearing loss not only represents a limit to audibility when speech is not amplified but may also serve as a perceptual "marker" for an increased likelihood of other (peripheral) processing deficits in these listeners that might have an impact on aided speech understanding. George et al. (2007) observed a significant correlation between a psychophysical measure of temporal resolution and speech-understanding performance in modulated noise, a finding also observed recently by Jin and Nelson (2006). In addition, Dubno et al. (2002b) observed excessive forward masking in older adults with near-normal hearing and also found negative correlations between forward-masked thresholds and speech-understanding in interrupted noise.

Humes et al. (2007) examined the performance of older listeners using two acoustic versions of "speeded speech." One version made use of conventional time-compressed monosyllabic words. As noted by Schneider et al. (2005), however, implementation of time compression can sometimes lead to inadvertent spectral distortion of the stimuli and it may be the spectral distortions that contribute to the performance declines of older adults for time-compressed speech. As a result, the other version of "speeded speech" used by Humes et al. (2007) was based on shortening the length of pauses between speech sounds. In this approach, brief but clearly articulated letters of the alphabet were presented that could be combined in sequence to spell various three-, four-, or five-letter words. The duration of the silent interval between each spoken letter was manipulated to vary the overall rate of presentation. The listener's task was to write down the word that had been spelled out acoustically. Another advantage of this form of speeded speech, in contrast to time compression, was that a direct visual analog of the auditory speeded-speech task could be implemented (i.e., speeded text recognition). In this study, there was again ~30% common variance between the auditory and visual analogs of "speeded speech" and this common variance was also negatively associated with age. Given the more complex nature of the task, these results could be interpreted as a general age-related cognitive deficit in the speed of processing. There was, however, an additional modality specific component and this was again related (inversely) to the amount of high-frequency hearing loss.

Additional study of commonalities across modalities is needed, but the work completed thus far is sufficient to at least urge caution when interpreting poor speech-understanding performance of some older adults on speech-based tests of "central-auditory processing" as pure modality-specific deficits. The use of sound as the stimulus, or spoken speech in particular, does not make the task an "auditory processing" task. In fact, significant amounts (~30%) of shared variance have been identified between auditory and visual analogs of some tasks thought to be auditory-specific tasks. Further evaluation of some of the visual analogs developed by Zekveld et al. (2007) and Humes et al. (2007) with older hearing-impaired listeners may result in the development of clinical procedures to help determine how much of an individual's speech-understanding deficit is due to a modality-specific auditory-processing problem or a general cognitive-processing problem. At the moment, however, there is no specific treatment for either form of higher-level processing deficit and both would likely constrain the immediate benefits provided by amplification designed to overcome the loss of audibility due to peripheral hearing loss.

As noted, dichotic measures of speech understanding represent another area often considered by audiologists to be within the domain of "central-auditory processing" but for which amodal cognitive factors could also play a role. Clearly, the hemispheric specialization for speech and language that gives rise to a right ear advantage in many dichotic listening tasks can be considered auditory specific, or at least speech specific (Kimura 1967; Berlin et al. 1973). However, there is also a general cognitive component to the processing of competing speech stimuli presented concurrently to each ear (e.g., Cherry 1953). Specifically, the ability to attend to a message in one ear or the other can impact performance. Several studies of dichotic speech understanding in older listeners, which also measured some aspect of general cognitive function, have reported significant correlations between these measures (Jerger et al. 1991; Hallgren et al. 2001; Humes 2005; Humes et al. 2006). Humes et al. (2006) measured the ability of young and older adults to identify words near the end of a sentence when a competing similar sentence was spoken by another talker. Various cues signaling the sentence to which the listener should attend were presented either immediately before each sentence pair (selective attention) or immediately after each sentence pair (divided attention). One type of cue, a lexical cue, was used in both a monaural and a dichotic condition. The expected right ear advantage was obtained when the results for the dichotic condition were analyzed. However, there were very strong and significant correlations ($r = 0.87$ and 0.94) between the monaural scores and dichotic scores for both the 10 YNH listeners and the 13 older adults with various degrees of hearing loss. (Stimuli were spectrally shaped in this study to minimize or eliminate the contributions of audibility loss.) Because there is no modality-specific or speech-specific interhemispheric advantage for the monaural condition, the strong correlations must reflect some other common denominator, perhaps cognitive in nature (attention). In fact, the measures of selective attention were found to be significantly correlated ($r = 0.6$) with digit-span measures from the older adults, a pattern observed frequently in studies of aging and cognition (Verhaeghen and De Meersman 1998a, b). Again, more work is needed in this area, but there is enough evidence to urge caution in the interpretation of results from dichotic speech-understanding measures. The use of speech does not make such measures "pure" measures of auditory processing and performance may be determined by other amodal cognitive (or linguistic) factors.

To summarize the review to this stage, the peripheral cochlear pathology commonly found in older adults is the primary factor underlying the speech-understanding problems of older adults. The restricted audibility of the speech stimulus, primarily due to the inaudibility of the higher frequencies, is the main contributor, especially for typical conversational listening conditions in quiet and in steady-state noise without amplification. Because the peripheral hearing loss in aging is sensorineural in nature, however, there may also be a reduction in nonlinearities and broadened tuning, and these peripheral factors may also contribute to the speech-understanding problems of older adults. In addition to these second-order peripheral effects, higher-level auditory and cognitive factors can also contribute to the speech-understanding problems of older adults, especially when speech is temporally degraded or presented in a temporally varying background. In fact, it has been argued that these higher-level factors are likely to play a more important role once the speech has been made audible through amplification (Humes 2007).

8.3 Amplification and Speech Understanding

According to the AI framework described previously, speech understanding will be optimal when the entire speech area (long-term average speech spectrum ± 15 dB from ~100 to 8,000 Hz) is audible. This theoretical objective is illustrated in Figure 8.10 for a hypothetical typical 80-year-old man. For simplicity, it is assumed that the upper end of the dynamic range (loudness discomfort level) for this listener is 95 dB SPL/Hz at all frequencies, the same as that for YNH listeners (Kamm et al. 1978). For the hypothetical case illustrated in Figure 8.10, an 80-year-old man with median hearing loss for that age has sufficient dynamic range to accommodate the full 30-dB range of the speech area (with speech in quiet), except at the very highest frequency (8,000 Hz). For reference purposes, the unaided long-term average speech spectrum is also shown in this panel and the difference between the aided

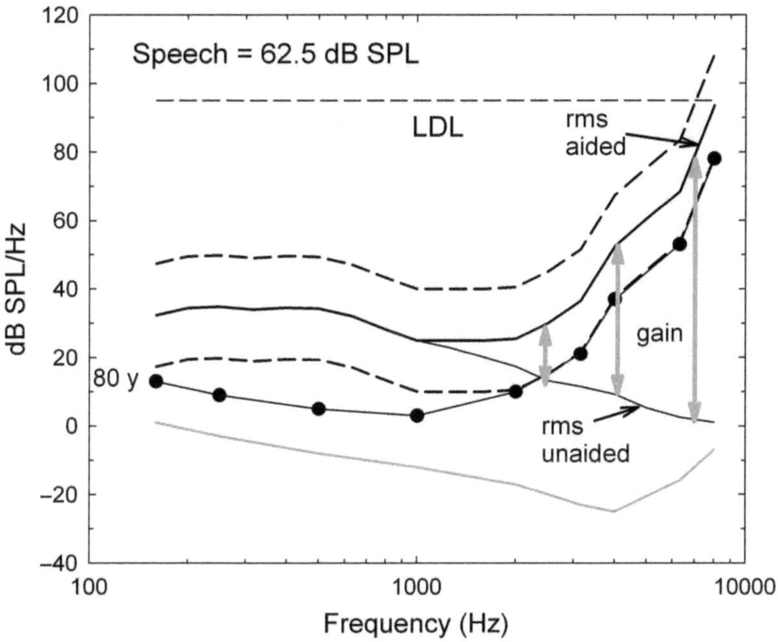

Fig. 8.10 Schematic illustration of the hearing aid gain needed to raise the rms level of unaided speech in quiet to a level 15 dB above the hearing thresholds of the 80-year-old man and below his loudness discomfort levels (LDLs).

and unaided long-term average spectra provides an indication of the gain at each frequency. The approach described here, one of making the full speech area just audible in aided listening, forms the basis of one of the most popular clinical hearing aid fitting algorithms, the desired sensation level (DSL) approach (Seewald et al. 1993, 2005; Cornelisse et al. 1995). It should be noted, however, that as long as the full speech area is audible and does not exceed discomfort at any frequency, there are many ways in which the speech may be amplified to yield equivalent speech-understanding performance (Horwitz et al. 1991; Van Buuren et al. 1995).

Of course, conversational-level speech is not the only input stimulus of importance to consider for the older adult with impaired hearing. The speech level, for example, may change with vocal effort or distance from the talker, with differing amounts of gain needed to position the amplified sound within the listener's dynamic range. In addition, many sounds of importance, such as environmental and warning sounds, span a wide range of levels and need to be amplified such that they are within the listener's dynamic range. Using the thresholds and loudness discomfort levels (LDLs) in Figure 8.10, Figure 8.11 illustrates the rationale behind the DSL input/output (I/O) method noted previously (Cornelisse et al. 1995), which positions aided stimuli within a listener's dynamic range. Given that this hypothetical listener's dynamic range is very narrow in the high frequencies (3,500-7,000 Hz), near normal in the low frequencies (100-1,000 Hz), and between these two

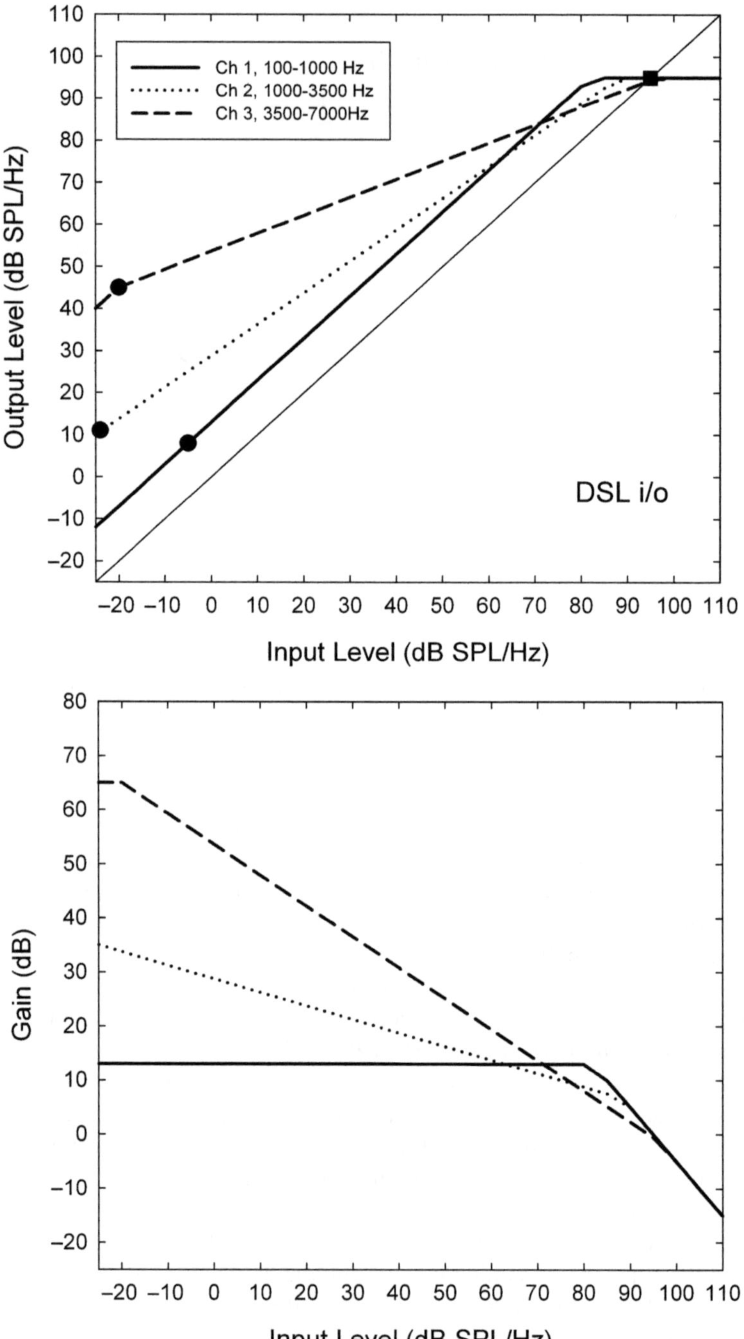

Fig. 8.11 Desired sensation level input/output (DSL I/O) hearing aid fitting approach. Top: output level as a function of the input level. Bottom: relative gain (output-inputs) as a function of input level.

extremes in the middle frequencies (1,000-3,500 Hz; Fig. 8.10), each of these frequency regions has been considered separately in Figure 8.11. In Figure 8.11, top, the average hearing thresholds (in dB SPL) of YNH adults for each frequency region are plotted against the corresponding hearing thresholds for the hypothetical 80 year old (circles; illustrated previously in Fig. 8.10). Thus, for each circle, the input level corresponding to hearing threshold for YNH adults is aligned with an output level generated by the hearing aid that corresponds to the hearing threshold of the older adult. In Figure 8.11, top, the square represents a similar alignment of the top end of the dynamic ranges of the YNH and older hearing-impaired listeners. Because, for simplicity, we've assumed that the LDL is the same at all frequencies and for both YNH and older hearing-impaired adults, only a single square at an I/O level of 95 dB SPL/Hz appears in Figure 8.11, top. In this implementation of the DSL approach, the circle representing the lower ends of the dynamic ranges for each frequency region are connected by a straight line to the square representing the upper end of the dynamic ranges.

The thin solid line along the positive-going diagonal in Figure 8.11, top, illustrates unity gain or output = input. Values above the diagonal reflect positive gain or amplification and functions running parallel to the diagonal represent linear increases in gain with input level. Linear I/O functions with slopes < 1, such as the 2 broken lines in Figure 8.11, top, represent decreasing gain with increasing input or amplitude compression. The dependence of gain on input level associated with each of three I/O functions in Figure 8.11, top, is illustrated in Figure 8.11, bottom. Because it is never desirable to have the output exceed the listener's LDL, peak clipping or output-limiting compression is typically implemented. In this case, the maximum output has been limited to 95 dB SPL/Hz.

Several things are noteworthy from the DSL illustrations in Figure 8.11. First, note that a considerable amount of gain (65 dB) is required in the higher frequencies for lower input levels. Before the advent and widespread of use of digital hearing aids in recent years, this amount of gain was very difficult to achieve without generating electroacoustic feedback. Although still challenging to obtain the full amount of gain needed to restore the entire speech area in older listeners with high-frequency hearing loss, one of the key advances in the transition to digital hearing aid circuitry was the development of sophisticated feedback-cancellation systems that significantly increased the amount of high-frequency gain that could be realized

Fig. 8.11 (continued) The spectrum was divided into three frequency regions or channels and the filled circles are the coordinates that correspond to the hearing thresholds of YNH adults, plotted along the input axis, and the corresponding hearing thresholds of an 80-year-old man, plotted along the output axis (from data in Figure 8.10). The filled circles basically align the low ends of the listeners' dynamic ranges in each frequency region or channel. The upper end of the dynamic range is based on the LDL and is assumed here to be constant over frequency and across listeners. As a result, the upper ends of the dynamic ranges are represented by a single point (filled square). Straight lines are then drawn from the low end (filled circles) to the upper end (filled square) of the dynamic range in each channel and represent the desired I/O function to map the wide dynamic range of the YNH adult to the narrower dynamic range of the older adult with impaired hearing.

in cases such as this (Dyrlund et al. 1994; Kates 1999). Because about two-thirds of hearing aid sales in the United States are to people over age 65 (Skafte 2000), most of whom have typical high-frequency sloping sensorineural hearing loss, resolution of the feedback problem is critical. With insufficient high-frequency gain, older listeners will never get beyond the channel width-restriction form of distortion noted above.

Second, note that the variations in gain with input level in Figure 8.11 are much more apparent in the higher frequencies than in the lower frequencies. In fact, rather than fitting a sloping line between the circles and square for the low frequencies in Figure 8.11, top, this is simply approximated by a constant gain of 13 dB up to LDL. Linear gain, however, would not be appropriate in the other two frequency regions. Instead, amplitude compression would be needed over a wide range of input levels, referred to as wide-dynamic-range compression, in the mid- and high-frequency regions. When different compression characteristics are desirable for different frequency regions, the spectrum is divided into compression channels for this purpose. Because the normal dynamic range is constant along the *x*-axis, as the listener's dynamic range narrows, the amount of compression needed to squeeze a wide range of inputs into a small range of outputs increases. Research has examined the number of compression channels needed for optimal aided speech understanding and, in general, four or fewer channels appear to be sufficient (Dillon 1996, 2001; Woods et al. 2006).

The I/O functions in Figure 8.11, top, were applied to the long-term average speech spectrum for input levels of 50- to 90-dB SPL, and the results obtained are illustrated in Figure 8.12. The thick solid line in Figure 8.12 represents positioning of the rms speech spectrum at 15 dB above threshold, which would be the minimum output needed to optimize the AI for speech in quiet. Note that, with the exception of the highest frequencies, the speech spectra have been amplified above this target while remaining comfortably (>15 dB) below LDL. Thus this approach, if realized in the wearer's ear, would come close to achieving the goal of making speech (in quiet) fully audible over a wide frequency range and for a wide range of speech levels while not being uncomfortably loud. Adjustments in the gain or frequency shaping of the highest frequency channel would lead to an even better match to this desired objective.

There are several other parameters used to specify the compression systems used in hearing aids, including the compression threshold (the input or output level at which compression begins), the compression ratio (the extent of compression expressed as the change in input, in decibels, divided by the change in output, in decibels), and temporal parameters, such as the time it takes for the compression to be activated (attack time) and the time it takes to deactivate (release time) (Dillon 1996, 2001). Values for each of these compression parameters are determined, in large part, by the wearer's dynamic range and considerations of the acoustical input (e.g., speech, music, noise).

Although Figure 8.12 presents only speech in quiet at several different presentation levels, it is important to note that if there is noise in the background, the noise will undergo equivalent amounts of amplification. Consider, for example, the

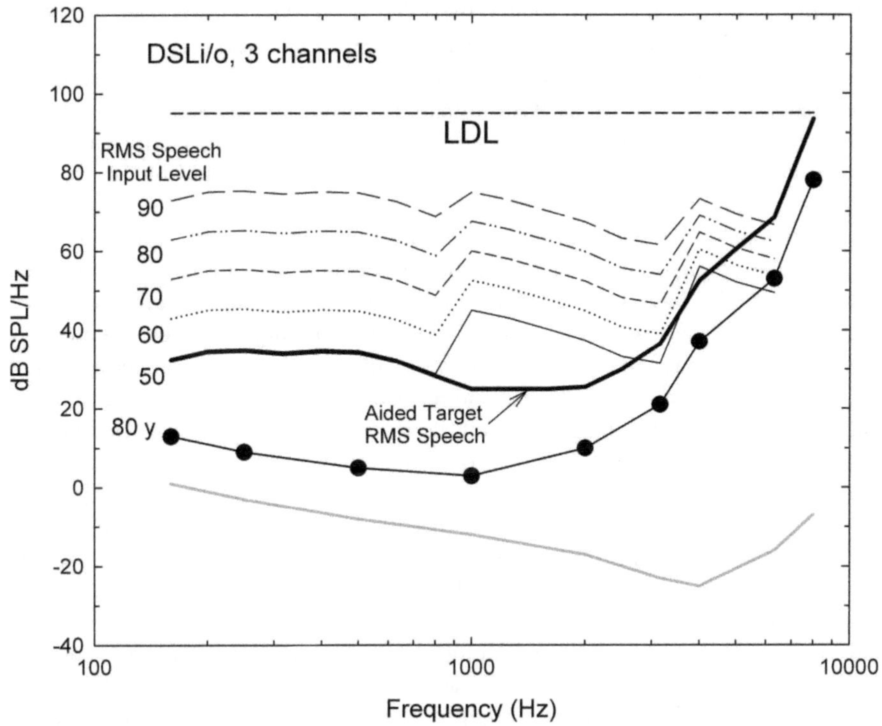

Fig. 8.12 Schematic illustration of the application of the gain derived from DSL I/O in Figure 8.11 to the speech spectrum (in quiet) when the speech varies from 50 to 90 dB SPL in 10-dB steps. The aided target corresponds to an aided rms long-term average speech spectrum that is 15 dB above hearing threshold in the region of hearing loss. Note that, with the exception of the highest frequencies and lower speech levels, the aided speech spectra are above the targeted minimum and below the LDL of the listener.

unaided case in which the noise level results in a 0-dB SNR and the noise has the same spectrum as the long-term spectrum of speech, as illustrated previously in the middle panel of Figure 8.2. In this unaided case, half of the 30-dB speech area is available to the normal-hearing listener at the input (Fig. 8.2, middle; AI = 0.5), but less than half of the speech area is available to the hearing-impaired listener (AI = 0.28; see Fig. 8.4, bottom, as an example). If the speech and noise undergo the same amplification (i.e., the same gain applied to the speech and noise) and the gain is designed to restore full audibility of the speech area in quiet, as illustrated in Figure 8.12, half of the speech area will be available to the hearing-impaired listener at the output. That is, an aided SNR of 0 dB (AI = 0.5), same as at the input, will be restored. Thus well-fit amplification, such as that in Figure 8.12, will, at best, restore the SNR to that of the input but not improve it. Because the 0-dB SNR has been restored over the full bandwidth for the hearing aid wearer, speech audibility has been improved relative to the unaided condition (from AI = 0.28 to AI = 0.5). In fact, if completely successful in restoring the full bandwidth of the speech (and noise) for the hearing aid wearer, the aided AI (0.5) for this listener will be the same as that of a normal-hearing listener without amplification (Fig. 8.2, middle). Given that the AI values for unaided listening in adults with normal hearing and aided listening in older adults with impaired hearing are the same (0.5), performance for these individuals should also be the same. (However, this argument ignores the slight level-dependent decreases in the AI that would likely occur for aided speech and noise.)

Do, in fact, older listeners with well-fit amplification of speech (and noise) perform like young adults under acoustically identical conditions? Several recent laboratory studies have been conducted in which speech understanding was measured in the presence of various competing stimuli, ranging from steady-state noise to a single competing talker, with the audibility of the speech and competing stimuli optimized through at least 4,000 Hz and acoustically identical (same overall levels and spectral shaping) listening conditions used for young and old alike (George et al. 2006, 2007; Humes et al. 2006, 2007; Amos and Humes 2007; Humes 2007; Horwitz et al. 2008). Consistently, if the study made use of an adaptive adjustment of SNR, a higher-than-normal SNR was needed by the older adults, even with the full bandwidth and input SNR restored. Interestingly, in a study of the same kind conducted with young hearing-impaired adults, a better-than-normal SNR was not needed when the audibility of the speech area had been restored fully (Van Tasell and Yanz 1987). For the studies with older adults that fixed the SNR and measured performance, group differences were observed in performance, with the older adults performing worse than the younger adults. Thus, in these studies, several of which were reviewed earlier in this chapter, restoring the audibility of the speech area, and the input SNR in the process, did not restore speech-understanding performance to normal (although, if data for unshaped speech were also obtained in a given study, performance improved when shaped). As was also noted, there could be second-order peripheral explanations (i.e., abnormal temporal resolution or frequency resolution) or higher-level central-auditory or cognitive explanations for this observation and additional research is needed to identify the nature of the underlying causes. Regardless of the specific underlying mechanism(s), the practical consequence is that many older adults need a better-than-normal SNR even when the audibility of the speech area has been restored with spectral shaping and amplification. This appears to be the case when the competition is more complex or meaningful, such as one or two competing talkers or fluctuating noise, and less so for simpler backgrounds such as steady-state noise. When speech understanding is assessed in quiet or in steady-state noise, on the other hand, results with spectral shaping have been consistent with the simple peripheral explanation in most cases, once audibility of the speech (and noise) has been fully restored (i.e., rms speech level at least 15 dB above threshold through at least 4,000 Hz). Specifically, older adults with impaired hearing perform like YNH adults in quiet or in steady-state noise when audibility has been restored. This does not appear to be the case, however, for temporally distorted speech, fluctuating noise competition, or speech competition.

Large-scale ($n > 100$) clinical studies of older hearing aid wearers also have observed less than optimal speech-understanding performance, especially when assessed in higher levels of background noise (Humes et al. 1999, 2001; Larson et al. 2000; Humes 2002, 2003; Shanks et al. 2002). As noted by Humes (2002), it is often the case that the audibility of the speech area has not been fully restored through 4,000 Hz in clinical studies of hearing aids, at least those using clinical fitting protocols and technology available at the time of the study. Even when this less than optimal audibility was accounted for through use of the AI framework, however, Humes (2002) still found many of the 171 older hearing aid wearers to be performing well below expected levels of performance with their hearing aids, especially in backgrounds of multitalker babble. In general, the less-than-optimal speech-understanding performance measured in these studies has been corroborated by parallel self-report measures of benefit, satisfaction, and use from the same hearing aid wearers (Larson et al. 2000; Humes et al. 2001; Humes 2003). In summary, many older adults with hearing loss appear to not only need a fully audible speech area for speech in quiet but also a better than normal SNR to optimize speech-understanding performance with amplification, especially in fluctuating noise or competing speech.

8.4 Improving the SNR

There have been many approaches to improving the SNR for older adults with impaired hearing, but most fall into one of two categories: (1) electroacoustic or technology approaches and (2) perceptual learning or training approaches. Regarding the first type of approach, the incorporation of sophisticated digital feedback-suppression circuitry in contemporary hearing aids has already been noted. This helps to increase the amplification in the high frequencies so that an otherwise restricted audible bandwidth won't require compensation in terms of a higher SNR. Directional microphones represent another technological approach to improving the SNR electroacoustically. Although initial evaluations of modern-day reincarnations of this technology suggested improvements in SNR of 7-8 dB (e.g., Valente et al. 1995), more recent evaluations suggest that the improvements are much smaller (2-3 dB) in typical listening conditions (Ricketts and Dittberner 2002) and that directional microphones are often not preferred by older hearing aid wearers (Walden et al. 2000; Surr et al. 2002; Cord et al. 2004). Noise-reduction algorithms have also been incorporated into many modern digital hearing aids, but the evidence supporting their benefit in listening situations that older adults typically find challenging, such as speech in the presence of other competing speech, is not compelling (Bentler 2005).

One could probably also place the trend to fit hearing aids to both ears of older adults (Strom 2006) in this category of technology-based solutions to improving the SNR. At a minimum, use of two well-fit hearing aids ensures that at least one ear will have the most favorable SNR in situations in which the target and competing speech stimuli originate from opposite sides of the head (Carhart 1965; Dirks and Wilson 1969; Bronkhorst and Plomp 1988, 1989, 1992; Festen and Plomp 1990).

Use of two hearing aids may also enable binaural processing that would lead to an improved perceptual SNR, as in binaural release from masking (Carhart et al. 1967), although there is some evidence that older listeners with hearing loss do not benefit from spatially separated sound sources to the same degree as younger listeners (Dubno et al. 2002a). With speech presented from a loudspeaker in front of the listener (0° azimuth) and noise presented from one side (90° azimuth), the level of the noise at higher frequencies is decreased in the ear away from the noise due to the effect of "head shadow." However, hearing-impaired listeners may not benefit from this improved SNR because speech information at higher frequencies that could be made audible remains inaudible due to high-frequency hearing loss. Because differences in audibility among listeners were accounted for in Dubno et al. (2002a), poorer performance by older listeners suggested an age-related deficit in the use of interaural difference cues to produce binaural advantages for speech recognition in noise. Dubno et al. (2008a) also found that spatial separation benefit was smaller than predicted by the AI framework for older adults with near-normal thresholds. In this experiment, binaural listening (speech at 0° azimuth, speech-shaped steady-state noise at 90° azimuth) was compared with monaural listening (speech and noise in the same orientation, but the ear closer to the noise was plugged). In this case, the only difference between monaural and binaural listening was the contribution of a second ear with an unfavorable SNR. Although a larger binaural listening benefit for speech in noise was predicted using the AI framework for older than younger adults, binaural benefit was observed for younger adults only. Thus, for younger listeners, the advantage of interaural difference cues provided by a second ear outweighed that ear's poorer SNR (an effect sometimes referred to as binaural "squelch").

To the extent that spatial benefit is improved by audibility of high-frequency amplitude-difference cues, providing amplification should improve spatial benefit by restoring these cues. However, differences between a listener's two hearing aids could reduce aided spatial benefit by altering these cues (e.g., Van den Bogaert et al. 2006). Ahlstrom et al. (2009) used the AI framework modified for aided sound-field conditions (aided audibility index; Souza and Turner 1999; Stelmachowicz et al. 2002) to predict hearing aid benefit and unaided and aided benefit of spatial separation of speech and multitalker babble for older hearing-impaired listeners wearing two hearing aids. Aided audibility indices provided a means to determine the extent to which listeners benefited from high-frequency audibility provided by the hearing aids. Speech recognition in noise improved with hearing aids, especially when speech and noise sources were separated in space. Thus, restoring mid- to high-frequency speech information with amplification allowed these older listeners to take advantage of the improved SNR in the higher frequencies provided by head shadow. Although aided performance was better than unaided, hearing aid benefit was less than predicted. In contrast, spatial benefit was measurable, better than predicted, and improved with hearing aids. Spatial benefit that is greater than predicted by the AI framework is expected because simple estimates of audibility do not take into account binaural advantages such as interaural level and time difference cues that improve speech recognition in noise. Measurable spatial benefit suggested that older hearing-impaired subjects were able to take advantage of these cues.

Another technological solution designed to improve the SNR for older adults with hearing aids involves moving the microphone closer to the desired sound source. These systems today typically make use of wireless connections between a remote microphone and the amplification device worn by the older individual with impaired hearing. The wireless connection is typically accomplished using FM or infrared carriers. These devices are often referred to as "assistive listening devices" when used with older adults. The amplification device may be a free-standing system and the only system worn by the older adult or it may be a system designed to be integrated into the personal hearing aids worn by the listener. In a typical listening environment with some amount of reverberation, as the distance between the remote microphone and the sound source is halved, the SNR is increased by 6 dB (following inverse square law behavior and assuming the distance traversed is entirely within the critical distance for the room). The biggest disadvantage to such systems is that they require a single sound source that is not varying over time, such as a speaker at a lecture or a loudspeaker on a television. Nonetheless, for some older adults with limited communication needs, these devices may be sufficient or they may complement the conventional hearing aid, which can be used in challenging listening situations.

In addition to technology-based approaches to improving the SNR for amplified speech in the case of older adults with hearing aids, various attempts have been made to train older adults to extract more information from existing acoustical SNRs. A complete review of all the approaches is beyond the scope of this chapter, but these approaches can generally be classified as being auditory based or visually based. Auditory-based approaches are commonly referred to as "auditory training" and may focus on improved discrimination, identification, or recognition of speech sounds, words, or sentences. In general, early attempts were adaptations of similar approaches developed for use with more severely hearing-impaired individuals and often focused on speech sounds in quiet. The study by Walden et al. (1981) represents one of the few successes for this approach when applied to adults with moderate high-frequency hearing loss. More recently, auditory training with older hearing-impaired adults have made use of word- or sentence-based speech materials and typically have focused on training in the presence of competing stimuli (Burk et al. 2006; Burk and Humes 2007). Another more comprehensive approach that focuses on auditory and cognitive training has been shown to improve speech-understanding in older hearing aid wearers (Sweetow and Sabes 2006). In general, the rationale behind these approaches is that older adult hearing aid candidates typically wait 10-15 years after first suspecting a hearing loss to seek treatment (Brooks 1979; Kyle et al. 1985). During this time, the hearing loss may have served as a passive means of eliminating competing background stimuli. In the process of restoring the audibility of speech with well-fit amplification, the audibility of competing background stimuli has also been restored. Training may be needed for the hearing aid wearer to relearn to extract the desired cues from the speech signal as well as to extract the speech signal from the competing background sounds.

Thus auditory training should occur with amplified speech signals immersed in competing background sounds. Much more work is needed, however, to demonstrate that such auditory training efforts lead to improved speech-understanding performance in older adults wearing hearing aids.

8.5 Summary

For unaided listening, the speech-understanding performance of older adults is determined primarily by the simplest peripheral factor: inaudibility of portions of the speech area due to the presence of high-frequency hearing loss. This is true regardless of the listening conditions (quiet, noise, reverberation, etc.). For temporally degraded speech or temporally varying competition, other factors may combine with the peripheral hearing loss to produce further declines in unaided performance. When the loss of audibility is compensated for by well-fit amplification, hearing loss is no longer the primary determiner of performance and the aided performance of older adults approaches that of younger listeners under acoustically equivalent conditions for quiet listening conditions and listening conditions with steady-state noise in the background. This is often not the case, however, when older adults listen to spectrally shaped speech that is temporally degraded or in a competing background that is temporally varying. This includes the very common listening condition of target speech produced by one talker with one to three competing talkers speaking in the background. For conditions such as these, audibility of the entire speech area must be restored with amplification and the SNR must be made better than normal for the aided performance of most older adults to approach that of their younger counterparts.

Acknowledgments This work was supported, in part, by National Institutes of Health Grants AG 008293 and AG 022334 from the National Institute on Aging (LEH) and Grants DC 00184 and DC 00422 (JRD) from the National Institute on Deafness and Other Communication Disorders. We also thank Jayne B. Ahlstrom for her feedback on earlier drafts of the chapter.

References

Ahlstrom JB, Horwitz AR, Dubno JR (2009) Spatial benefit of bilateral hearing aids. Ear Hear 30:203–218.

Amos NE, Humes LE (2007) Contribution of high frequencies to speech recognition in quiet and noise in listeners with varying degrees of high-frequency sensorineural hearing loss. J Speech Lang Hear Res 50:819–834.

ANSI (1969) ANSI S3.5–1969, American National Standard Methods for the Calculation of the Articulation Index. New York: American National Standards Institute.

ANSI (1997) ANSI S3.5–1997, American National Standard Methods for the Calculation of the Speech Intelligibility Index. New York: American National Standards Institute.

Baltes PB, Lindenberger U (1997) Emergence of a powerful connection between sensory and cognitive functions across the adult life span: a new window to the study of cognitive aging? Psychol Aging 12:12–21.

Bentler RA (2005) Effectiveness of directional microphones and noise reduction schemes in hearing aids: a systematic review of evidence. J Am Acad Audiol 16:477–488.
Berlin CI, Lowe-Bell SS, Cullen JK Jr, Thompson CL (1973) Dichotic speech perception: an interpretation of right-ear advantage and temporal offset effects. J Acoust Soc Am 53:699–709.
Bilger RC, Nuetzel MJ, Rabinowitz WM, Rzeckowski C (1984) Standardization of a test of speech perception in noise. J Speech Hear Res 27:32–48.
Bode DL, Carhart R (1974) Stability and accuracy of adaptive tests of speech discrimination scores. J Acoust Soc Am 56:963–970.
Bronkhorst AW, Plomp R (1988) The effect of head-induced interaural time and level differences on speech intelligibility in noise. J Acoust Soc Am 83:1508–1516.
Bronkhorst AW, Plomp R (1989) Binaural speech intelligibility in noise for hearing-impaired listeners. J Acoust Soc Am 86:1374–1383.
Bronkhorst AW, Plomp R (1992) Effect of multiple speechlike maskers on binaural speech recognition in normal and impaired hearing. J Acoust Soc Am 92:3132–3139.
Brooks DN (1979) Hearing aid candidates—some relevant features. Br J Audiol 13:81–84.
Burk MH, Humes LE, Amos NE, Strauser LE (2006). Effect of training on word-recognition performance in noise for young normal-hearing and older hearing-impaired listeners. Ear Hear 27:263–278.
Burk MH, Humes LE (2008). Effects of long-term training on aided speech-recognition performance in noise in older adults. J Speech Lang Hear Res 51:759–771.
Cacace AT, McFarland DJ (1998) Central auditory processing disorder in school-aged children: a critical review. J Speech Lang Hear Res 41:335–373.
Cacace AT, McFarland DJ (2005) The importance of modality specificity in diagnosing central auditory processing disorder. Am J Audiol 14:112–123.
Carhart R (1965) Monaural and binaural discrimination against competing sentences. Int J Audiol 4:5–10.
Carhart R, Tillman TW, Johnson KR (1967) Release of masking for speech through interaural time delay. J Acoust Soc Am 42:124–138.
Cherry EC (1953) Some experiments on the recognition of speech, with one and with two ears. J Acoust Soc Am 25:975–979.
Clopper CG, Pisoni DB, Tierney AT (2006) Effects of open-set and closed-set task demands on spoken word recognition. J Am Acad Audiol 17:331–349.
Committee on Hearing, Bioacoustics, and Biomechanics (CHABA) (1988) Speech understanding and aging. J Acoust Soc Am 83:859–895.
Cooper JC Jr, Gates GA (1991) Hearing in the elderly—the Framingham cohort, 1983–1985: Part II. Prevalence of central auditory processing disorders. Ear Hear 12:304–311.
Cord MT, Surr RK, Walden BE, Dyrlund O (2004) Relationship between laboratory measures of directional advantage and everyday success with directional microphone hearing aids. J Am Acad Audiol 15:353–364.
Cornelisse LE, Seewald RC, Jamieson DG (1995) The input/output formula: a theoretical approach to the fitting of personal amplification devices. J Acoust Soc Am 97:1854–1864.
Dillon H (1996) Compression? Yes, but for low or high frequencies, for low or high intensities, and with what response times? Ear Hear 17:287–307.
Dillon H (2001) Hearing Aids. New York: Thieme.
Dirks DD, Bower D (1969) Masking effects of speech competing messages. J Speech Hear Res 12:229–245.
Dirks DD, Wilson RH (1969) The effect of spatially separated sound sources on speech intelligibility. J Speech Hear Res 12:5–38.
Dirks DD, Morgan DE, Dubno JR (1982) A procedure for quantifying the effects of noise on speech recognition. J Speech Hear Disord 47:114–123.
Dirks DD, Bell TS, Rossman RN, Kincaid GE (1986) Articulation index predictions of contextually dependent words. J Acoust Soc Am 80:82–92.
Divenyi PL, Haupt KM (1997a) Audiological correlates of speech understanding in elderly listeners with mild-to-moderate hearing loss. I. Age and lateral asymmetry effects. Ear Hear 18:42–61.

Divenyi PL, Haupt KM (1997b) Audiological correlates of speech understanding in elderly listeners with mild-to-moderate hearing loss. II. Correlational analysis. Ear Hear 18:100–113.

Divenyi PL, Haupt KM (1997c) Audiological correlates of speech understanding in elderly listeners with mild-to-moderate hearing loss. III. Factor representation. Ear Hear 18:189–201.

Divenyi PL, Stark PB, Haupt KM (2005) Decline of speech understanding and auditory thresholds in the elderly. J Acoust Soc Am 118:1089–1100.

Dubno JR, Ahlstrom JB (1995a) Masked thresholds and consonant recognition in low-pass maskers for hearing-impaired and normal-hearing listeners. J Acoust Soc Am 97:2430–2441.

Dubno JR, Ahlstrom JB (1995b) Growth of low-pass masking of pure tones and speech for hearing-impaired and normal-hearing listeners. J Acoust Soc Am 98:3113–3124.

Dubno JR, Dirks DD (1993) Factors affecting performance on psychoacoustic and speech-recognition tasks in the presence of hearing loss. In: Studebaker GA, Hochberg I (eds) Acoustical Factors Affecting Hearing-Aid Performance. Boston: Allyn & Bacon, pp. 235–253.

Dubno JR, Schaefer AB (1992) Comparison of frequency selectivity and consonant recognition among hearing-impaired and masked-normal listeners. J Acoust Soc Am 91:2110–2121.

Dubno JR, Schaefer AB (1995) Frequency selectivity and consonant recognition for hearing-impaired and normal-hearing listeners with equivalent masked thresholds. J Acoust Soc Am 97:1165–1174.

Dubno JR, Dirks DD, Morgan DE (1984) Effects of age and mild hearing loss on speech recognition. J Acoust Soc Am 76:87–96.

Dubno JR, Lee FS, Matthews LJ, Mills JH. (1997) Age-related and gender-related changes in monaural speech recognition. J Speech Hear Res 40:444–452.

Dubno JR, Ahlstrom JB, Horwitz AR (2000) Use of context by young and aged persons with normal hearing. J Acoust Soc Am 107:538–546

Dubno JR, Ahlstrom JB, Horwitz AR (2002a) Spectral contributions to the benefit from spatial separation of speech and noise. J Speech Lang Hear Res 45:1297–1310.

Dubno JR, Horwitz AR, Ahlstrom JB (2002b) Benefit of modulated maskers for speech recognition by younger and older adults with normal hearing. J Acoust Soc Am 111:2897–2907.

Dubno JR, Horwitz AR, Ahlstrom JB (2003) Recovery from prior stimulation: masking of speech by interrupted noise for younger and older adults with normal hearing. J Acoust Soc Am 113:2084–2094.

Dubno JR, Horwitz AR, Ahlstrom JB (2005a) Word recognition in noise at higher-than-normal levels: decreases in scores and increases in masking. J Acoust Soc Am 118:914–922.

Dubno JR, Horwitz AR, Ahlstrom JB (2005b) Recognition of filtered words in noise at higher-than-normal levels: decreases in scores with and without increases in masking. J Acoust Soc Am 118:923–933.

Dubno JR, Horwitz AR, Ahlstrom JB. (2006) Spectral and threshold effects on recognition of speech at higher-than-normal levels. J Acoust Soc Am 120:310–320.

Dubno JR, Horwitz AR, Ahlstrom JB (2007) Estimates of basilar-membrane nonlinearity effects on masking of tones and speech. Ear Hear 28:2–17.

Dubno JR, Ahlstrom JB, Horwitz AR (2008a) Binaural advantage for younger and older adults with normal hearing. J Speech Lang Hear Res 51:539–556.

Dubno JR, Lee FS, Matthews LJ, Ahlstrom JB, Horwitz AR, Mills JH (2008b). Longitudinal changes in speech recognition in older persons. J Acoust Soc Am 123:462–475.

Dyrlund O, Hennignsen LB, Bisgaard N, Jensen JH (1994) Digital feedback suppression (DFS): characterization of feedback-margin improvements in a DFS hearing instrument. Scand Audiol 23:135–138.

Fabry DA, Van Tasell DJ (1986) Masked and filtered simulation of hearing loss: effects on consonant recognition. J Speech Hear Res 29:170–178.

Festen JM, Plomp R (1990) Effects of fluctuating noise and interfering speech on the speech-reception threshold for impaired and normal hearing. J Acoust Soc Am 88:1725–1736.

Fitzgibbons PJ, Gordon-Salant S (1995) Age effects on duration discrimination with simple and complex stimuli. J Acoust Soc Am 98:3140–3145.

Fitzgibbons PJ, Gordon-Salant S (1998) Auditory temporal order perception in younger and older adults. J Speech Lang Hear Res 41:1052–1060.

Fitzgibbons PJ, Gordon-Salant S (2004) Age effects on discrimination of timing in auditory sequences. J Acoust Soc Am 116:1126–1134.

Fitzgibbons PJ, Gordon-Salant S (2006) Effects of age and sequence presentation rate on temporal order recognition. J Acoust Soc Am 120:991–999.

Fletcher H (1953) Speech and Hearing in Communication. New York: Van Nostrand.

Fletcher H, Galt RH (1950) The perception of speech and its relation to telephony. J Acoust Soc Am 22:89–151.

French NR, Steinberg JC (1947) Factors governing the intelligibility of speech sounds. J Acoust Soc Am 19:90–119.

Gates GA, Cooper JC Jr, Kannel WB, Miller NJ (1990) Hearing in the elderly: the Framingham cohort, 1983–1985. Part I. Basic audiometric test results. Ear Hear 11:247–256.

George ELJ, Festen JM, Houtgast T (2006) Factors affecting masking release for speech in modulated noise for normal-hearing and hearing-impaired listeners. J Acoust Soc Am 120:2295–2311.

George ELJ, Zekveld AA, Kramer SE, Goverts ST, Festen JM, Houtgast T (2007) Auditory and nonauditory factors affecting speech reception in noise by older listeners. J Acoust Soc Am 121:2362–2375.

Gordon-Salant S, Fitzgibbons PJ (1993) Temporal factors and speech recognition performance in young and elderly listeners. J Speech Hear Res 36:1276–1285.

Gordon-Salant S, Fitzgibbons PJ (1995) Recognition of multiply degraded speech by young and elderly listeners. J Speech Hear Res 38:1150–1156.

Gordon-Salant S, Fitzgibbons PJ (1999) Profile of auditory temporal processing in older listeners. J Speech Lang Hear Res 42:300–311.

Gordon-Salant S, Fitzgibbons PJ (2001) Sources of age-related recognition difficulty for time-compressed speech. J Speech Lang Hear Res 44:709–719.

Gordon-Salant S, Fitzgibbons PJ (2004) Effects of stimulus and noise rate variability on speech perception by younger and older adults. J Acoust Soc Am 115:1808–1817.

Hallgren M, Larsby B, Lyxell B, Arlinger S (2001) Cognitive effects in dichotic speech testing in elderly persons. Ear Hear 22:120–129.

Halling DC, Humes LE (2000) Factors affecting the recognition of reverberant speech by elderly listeners. J Speech Lang Hear Res 43:414–431.

Helfer KS (1992) Aging and the binaural advantage in reverberation and noise. J Speech Hear Res 35:1394–1401.

Helfer KS, Wilber LA (1990) Hearing loss, aging, and speech perception in reverberation and noise. J Speech Lang Hear Res 33:149–155.

Horwitz AR, Turner CW, Fabry DA (1991) Effects of different frequency response strategies upon recognition and preference for audible speech stimuli. J Speech Hear Res 34:1185–1196.

Horwitz AR, Dubno JR, Ahlstrom JB (2002) Recognition of low-pass-filtered consonants in noise with normal and impaired high-frequency hearing. J Acoust Soc Am 11:409–416.

Horwitz AR, Ahlstrom JB, Dubno JR (2007) Speech recognition in noise: estimating effects of compressive nonlinearities in the basilar-membrane response. Ear Hear 28:682–693.

Horwitz AR, Ahlstrom JB, Dubno JR (2008) Factors affecting the benefits of high-frequency amplification. J Speech Lang Hear Res 51:798–813.

Houtgast T, Steeneken HJM (1985) A review of the MTF-concept in room acoustics. J Acoust Soc Am 77:1069–1077.

Humes LE (1991) Understanding the speech-understanding problems of the hearing impaired. J Amer Acad Audiol 2:59–70.

Humes LE (1996) Speech understanding in the elderly. J Am Acad Audiol 7:161–167.

Humes LE (2002) Factors underlying the speech-recognition performance of elderly hearing-aid wearers. J Acoust Soc Am 112:1112–1132.

Humes LE (2003) Modeling and predicting hearing-aid outcome. Trends Amplif 7:41–75.

Humes LE (2005) Do 'auditory processing' tests measure auditory processing in the elderly? Ear Hear 26:109–119.
Humes LE (2007) The contributions of audibility and cognitive factors to the benefit provided by amplified speech to older adults. J Am Acad Audiol 18:590–603.
Humes LE (2008) Issues in the assessment of auditory processing in older adults. In: Cacace AT, McFarland DJ (eds) Controversies in Central Auditory Processing Disorder. San Diego, CA: Plural Publishing, pp. 21–150.
Humes LE, Christopherson L (1991) Speech-identification difficulties of the hearing-impaired elderly: the contributions of auditory-processing deficits. J Speech Hear Res 34:686–693.
Humes LE, Floyd SS (2005) Measures of working memory, sequence learning, and speech recognition in the elderly. J Speech Lang Hear Res 48:224–235.
Humes LE, Roberts L (1990) Speech-recognition difficulties of hearing-impaired elderly: the contributions of audibility. J Speech Hear Res 33:726–735.
Humes LE, Wilson DL (2003) An examination of the changes in hearing-aid performance and benefit in the elderly over a 3-year period of hearing-aid use. J Speech Lang Hear Res 46:137–145.
Humes LE, Nelson KJ, Pisoni DB (1991) Recognition of synthetic speech by hearing-impaired elderly listeners. J Speech Hear Res 34:1180–1184.
Humes LE, Christopherson LA, Cokely CG (1992) Central auditory processing disorders in the elderly: fact or fiction? In: Katz J, Stecker N, Henderson D (eds) Central Auditory Processing: A Transdisciplinary View. Philadelphia: BC Decker, pp. 41–150.
Humes LE, Nelson KJ, Pisoni DB, Lively SE (1993) Effects of age on serial recall of natural and synthetic speech. J Speech Hear Res 36:634–639.
Humes LE, Watson BU, Christensen LA, Cokely CA, Halling DA, Lee L (1994) Factors associated with individual differences in clinical measures of speech recognition among the elderly. J Speech Hear Res 37:465–474.
Humes LE, Coughlin M, Talley L (1996) Evaluation of the use of a new compact disc for auditory perceptual assessment in the elderly. J Am Acad Audiol 7:419–427.
Humes LE, Christensen LA, Bess FH, Hedley-Williams A, Bentler R (1999) A comparison of the aided performance and benefit provided by a linear and a two-channel wide-dynamic-range-compression hearing aid. J Speech Lang Hear Res 42:65–79.
Humes LE, Garner CB, Wilson DL, Barlow NN (2001) Hearing-aid outcome measures following one month of hearing aid use by the elderly. J Speech Lang Hear Res 44:469–486.
Humes LE, Wilson DL, Barlow NN, Garner CB (2002) Measures of hearing-aid benefit following 1 or 2 years of hearing-aid use by older adults. J Speech Lang Hear Res 45:772–782.
Humes LE, Lee JH, Coughlin MP (2006) Auditory measures of selective and divided attention in young and older adults using single-talker competition. J Acoust Soc Am 120:2926–2937.
Humes LE, Burk MH, Coughlin MP, Busey TA, Strauser LE (2007) Auditory speech recognition and visual text recognition in younger and older adults: similarities and differences between modalities and the effects of presentation rate. J Speech Lang Hear Res 50:283–303.
ISO (2000) ISO-7029, Acoustics-Statistical Distribution of Hearing Tthresholds as a Function of Age. Basel, Switzerland: International Standards Organization.
Jerger J, Chmiel R (1997) Factor analytic structure of auditory impairment in elderly persons. J Am Acad Audiol 8:269–276.
Jerger J, Jerger S, Oliver T, Pirozzolo F (1989) Speech understanding in the elderly. Ear Hear 10:79–89.
Jerger J, Jerger S, Pirozzolo F (1991) Correlational analysis of speech audiometric scores, hearing loss, age and cognitive abilities in the elderly. Ear Hear 12:103–109.
Jin S-H, Nelson PB (2006) Speech perception in gated noise: the effects of temporal resolution. J Acoust Soc Am 119:3097–3108.
Kalikow DN, Stevens KN, Elliott LL (1977) Development of a test of speech intelligibility in noise using test material with controlled word predictability. J Acoust Soc Am 61:1337–1351.
Kamm CA, Dirks DD, Mickey MR (1978) Effects of sensorineural hearing loss on loudness discomfort level and most comfortable loudness. J Speech Hear Res 21:668–681.

Kamm CA, Dirks DD, Bell TS (1985) Speech recognition and the Articulation Index for normal and hearing-impaired listeners. J Acoust Soc Am 77:281–288.
Kates J (1999) Constrained adaptation for feedback cancellation in hearing aids. J Acoust Soc Am 106:1010–1019.
Killion MC, Niquette PA, Gudmundsen GI, Revit LJ, Banerjee S (2004) Development of a quick speech-in-noise test for measuring signal-to-noise ratio loss in normal-hearing and hearing-impaired listeners. J Acoust Soc Am 116:2395–2405.
Kimura D (1967) Functional asymmetry of the brain in dichotic listening. Cortex 3:163–178.
Kyle JG, Jones LG, Wood PL (1985) Adjustment to acquired hearing loss: a working model. In: Orlans H (ed) Adjustment to Hearing Loss. San Diego, CA: College-Hill Press, pp.119–138.
Larson VD, Williams DW, Henderson WG, Luethke LE, Beck LB, Noffsinger D, Wilson RH, Dobie RA, Haskell GB, Bratt GW, Shanks JE, Stelmachowicz P, Studebaker GA, Boysen AE, Donahue A, Canalis R, Fausti SA, Rappaport BZ (2000) Efficacy of 3 commonly used hearing aid circuits: a crossover trial. J Am Med Assoc 284:1806–1813.
Lee LW, Humes LE (1993) Evaluating a speech-reception threshold model for hearing-impaired listeners. J Acoust Soc Am 93:2879–2885.
Levitt H, Rabiner LR (1967) Predicting binaural gain in intelligibility and release from masking for speech. J Acoust Soc Am 42:820–829.
Lindenberger U, Baltes PB (1994) Sensory functioning and intelligence in old age: a strong connection. Psychol Aging 9:339–355.
Lopez OL, Jagust WJ, DeKosky ST, Becker JT, Fitzpatrick A, Dulberg C, Breitner J, Lyketsos C, Jones B, Kawas C, Carlson M, Kuller LH (2003) Prevalence and classification of mild cognitive impairment in the Cardiovascular Health Study Cognition Study: part I. Arch Neurol 60:1385–1389.
Martin JS, Jerger JF (2005) Some effects of aging on central auditory processing. J Rehabil Res Dev 42:25–44.
McDowd JM, Shaw RJ (2000) Attention and aging: a functional perspective. In: Craik FIM, Salthouse TA (eds) The Handbook of Aging and Cognition, 2nd Ed. Mahwah, NJ: Erlbaum, pp. 221–292.
McFarland DJ, Cacace AT (1995) Modality specificity as a criterion for diagnosing central auditory processing disorders. Am J Audiol 4:36–48.
Miller GA, Heise GA, Lichten W (1951) The intelligibility of speech as a function of the context of the text materials. J Exp Psychol 41:329–335.
Nilsson M, Soli S, Sullivan JA (1994) Development of the Hearing In Noise Test for the measurement of speech reception thresholds in quiet and in noise. J Acoust Soc Am 94:1085–1099.
Pavlovic CV (1984) Use of the articulation index for assessing residual auditory function in listeners with sensorineural hearing impairment. J Acoust Soc Am 75:1253–1258.
Pavlovic CV, Studebaker GA, Sherbecoe RL (1986) An articulation index based procedure for predicting the speech recognition performance of hearing-impaired individuals. J Acoust Soc Am 80:50–57.
Pichora-Fuller MK (2003) Cognitive aging and auditory information processing. Int J Audiol 42, Suppl 2:S26-S32.
Pichora-Fuller MK, Singh G (2006) Effects of age on auditory and cognitive processing: implications for hearing aid fitting and audiological rehabilitation. Trends Amplif 10:29–59.
Pichora-Fuller MK, Schneider BA, Daneman M (1995). How young and old listen to and remember speech in noise. J Acoust Soc Am 97:593–608.
Plomp R (1978) Auditory handicap of hearing impairment and the limited benefit of hearing aids. J Acoust Soc Am 63:533–549.
Plomp R (1986) A signal-to-noise ratio model for the speech-reception threshold of the hearing impaired. J Speech Hear Res 29:146–154.
Plomp R, Mimpen AM (1979a) Improving the reliability of testing the speech reception threshold for sentences. Audiology 18:43–52.
Plomp R, Mimpen AM (1979b) Speech-reception threshold for sentences as a function of age and noise level. J Acoust Soc Am 66:1333–1342.

Portet F, Ousset PJ, Visser PJ, Frisoni GB, Nobili F, Scheltens P, Vellas B, Touchon J (2006) Mild cognitive impairment (MCI) in medical practice: a critical review of the concept and new diagnostic procedure. Report of the MCI Working Group of the European Consortium on Alzheimer's Disease. J Neurol Neurosurg Psychiatry 77:714–718.
Rabbitt P (1968) Channel capacity, intelligibility and immediate memory. Q J Exp Psychol 20:241–248.
Rhebergen KS, Versfeld NJ (2005) A Speech Intelligibility Index-based approach to predict the speech reception threshold for sentences in fluctuating noise for normal-hearing listeners. J Acoust Soc Am 117:2181–2192.
Rhebergen KS, Versfeld NJ, Dreschler WA (2006) Extended speech intelligibility index for the prediction of the speech reception threshold in fluctuating noise. J Acoust Soc Am 106:3988–3997.
Ricketts TA, Dittberner AB (2002) Directional amplification for improved signal-to-noise ratio: strategies, measurements, and limitations. In: Valente M (ed) Hearing Aids: Standards, Options, and Limitations, 2nd Ed. New York: Thieme, pp. 274–346.
Rogers WA (2000) Attention and aging. In: Park DC, Schwarz N (eds) Cognitive Aging: A Primer. Philadelphia, PA: Psychology Press, pp. 57–73.
Roup C, Wiley T, Wilson R (2006) Dichotic word recognition in young and older adults. J Am Acad Audiol 17:230–240.
Salthouse TA (1985) A Theory of Cognitive Aging. Amsterdam: North-Holland.
Salthouse TA (1991) Theoretical Perspectives on Cognitive Aging. Hillsdale, NJ: Lawrence Erlbaum Associates.
Salthouse TA (2000) Aging and measures of processing speed. Biol Psychol 54:35–54.
Schneider BA, Pichora-Fuller MK, Kowalchuk D, Lamb M (1994) Gap detection and the precedence effect in young and old adults. J Acoust Soc Am 95:980–991.
Schneider BA, Pichora-Fuller MK (2000) Implications of perceptual processing for cognitive aging research. In: Craik FIM, Salthouse TA (eds) The Handbook of Aging and Cognition, 2nd Ed. New York: Lawrence Erlbaum Associates, pp. 155–220.
Schneider BA, Daneman M, Murphy DR (2005) Speech comprehension difficulties in older adults: cognitive slowing or age-related changes in hearing? Psychol Aging 20:261–271.
Seewald RC, Ramji KV, Sinclair ST, Moodie KS, Jamieson DG (1993) Computer-assisted implementation of the Desired Sensation Level method for electroacoustic selection and fitting in children: version 3.1. User's Manual. The University of Western Ontario, London, Ontario, Canada.
Seewald RC, Moodie S, Scollie S, Bagatto M (2005) The DSL method for pediatric hearing instrument fitting: historical perspective and current issues. Trends Amplif 9:145–157.
Shanks JE, Wilson RH, Larson V, Williams D (2002) Speech recognition performance of patients with sensorineural hearing loss under unaided and aided conditions using linear and compression hearing aids. Ear Hear 23:280–290.
Skafte MD (2000) The 1999 hearing instrument market—the dispenser's perspective. Hear Rev 7:8–40.
Smoorenburg GF (1992) Speech reception in quiet and in noisy conditions by individuals with noise-induced hearing loss in relation to their tone audiogram. J Acoust Soc Am 91:421–437.
Snell KB, Frisina DR (2000) Relationships among age-related differences in gap detection and word recognition. J Acoust Soc Am 107:1615–1626.
Souza PE, Turner CW (1994) Masking of speech in young and elderly listeners with hearing loss. J Speech Hear Res 37:655–661.
Souza PE, Turner CW (1999) Quantifying the contribution of audibility to recognition of compression-amplified speech. Ear Hear 20:12–20.
Steeneken HJM, Houtgast T (1980) A physical method for measuring speech-transmission quality. J Acoust Soc Am 67:318–326.
Stelmachowicz PG, Pittman AL, Hoover BM, Lewis DE (2002) Aided perception of /s/ and /z/ by hearing-impaired children. Ear Hear 23:316–324.
Strom KE (2006) The HR 2006 dispenser survey. Hear Rev 13(6):13–39.
Strouse A, Ashmead DH, Ohde RN, Granthan DW (1998) Temporal processing in the aging auditory system. J Acoust Soc Am 104:2385–2399.

Studebaker GA, Sherbecoe RL, McDaniel DM, Gwaltney CA (1999) Monosyllabic word recognition at higher-than-normal speech and noise levels. J Acoust Soc Am 105:2431–2444.
Summers V, Cord MT (2007) Intelligibility of speech in noise at high presentation levels: effects of hearing loss and frequency region. J Acoust Soc Am 122:1130–1137.
Surprenant AM (2007) Effects of noise on identification and serial recall of nonsense syllables in older and younger adults. Neuropsychol Dev Cogn B Aging Neuropsychol Cogn 14:126–143.
Surr RK, Walden BE, Cord MT, Olsen L (2002) Influence of environmental factors on hearing aid microphone preference. J Am Acad Audiol 13:308–322.
Sweetow RW, Sabes JH (2006) The need for and development of an adaptive listening and communication enhancement (LACE) program. J Am Acad Audiol 17:538–558.
Valente M, Fabry DA, Potts LG (1995). Recognition of speech in noise with hearing aids using dual microphones. J Am Acad Audiol 6:440–449.
van Buuren RA, Festen JM, Plomp R (1995) Evaluation of a wide range of amplitude-frequency responses for the hearing impaired. J Speech Hear Res 38:211–221.
Van den Bogaert T, Klasen TJ, Moonen M, Van Deun L, Wouters J (2006) Horizontal localization with bilateral hearing aids: without is better than with. J Acoust Soc Am 119:515–526.
van Rooij JCGM, Plomp R (1990) Auditive and cognitive factors in speech perception by elderly listeners. II. Multivariate analyses. J Acoust Soc Am 88:2611–2624.
van Rooij JCGM, Plomp R (1992) Auditive and cognitive factors in speech perception by elderly listeners. III. Additional data and final discussion. J Acoust Soc Am 91:1028–1033.
van Rooij JCGM, Plomp R, Orlebeke JF (1989) Auditive and cognitive factors in speech perception by elderly listeners. I. Development of test battery. J Acoust Soc Am 86:1294–1309.
Van Tasell DJ, Yanz JL (1987) Speech recognition threshold in noise: effects of hearing loss, frequency response, and speech materials. J Speech Hear Res 30:377–386.
Verhaeghen P, De Meersman L (1998a) Aging and the negative priming effect: a meta-analysis. Psychol Aging 13:1–9.
Verhaeghen P, De Meersman L (1998b) Aging and the Stroop effect: a meta-analysis. Psychol Aging 13:120–126.
Versfeld NJ, Dreschler WA (2002) The relationship between the intelligibility of time-compressed speech and speech in noise in young and elderly listeners. J Acoust Soc Am 111:401–408.
Walden BE, Erdman S, Montgomery A, Schwartz D, Prosek R (1981) Some effects of training on speech recognition by hearing-impaired adults. J Speech Hear Res 24:207–216.
Walden BE, Surr RK, Cord MT, Edwards B, Olson L. (2000) Comparison of benefits provided by different hearing aid technologies. J Am Acad Audiol 11:540–560.
Wiley TL, Cruickshanks KJ, Nondahl DM, Tweed TS, Klein R, Klein BEK (1998) Aging and word recognition in competing message. J Am Acad Audiol 9:191–198.
Wingfield A (1996) Cognitive factors in auditory performance: context, speed of processing, and constraints on memory. J Am Acad Audiol 7:175–182.
Wingfield A, Tun PA (2001) Spoken language comprehension in older adults: interactions between sensory and cognitive change in normal aging. Semin Hear 22:287–301.
Wingfield A, Poon LW, Lombardi L, Lowe D (1985) Speed of processing in normal aging: effects of speech rate, linguistic structure, and processing time. J Gerontol 40:579–585.
Wingfield A, Tun PA, Koh CK, Rosen MJ (1999) Regaining lost time: adult aging and the effect of time restoration on recall of time-compressed speech. Psychol Aging 14:380–389.
Woods WS, Van Tasell DJ, Rickert ME, Trine TD (2006) SII and fit-to-target analysis of compression system performance as a function of number of compression channels. Int J Audiol 45:630–644.
Zekveld AA, George ELJ, Kramer SE, Goverts ST, Houtgast T (2007) The development of the Text Reception Threshold test: a visual analogue of the Speech Reception Threshold test. J Speech Lang Hear Res 50:576–584.
Zurek PM (1993) Binaural advantages and directional effects in speech intelligibility. In: Studebaker GA, Hochberg I (eds) Acoustical Factors Affecting Hearing Aid Performance, 2nd Ed. Needham Heights, MA: Allyn & Bacon, pp. 255–276.

Studebaker GA, Sherbecoe RL, McDaniel DM, Gwaltney CA (1999) Monosyllabic word recognition at higher than normal speech and noise levels. J Acoust Soc Am [illegible]

[illegible]

Chapter 9
Epidemiology of Age-Related Hearing Impairment

Karen J. Cruickshanks, Weihai Zhan, and Wenjun Zhong

9.1 Introduction

Hearing loss in older adults is an important problem affecting communication and quality of life, yet few epidemiological studies have been conducted to evaluate the magnitude of the problem, identify modifiable factors contributing to the risk of hearing impairment, or test potential prevention strategies. The purpose of this chapter is to summarize the available epidemiological data about the prevalence and incidence of age-related hearing loss (ARHL) and its risk factors. Although prevalent cases of hearing impairment in older adults may represent a wide range of causes or disorders, prospective studies of older adults that can identify new cases occurring at older ages offer great promise for identifying modifiable factors for ARHL. Longitudinal epidemiological studies can identify ways to reduce the burden of hearing loss among older adults.

Although most hearing scientists use the term ARHL to imply hearing loss caused by aging, it is likely that hearing loss that develops in humans at older ages has a multifactorial etiology and does not represent an inevitable consequence of getting older. Dementia, cardiovascular disease, and even prostate cancer once were viewed as normal aging processes but now are considered preventable disorders; it is likely that one day ARHL also will be accepted as a preventable disorder. In this chapter, the evidence that ARHL is a potentially preventable health problem is reviewed.

K.J. Cruickshanks (✉)
Department of Ophthalmology and Visual Sciences,
University of Wisconsin, School of Medicine and Public Health,
and
Department of Population Health Science, University of Wisconsin,
School of Medicine and Public Health
e-mail: cruickshanks@episense.wisc.edu

W. Zhan and W. Zhong
Department of Ophthalmology and Visual Sciences,
University of Wisconsin, School of Medicine and Public Health
e.mail: wzhan@wisc.edu; wzhong@wisc.edu

S. Gordon-Salant et al. (eds.), *The Aging Auditory System*,
Springer Handbook of Auditory Research 34,
DOI 10.1007/978-1-4419-0993-0_9,

9.2 Classification/Definition of ARHL

The first challenge for reviewing the epidemiology of age-related hearing impairment is the lack of a standardized definition of hearing impairment. Some studies used self-reported hearing loss, assessed through a variety of questionnaires, although it has been shown that self-report results in significant misclassification because many people with measurable impairments do not report hearing problems (Ries 1994; Nondahl 1998). In addition, the variety of questions used further hampers comparisons of the prevalence across study populations.

Studies relying on audiometric measures also differed in the definitions of hearing impairment used because of averaging different combinations of frequencies or ears to arrive at a person-level measure or employing different criteria for the magnitude of the threshold or pure-tone average required to be classified as hearing impaired (Table 9.1). Some studies used automated audiometry, but most studies employed conventional manual bracketing procedures for threshold determinations; some studies included bone conduction testing but most did not. Differences in equipment (including transducers), frequency of calibration, training and certification of testers,

Table 9.1 Selected Studies of the Prevalence of Hearing Impairment.

Author/ Study	Sample	Hearing Loss Definition	Prevalence
Ries 1985 National Health and Nutrition Examination Survey 1974-1975, USA	Nationwide probability sample, *n* = 6,805, 25-74 years	Better ear $PTA_{0.5,1,2} \geq$ 26-dB HL	8%
Moscicki et al. 1985 Framingham Heart Study 1978-1979, MA, USA	*n* = 2,293, 57-89 years.	Better ear $PTA_{0.5,1,2,4} >$ 25-dB HL	47%
Davis 1989 British National Study on Hearing 1980-1986, UK	*n* = 2,708, 17-80 years	$PTA_{0.5,1,2,4} \geq$ 25-dB HL	16.1% better ear 26.1% worse ear
Cruickshanks et al. 1998b Epidemiology of Hearing Loss Study 1993-1995, WI, USA	Population-based, *n* = 3,753, 48-92 years	Either ear $PTA_{0.5,1,2,4}$ >25-dB HL	45.9%
Borchgrevink et al. 2005 Nord-Trøndelag (NT) Norway Audiometric Survey 1996-1998, NT, Norway	n = 50,723, 20-101 years	$PTA_{0.5,1,2,4}$ ≥25-dB HL	18.8% better ear 27.2% worse ear
Chia et al. 2007 Blue Mountains Hearing Study 1997-2000, Sydney, Australia	Population-based, n = 2,431, 49+ years	$PTA_{0.5,1,2,4}$ >25-dB HL.	Unilateral: 13.3% Bilateral: 31.3% Combined: 44.6%
Agrawal et al. 2008 National Health and Nutrition Examination Survey 1999-2004, USA	Nationwide probability sample, n = 5,742, 20-69 years	Either ear $PTA_{0.5,1,2,4}$ ≥25-dB HL	15.7%

PTA, pure-tone average; HL, hearing level.

and other quality assurance methods that may affect measurement variability complicate the task of comparing results.

Most commonly in epidemiological studies, a pure-tone average of thresholds at 500, 1,000, 2,000, and 4,000 Hz that is >25-dB hearing level (HL) has been used to define hearing impairment, with studies differing between usage of either (worse) ear or better ear approaches. Defining hearing impairment as a hearing loss affecting one or both ears is consistent with definitions in other chronic disorders affecting paired organs such as lung, kidney, or breast cancer, where the interest is in whether the person is affected or not. Although functionally the person may derive benefit from the remaining organ, reducing the handicap associated with the condition, the disorder is present once a single organ is affected. In age-related hearing impairment, most losses are symmetrical and affect both ears, but due to measurement error and the use of cut-points, they may appear to be unilateral at one point in time. Discounting unilateral losses (both age related and from other causes such as trauma) will result in lower prevalence estimates of hearing impairment in populations. Recent studies in patients receiving cochlear implants reinforce the importance of two-ear hearing for optimal function (Brown and Balkany 2007).

Audiogram shape has been suggested as a way to subclassify hearing impairments into distinct etiologic pathways such as strial or metabolic, neural, or sensory, but these systems have not been validated by population level data (Schuknecht 1964) (see Schmiedt, Chapter 2, for a further discussion of these forms of presbycusis and their associated audiograms). Indeed, in one large population-based cohort study of >3,900 older adults with a high prevalence of hearing impairment, only one participant had a flat hearing impairment proposed by Schuknecht (1964) to suggest strial presbycusis (Cruickshanks, unpublished data). In Chapter 2, Schmiedt suggested that strial changes in the cochlea may produce a sloping hearing loss pattern, the more typical pattern of hearing loss seen in population-based studies of older adults (Cruickshanks et al. 1998b). Threshold testing with bone and air conduction may help to distinguish conductive from sensorineural hearing losses, but even then there is variation in the definition of a conductive hearing loss. Few longitudinal studies of hearing in adults have been conducted so the natural history of small air-bone gaps at individual frequencies is unknown. Although audiometric thresholds provide a reliable and repeatable measure of hearing sensitivity, there remains a need for standardized definitions of hearing impairment to facilitate comparisons across populations and the search for etiologic pathways. For the purposes of this review, studies using audiometric measures are included when available, and when possible, the definition of ARHL employed is noted.

9.3 Evaluating the Evidence for Causal Associations: Identifying Risk Factors

In the field of hearing science, many advances in understanding human age-related auditory pathology have been derived from studies using animal models because of the difficulty of studying intact auditory systems in humans and ethical limitations for experimental designs in humans (see Schmiedt, Chapter 2). In addition, human pathology studies and small cross-sectional studies of humans have contributed to ideas about key mechanisms of age-related changes. These important lines of evidence generate hypotheses that should be tested through longitudinal studies in humans using analytical observational or randomized controlled trial designs. Epidemiological data are important components of the evidence considered by systematic systems evaluating causal factors. As reviewed by Evans (1995), many current guidelines have their origins in the Henle-Koch postulates, a systematic approach for determining if a virus or strain of bacteria was the cause of an infectious disease. Although many variations exist, these criteria for causal associations are frequently summarized as consistency, strength, specificity, temporal relationship, and coherence of the association (Table 9.2) (Evans 1995). Combined with systems weighing the quality of the studies including strengths and weaknesses of the design, measurement methods, analytic approach, attempts to control for potential

Table 9.2 Criteria for Causation.

Factor	Definition*	Comments
Consistency	Close conformity between findings in different samples, strata, or populations; at different times; in different circumstances; or in studies conducted by different methods or different investigators	Replication suggests that the finding is not due to chance, inherent limitations of one study design, or selection factors. Animal models and small experimental human studies contribute, but stronger evidence is from population-based prospective analytical studies and randomized, controlled trials.
Strength	The size of the risk as measured by appropriate statistical tests.	Although larger effects may be more stable and reproducible, small effects may be clinically important. Greater precision of the estimates is reflected in small confidence intervals. Evidence of a dose-response relationship adds to the evidence.
Specificity	A single [factor] produces a specific effect.	This is often not the case in chronic diseases. For example, cigarette smoking does not just cause lung cancer but cardiovascular disease as well.
Temporal Relationship	Exposure precedes the outcome. This is the only absolutely essential criterion.	Cross-sectional studies are not sufficient because the onset of disease can alter exposures and biomarkers. Prospective designs are required.
Coherence	The association should be compatible with existing theory and knowledge.	A biologically plausible explanation for the association based on clinical experience and current scientific knowledge.

Adapted from Evans 1995. *Adapted from Last 2001.

confounders, sample selection, and completeness of follow-up, these guidelines have been used widely in epidemiology and health policy decision-making. However, audiological research has rarely incorporated epidemiological methodologies, and few epidemiologists have engaged in hearing research, leading to significant gaps in the body of evidence critical for determining causal pathways in humans and developing prevention strategies. For the purposes of this chapter, evidence for risk factors from epidemiological studies in humans is reviewed because the evidence from animal models and small cross-sectional studies are discussed in other chapters.

9.4 Descriptive Epidemiology

9.4.1 Prevalence

A number of epidemiological studies of the prevalence of ARHL have been conducted (Table 9.1) using a pure-tone average (PTA)-based definition of ARHL, although cut-points and other details may vary. Hearing was tested as part of the 1974-75 National Health and Nutrition Examination Survey (NHANES), a nationally representative sample of the United States population aged 25-74 years (Ries 1985). In this study, 29% of adults aged 65-74 years and 8% of adults aged 25-74 years had hearing impairments in the better ear. In the Framingham Heart Study, hearing was first measured in 1978-79; the prevalence of ARHL among adults aged 57-89 years was 47% based on PTA from 0.5 to 4 kHz >25-dB HL (Moscicki et al. 1985). In the population-based Epidemiology of Hearing Loss Study (EHLS) of adults aged 48-92 years, the prevalence of hearing impairment was 45.9% (Cruickshanks et al. 1998b). The Australian Blue Mountains Hearing Study was similar to the EHLS in study design and measurement methods and reported a prevalence of 44.6% among adults aged 49 years or older (Chia et al. 2007). Recent reports from NHANES 1999-2004 indicated that the prevalence of hearing impairment among adults 60-69 years was 49% (Agrawal et al. 2008). These studies demonstrate that ARHL is a common disorder among older adults. Direct comparisons of geographic and temporal variations are difficult because of the methodological differences, including differences in the distributions of age and gender, but the studies consistently report a high prevalence of ARHL among older adults.

9.4.2 Incidence

Few studies have followed cohorts over time to determine the incidence of ARHL. Prevalence studies, as summarized in Section 4.1, count the number of people with a disorder at a point in time. Incidence studies require studying people who are initially healthy to determine the incidence rate or number of new events that develop in a given population at risk during a specified period of time. These prospective studies are used to measure the risk of developing a health problem and identify factors that predict the onset of that disorder.

The Framingham Heart Study tested hearing in the same participants six years later (Gates and Cooper 1991). They reported the incidence of hearing impairment was 13.7% in left ears and 8.4% in right ears but did not provide person-level estimates. A short-term pilot study in Great Britain and Denmark reported a two-year incidence rate of 12% (Davis et al.1991). The Baltimore Longitudinal Study of Aging (BLSA) measured hearing repeatedly in men without a history of noise exposure or middle ear disease (Brant et al. 1996). Among men aged 60-69 years at baseline, 13% developed ARHL, with a mean follow-up time of 7.8 years. The incidence was slightly higher (17%) among older men (aged 70+ years). In the EHLS, among adults aged 48-92 years at the baseline, the 5-year incidence of hearing impairment was 21% and the cumulative 10-year incidence was 37.2% (Cruickshanks et al. 2003). All of these studies primarily studied non-Hispanic white (NHW) subjects. There are no reliable estimates of the incidence of ARHL in African American, Latino, or other racial/ethnic groups. Nonetheless, the incidence of hearing impairment among older NHW adults in each study was high. Using the EHLS data and projecting to the US population, it is estimated that there will be 69 million adults aged 45 years or older with hearing impairment by 2030. Because <15% of people with hearing impairment are currently treated, this figure represents a large unmet future need for hearing health care; 58 million of these individuals would likely be untreated if current trends continue (Popelka et al.1998; Cruickshanks et al. 2003).

9.4.3 Age, Gender, Ethnicity/Race, Geographic Location, and Temporal Trends

It is well established that age and male gender are associated with the prevalence of ARHL. In the EHLS, the risk of developing ARHL increased with age (the odds of having ARHL were 81% higher for every 5 years of age) and men were more than twice as likely to have ARHL as women (Cruickshanks et al. 2003). Figure 9.1 illustrates the 5-year incidence of ARHL in the EHLS by age at the baseline examination and gender. The incidence was higher for men than women in each age group, although the gender difference was small among the oldest group, perhaps due to the small sample sizes in the oldest age group (22 adults aged 80-92 years without hearing impairment at baseline).

As noted above, there are no incidence data for minority populations. However, the NHANES data suggest that the prevalence of ARHL is lower among black non-Hispanic participants than NHW subjects (8.2 vs. 18%, aged 20-69 years) (Agrawal et al. 2008). The Health, Aging, and Body Composition Study (Health ABC) found ARHL was more common among white than black participants (Helzner et al. 2005).

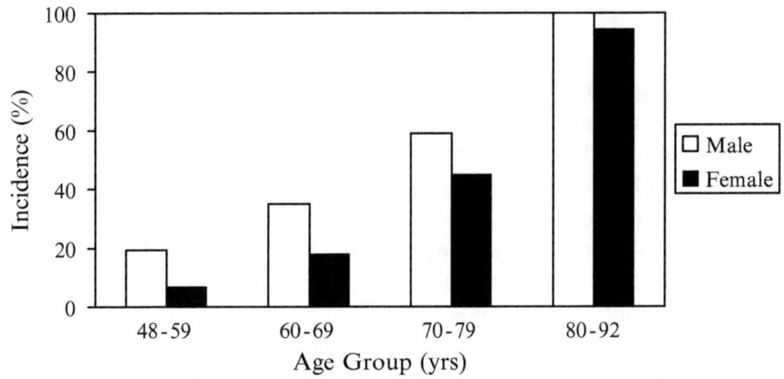

Fig. 9.1. The 5-year incidence of age-related hearing loss by age and gender: The Epidemiology of Hearing Loss Study, Beaver Dam, WI. (From data published in table form in Cruickshanks et al. 2003.)

In the Hispanic Health and Nutrition Examination Survey, the prevalence of hearing impairment appeared to differ by Latino subgroups (Lee et al. 1991). The ongoing Hispanic Community Health Study of 16,000 Latinos from different origins (http://www.cscc.unc.edu/hchs/about.php) and an ongoing study of hearing in African-American participants in the Jackson Heart Study will provide important information about the prevalence of hearing impairments among minority groups, but incidence data are not likely to be available in the near future.

Early epidemiological studies by Rosen et al. (1962) demonstrated that hearing was better among Africans living in the Sudan than expected based on hearing loss patterns in industrial societies (also Rosen and Olin 1965). In this study, older African men thought to be 60-69 years of age had median hearing thresholds <20-dB HL through 6 kHz. In contrast, NHW men of similar ages participating in the Framingham Heart Study and the EHLS had average hearing thresholds >25-dB HL at 2 kHz and above (Gates et al. 1990; Cruickshanks et al. 1998b).

Doll and Peto (1981) developed an important model for evidence that a disorder has environmental or modifiable causes. They suggested that disorders are likely to be preventable when (1) the incidence is different between countries or geographic regions (geographic variation), (2) some groups of people are more or less likely to have the disorder than others, (3) rates of disease change over time (temporal trends), and (4) risk changes as people move from areas with low rates to areas with high rates or vice versa (rates change with migration). There have been too few longitudinal studies with comparable methods and populations to permit conclusions about the patterns of ARHL by race, ethnicity, and geographic location and there are no published data on temporal trends or migration effects. Fundamental epidemiological evidence is still needed to understand the patterns of ARHL across countries, geographic regions, race/ethnicity, and time to help identify modifiable factors contributing to the etiology of ARHL. These types of data have been useful in understanding cancer and hold great promise for reducing the burden of hearing impairment in aging populations (Doll and Peto 1981).

9.5 Risk Factors

There is a growing body of evidence from cross-sectional epidemiological studies of ARHL that modifiable factors are associated with the prevalence of ARHL. However, longitudinal data are needed because cross-sectional data cannot establish that the exposures to potential risk factors preceded the development of the disorder (temporal sequence; see Table 9.2). Identifying etiologic factors may require studying younger adults for many years because the changes in the auditory system that contribute to ARHL may occur over a long period of time. ARHL may develop slowly through pathological changes affecting multiple components of the auditory system (Willott 1991; Jennings and Jones 2001; Schmiedt, Chapter 2; Canlon, Illing and Walton, Chapter 3). Aging changes occur primarily within the cochlea and may affect the central auditory system as described by Schmiedt (Chapter 2) and Canlon, Illing and Walton (Chapter 3). Although age-related changes in cellular function (i.e., oxidative stress) may lead to decrements in hearing function, it is also likely that other factors such as genes, trauma (e.g., noise, vibration), direct toxic effects of medications and smoking, and ischemia contribute to auditory dysfunction in aging (Van Eyken et al. 2007a).

9.5.1 Genetic Factors

Epidemiological studies have found that ARHL clusters within families, with estimates of the heritability ranging from 25 to 75% (Karlsson et al. 1997; Gates et al. 1999; Christensen et al. 2001; Viljanen et al. 2007). Some of the variability in these estimates may be due to differences in the age and gender distributions of the subjects studied. Although these data suggest that ARHL is partially a genetic disorder, the genes involved are not known. Many studies of nonsyndromic forms of hearing loss have been completed, but there have been few genomewide linkage studies of ARHL. The Framingham Study found several regions suggestive of linkage: chromosomes 10q, 11q13.5, and 18q (DeStefano et al. 2003). Huyghe et al. (2008) detected significant linkage on chromosome 8 in a study of 200 sibships. A multicenter European study found only one gene (grainyhead-like 2 [*GRHL2*]) significantly associated with ARHL (Van Laer et al. 2008). Candidate gene associations studies of ARHL have also reported positive associations for *N*-acetyltransferase 2 (*NAT2*), glutathione S-transferase M1 (*GSTM1*), glutathione S-transferase T1 (*GSTT1*), and apolipoprotein E (*APOE*) (Unal et al. 2005; O'Grady et al. 2007; Van Eyken et al. 2007b; Van Laer et al. 2008).

In the only genomewide association study published to date, Huyghe et al. (2008) tested 169,154 single-nucleotide polymorphisms (SNPs) and found no variants that reached genomewide significance for hearing impairment. Thus although it is likely genetic factors contribute significantly to the risk of ARHL, the genes responsible are unknown. Large-scale collaborative studies are needed to identify the genetic markers for ARHL.

9.5.2 Cardiovascular Disease and Cardiovascular Disease Risk Factors

Rosen suggested that the better hearing he observed in Mabaan men living in the Sudan region of Africa might be due to less noise exposure and healthier lifestyles (Rosen et al. 1962; Rosen and Olin 1965). These people had little heart disease, hypertension, or diabetes; consumed diets high in fruits and fiber; did not smoke; and were very physically active. Results from recent epidemiological studies have been conflicting. In Framingham and the Health ABC Study, cardiovascular disease (CVD) and hearing loss were cross-sectionally associated (Gates et al. 1993; Helzner et al. 2005). There was no association between blood pressure and hearing loss. The BLSA found systolic blood pressure was associated with the incidence of hearing loss in men (Brant et al. 1996). Some population-level studies have shown that the prevalence of hearing loss is higher in areas with high rates of CVD compared with areas with low rates of CVD (Rosen 1966; Rosen et al. 1970).

Results from some animal and histopathological studies have demonstrated that atherosclerosis can affect blood vessels supplying the inner ear and that a reduction in blood supply may lead to degeneration of hair cells and other components of the auditory system (Johnsson and Hawkins 1972; Willott 1991; Jennings and Jones 2001). A case-control study of women found the odds of having a hearing impairment were eightfold higher among the cases with ischemic heart disease compared with the control subjects with normal coronary arteries (Susmano and Rosenbush 1988). In the EHLS, history of CVD was associated with prevalence of ARHL and with cochlear function in women but not men (Cruickshanks et al. 1996; Torre et al. 2005).

Cigarette smoking was not associated with prevalent ARHL in the Framingham Heart Study or the BLSA (Gates et al. 1993; Brant et al. 1996). In the EHLS, current smokers were more likely to have a hearing impairment than nonsmokers (Cruickshanks et al. 1998a). A recent multicenter study in Europe and the NHANES reported associations between smoking and ARHL (Agrawal et al. 2008; Fransen et al. 2008). In the Health ABC Study, the odds of hearing impairment were higher among smokers than nonsmokers (Helzner et al. 2005). These findings are consistent with earlier clinical studies, but prospective data are needed to confirm them.

In the EHLS, moderate alcohol consumption also was associated with ARHL and was recently reported in the European study, providing weak evidence of a protective effect (Popelka et al. 2000; Fransen et al. 2008). Similarly, socioeconomic status (income or education), a known CVD risk factor, also has been associated with ARHL in several studies, with higher socioeconomic status having a modest protective effect (Cruickshanks et al. 2003; Helzner et al. 2005; Agrawal et al. 2008). Participants in the EHLS with less than a high school education were more than twice as likely to develop a hearing impairment as participants with a college education (Cruickshanks et al. 2003).

Diabetes and CVD share many risk factors, and Type 2 diabetes has been found to be associated with ARHL in the EHLS, adjusting for age (Dalton et al. 1998). Diabetes also was associated with hearing impairment in the Health ABC Study and, more recently, in the NHANES (Helzner et al. 2005; Bainbridge et al. 2008). These studies suggest that diabetes and ARHL may share common risk factors or similar underlying etiologic pathways.

Although strong evidence is not available because few longitudinal studies have been conducted, taken together, these studies suggest that cardiovascular risk factors may play important roles in ARHL.

9.5.3 Noise

It is well known that occupational exposure to loud noise can damage hearing over many years of exposure. However, the role of noise exposure in the etiology of ARHL is unclear. The onset of ARHL is most frequent over age 60 when most occupational exposure to noise has ceased due to retirement. In the EHLS, leisure noise exposure to loud hobbies such as woodworking was associated with the prevalence of ARHL (Dalton et al. 2001). In addition, occupation, a marker of socioeconomic status, as well as potential noise exposure was associated with ARHL (Cruickshanks et al. 1998b). The 5-year incidence of ARHL was almost twice as high among people employed in production, operations, and fabrications than people employed in management and professional occupations (Cruickshanks et al. 2003). Understanding the contribution of noise exposure to ARHL will be difficult given the challenges of measuring past and current noise exposure and disentangling the potential confounding effects of socioeconomic status, lifestyle, and health factors.

9.5.4 Solvents and Other Environmental Chemical Exposures

There is good evidence from animal models that some chemicals, particularly solvents such as toluene and styrene, may damage the auditory pathway (Fuente and McPherson 2006; Morata and Campo 2002). In humans, the best evidence comes from cross-sectional studies in occupationally exposed middle-aged workers (Morata et al. 2002; Moshe et al. 2002; Sliwinska-Kowalska et al. 2003). These studies have been limited by small samples sizes, analyses inadequately controlling for other potential confounding factors, methodological challenges in measuring exposures, abbreviated audiometric threshold testing, differences in definitions of hearing impairment, and their cross-sectional designs.

The Copenhagen Male Study, an epidemiological study of 3,284 men employed in 14 private or public companies in 1971 evaluated the relationship between self-reported occupational exposure to solvents and self-reported hearing loss (Jacobsen et al. 1993). Solvent exposure was associated with self-reported hearing problems in models adjusting for noise and family history (Jacobsen et al. 1993). A study of two birth cohorts from Göteborg, Sweden, found no correlation between reported exposure to industrial chemicals and hearing thresholds at ages 70-85 years, but the numbers of exposed individuals were very small, thereby limiting the power to detect an association (Rosenhall et al. 1993). One paper from the Second National Health and Nutrition Examination Survey and one from the Hispanic Health and Nutrition Survey evaluated the association between blood lead levels and hearing thresholds (Schwartz and Otto 1987; 1991). Both studies reported higher hearing thresholds among people with higher blood lead levels.

There are few, if any, large studies evaluating chemical exposures and the development of hearing impairment in older adults. Because there is good evidence of the potential hazard, measures of solvent and chemical exposures such as mercury should be included in future epidemiological studies to measure the relative contribution of such exposures over the lifetime to hearing impairments. Low-dose exposures are difficult to measure because they may accrue in many occupational settings; during hobby and recreational activities; through exposure to ordinary household products, paints, adhesives, and other building materials; or through water and food sources such as fish contaminated with mercury or other chemicals. Approaches developed for environmental epidemiological studies of indoor air pollution from volatile organic solvents and other emissions as well as studies of contaminated fish and water could be adapted to begin to address this important question (Turyk et al. 2006; Lin et al. 2008).

9.5.5 Radiation and Medications

There is a wealth of clinical studies documenting the ototoxic effects of radiation and chemotherapy, particularly with cisplatin-C (Van Eyken et al. 2007a). However, there are no population-based estimates of their importance. Few participants in population-based studies would have these exposures due to the low prevalence of cancer survivors, making precise estimates of the hearing-health effects difficult to obtain and suggesting that these exposures do not contribute substantially to the overall prevalence of ARHL. Whether lower-dose radiation such as exposure received during dental exams or head imaging studies is associated with age-related hearing impairment remains unknown.

Older adults often take numerous mediations to prevent or treat other health conditions, making medication effects on hearing an important issue for epidemiological studies of hearing. To date, few studies have included data on medication usage. Several small cross-sectional studies have found differences in hearing associated with the use of salicylates (better hearing thresholds among users compared with nonusers), antihistamines, β-adrenergic agents, and calcium-channel blockers, although results have been inconsistent (Lee et al. 1998; Mills et al. 1999; Van Eyken et al. 2007a). The Health ABC Study reported a protective effect of salicylates (Helzner et al. 2005).

9.5.6 *Vitamins B_{12} and D and Folate*

Previously, studies have found suggestive evidence that people with hearing loss had lower serum levels of vitamin B_{12}, vitamin D, and/or folate than people with normal hearing (Houston et al. 1999; Berner et al. 2000; Durga et al. 2006). Recently, results from a randomized controlled trial of long-term folic supplementation in 728 adults ages 50-70 years of age were reported (Durga et al. 2007). Participants receiving 800 μg/day of folic acid had smaller declines in hearing ($PTA_{0.5-2}$, both ears) during a 3-year follow-up period than subjects randomized to placebo (1- vs. 1.7-dB HL, respectively) (Durga et al. 2007). Numerous selection criteria may limit the generalizablity of the study results and the study was conducted in a country where foods are not fortified with folate, further limiting the findings. Nevertheless, these results, if replicated, are exciting because they suggest that it may be possible to reduce the deterioration of hearing in aging populations.

9.6 Summary

Epidemiological studies have demonstrated that ARHL is a common health problem for older adults. The EHLS data suggest that ~1 in 25 older adults develops a hearing impairment each year and that 1 in 10 people with hearing impairment experience deterioration in their hearing. Doll and Peto (1981) proposed that differences in incidence between geographic areas, changes in incidence with migration, changes in incidence over time, and identifying risk factors provide strong evidence that a disease is preventable. The lack of prospective data for ARHL severely limits conclusions that can be drawn about its etiology and hampers identifying effective preventive strategies. The data summarized in this chapter do not meet the established standards for causality (Table 9.2). Nonetheless, the limited epidemiological data available, combined with smaller clinical and animal studies, suggest that ARHL is a preventable disorder. Rosen's early work suggests that risk may vary by geographic location and it may be possible to lower rates of ARHL in developed nations to those experienced in other regions (Rosen et al. 1962; Rosen and Olin 1965; Rosen 1966; Rosen et al. 1970). Many epidemiological studies have identified factors associated with prevalent cases of ARHL, although incidence studies are needed.

Multidisciplinary research on the epidemiology of ARHL has the potential to provide important insights into the causes of ARHL. If vascular health or cardiovascular risk factors contribute to the risk of ARHL, medications and preventive strategies meant to reduce the risk of CVD may provide ancillary auditory protective benefits and help reduce the burden of hearing impairment for the aging baby boomer generation.

References

Agrawal Y, Platz EA, Niparko JK (2008) Prevalence of hearing loss and differences by demographic characteristics among US adults. Arch Intern Med 168:1522–1530.

Bainbridge KE, Hoffmann HJ, Cowie CC (2008) Diabetes and hearing impairment in the United States: audiometric evidence from the National Health and Nutrition Exam Survey, 1999 to 2004. Ann Intern Med 149:1–10.

Berner B, Odum L, Parving A (2000) Age-related hearing impairment and B vitamin status. Acta Otolaryngol 120:633–637.

Borchgrevink HM, Tambs K, Hoffman HJ (2005) The Nord-Trondelag Norway audiometric survey 1996–98: unscreened thresholds and prevalence of hearing impairment for adults >20 years. Noise Health 7:1–15.

Brant LJ, Gordon-Salant S, Pearson JD, Klein LL, Morrell CH, Metter EJ, Fozard JL (1996) Risk factors related to age-associated hearing loss in the speech frequencies. J Am Acad Audiol 7:152–160.

Brown KD, Balkany TJ (2007) Benefits of bilateral cochlear implantation: a review. Curr Opin Otolaryngol Head Neck Surg 15:315–318.

Chia EM, Wang JJ, Rochtchina E, Cumming RR, Newall P, Mitchell P (2007) Hearing impairment and health-related quality of life: the Blue Mountains Hearing Study. Ear Hear 28:187–195.

Christensen K, Frederiksen H, Hoffman H (2001) Genetic and environmental influences on self-reported reduced hearing in the old and oldest old. J Am Geriatr Soc 49:1512–1517.

Cruickshanks KJ, Nondahl DM, Klein R, Klein BEK (1996) Sex, cardiovascular disease and hearing loss: the Epidemiology of Hearing Loss Study. Am J Epidemiol 143, Suppl 11:S65.

Cruickshanks KJ, Klein R, Klein BEK, Wiley TL, Nondahl DM, Tweed TS (1998a) Cigarette smoking and hearing loss. The Epidemiology of Hearing Loss Study. J Am Med Assoc 279:1715–1719.

Cruickshanks KJ, Wiley TL, Tweed TS, Klein BE, Klein R, Mares-Perlman JA, Nondahl DM (1998b) Prevalence of hearing loss in older adults in Beaver Dam, Wisconsin. Am J Epidemiol 148:879–886.

Cruickshanks KJ, Tweed TS, Wiley TL, Klein BE, Klein R, Chappell R, Nondahl DM, Dalton DS (2003) The 5-year incidence and progression of hearing loss: the Epidemiology of Hearing Loss Study. Arch Otolaryngol Head Neck Surg 129:1041–1046.

Dalton DS, Cruickshanks KJ, Klein R, Klein BE, Wiley TL (1998) Association of diabetes and hearing loss. Diabetes Care 21:1540–1544.

Dalton DS, Cruickshanks KJ, Wiley TL, Klein BEK, Klein R, Tweed TS (2001) The association of leisure time noise exposure and hearing loss. Audiology 40:1–9.

Davis AC (1989) The prevalence of hearing impairment and reported disability among adults in Great Britain. Int J Epidemiol 18:911–917.

Davis AC, Ostri B, Parving A (1991) Longitudinal study of hearing. Acta Otolaryngol Suppl 476:12–22.

DeStefano AL, Gates GA, Heard-Costa N, Myers RH, Baldwin CT (2003) Genomewide linkage analysis to presbycusis in the Framingham Heart Study. Arch Otolaryngol Head Neck Surg 129:285–289.

Doll R, Peto R (1981) The Causes of Cancer. New York: Oxford Press.

Durga J, Anteunis LJ, Schouten EG, Bots ML, Kok FJ, Verhoef P (2006) Association of folate with hearing is dependent on the 5.10-methylenetetrahdyrofolate reductase 677C → T Mutation. Neurobiol Aging 27:482–489.

Durga J, Verhoef P, Anteunis LJ, Schouten E, Kok FJ (2007) Effects of folic acid supplementation on hearing in older adults. Ann Intern Med 146:1–9.

Evans AS (1995) Causation and disease: a chronological journey. The Thomas Parran Lecture. Am J Epidemiol 142:1126–1135.

Fransen E, Topsakal V, Hendrickx JJ, Van Laer L, Huyghe JR, Van Eyken E, Lemkens N, Hannula S, Mäki-Torkko E, Jensen M, Demeester K, Tropitzsch A, Bonaconsa A, Mazzoli M, Espeso

A, Verbruggen K, Huyghe J, Huygen PL, Kunst S, Manninen M, Diaz-Lacava A, Steffens M, Wienker TF, Pyykkö I, Cremers CW, Kremer H, Dhooge I, Stephens D, Orzan E, Pfister M, Bille M, Parving A, Sorri M, Van de Heyning P, Van Camp G (2008) Occupational noise, smoking, and a high body mass index are risk factors for age-related hearing impairment and moderate alcohol consumption is protective: a European population-based multicenter study. J Assoc Res Otolaryngol 9:264–276.

Fuente A, McPherson B (2006) Organic solvents and hearing loss: the challenge for audiology. Int J Audiol 45:367–381.

Gates GA, Cooper JC (1991) Incidence of hearing decline in the elderly. Acta Otolaryngol 111: 240–248.

Gates GA, Cooper JC Jr, Kannel WB, Miller NJ (1990) Hearing in the elderly: the Framingham cohort, 1983–1985. Part I. Basic audiometric test results. Ear Hear 11:247–256.

Gates GA, Cobb JL, D'Agostino RB, Wolf PA. (1993) The relation of hearing in the elderly to the presence of cardiovascular disease and cardiovascular risk factors. Arch Otolaryngol Head Neck Surg 119:156–161.

Gates G, Couropmitree N, Myers R (1999) Genetic associations in age-related hearing thresholds. Arch Otolaryngol Head Neck Surg 125:654–659.

Helzner EP, Cauley JA, Pratt SR, Wisniewski SR, Zmuda JM, Talbott EO, de Rekeneire N, Harris TB, Rubin SM, Simonsick EM, Tylavsky FA, Newman AB (2005) Race and sex differences in age-related hearing loss: the Health, Aging, and Body Composition Study. J Am Geriatr Soc 53:2119–2127.

Houston DK, Johnsson MA, Nozza RJ, Gunter EW, Shea KJ, Cutler GM, Edmonds JT (1999) Age-related hearing loss, vitamin B-12, and folate in elderly women. Am J Clin Nutr 69:564–571.

Huyghe JR, Van Laer L, Hendrickx JJ, Fransen E, Demeester K, Topsakal V, Kunst S, Manninen M, Jensen M, Bonaconsa A, Mazzoli M, Baur M, Hannula S, Mäki-Torkko E, Espeso A, Van Eyken E, Flaquer A, Becker C, Stephens D, Sorri M, Orzan E, Bille M, Parving A, Pyykkö I, Cremers CW, Kremer H, Van de Heyning PH, Wienker TF, Nürnberg P, Pfister M, Van Camp G (2008) Genome-wide SNP-based linkage scan identifies a locus on 8q24 for an age-related hearing impairment trait. Am J Hum Genet 83:401–407.

Jacobsen P, Hein HO, Suadicani P, Parving A, Gyntelberg F (1993) Mixed solvent exposure and hearing impairment: an epidemiological study of 3284 men. The Copenhagen male study. Occup Med 43:180–184.

Jennings CR, Jones NS (2001) Presbyacusis. J Laryngol Otol 115:171–178.

Johnsson LG, Hawkins JE (1972) Vascular changes in the human inner ear associated with aging. Ann Otol 81:364–376.

Karlsson KK, Harris JR, Svartengren M (1997) Description and primary results from an audiometric study of male twins. Ear Hear 18:114–120.

Last JM (2001) A Dictionary of Epidemiology, 4th Ed. New York: Oxford University Press.

Lee DJ, Carlson DL, Lee HM, Ray LA, Markides KS (1991) Hearing loss and hearing aid use in Hispanic adults: results from the Hispanic Health and Nutrition Examination Survey. Am J Public Health 8:1471–1474.

Lee FS, Matthews LJ, Mills JH, Dubno JR, Adkins WY (1998) Gender-specific effects of medicinal drugs on hearing levels of older persons. Otolaryngol Head Neck Surg 118:221–227.

Lin YS, Egeghy PP, Rappaport SM (2008) Relationships between levels of volatile organic compounds in air and blood from the general population. J Expo Sci Environ Epidemiol 18:421–429.

Mills JH, Matthews LJ, Lee FS, Dubno JR, Schulte BA, Weber PC (1999) Gender-specific effects of drugs on hearing levels of older persons. Ann NY Acad Sci 884:381–388.

Morata TC, Campo P (2002) Ototoxic effects of styrene alone or in concert with other agents:aA review. Noise Health 4:15–24.

Morata TC, Johnson AC, Nylen P, Svensson EB, Cheng J, Krieg EF, Lindblad AC, Ernstgård L (2002) Audiometric findings in workers exposed to low levels of styrene and noise. J Occup Environ Med 44:806–814.

Mościcki EK, Elkins EF, Baum HM, McNamara PM (1985) Hearing loss in the elderly: an epidemiologic study of the Framingham Heart Study cohort. Ear Hear 6:184–190.

Moshe S, Frenkel A, Hager M, Skulsky M, Skulsky M, Sulkis J, Himelfarbe M (2002) Effects of occupational exposure to mercury or chlorinated hydrocarbons on the auditory pathway. Noise Health 4:71–77.

Nondahl DM, Cruickshanks KJ, Wiley TL, Tweed TS, Ershler W, Klein R, Klein BEK (1998) Accuracy of self-reported hearing loss. Audiology 37:295–301.

O'Grady G, Boyles AL, Speer M, DeRuyter F, Strittmatter W, Worley G. (2007) Apolipoprotein E alleles and sensorineural hearing loss. Int J Audiol 46:183–186.

Popelka MM, Cruickshanks KJ, Wiley TL, Tweed TS, Klein BEK, Klein R (1998) Low prevalence of hearing aid use among older adults with hearing loss: the Epidemiology of Hearing Loss Study. J Am Geriatr Soc 46:1075–1078.

Popelka MM, Cruickshanks KJ, Wiley TL, Tweed TS, Klein BE, Klein R, Nondahl DM (2000) Moderate alcohol consumption and hearing loss: a protective effect. J Am Geriatr 48:1273–1278

Ries PW (1985) The demography of hearing loss. In: Orlans H (ed) Adjustments to Adult Hearing Loss. San Diego, CA: College Hill Press, pp. 3–20.

Ries PW (1994) Prevalence and characteristics of persons with hearing trouble: United States, 1990–91. Vital Health Stat 10, p. 1–75.

Rosen S (1966) Hearing studies in selected urban and rural populations. Trans NY Acad Sci 29:9–21.

Rosen S, Olin P (1965) Hearing loss and coronary heart disease. Arch Otolaryngol 82:236–243.

Rosen S, Bergman M, Plester D, El-mofty A, Satti MH (1962) Presbycusis study of a relatively noise-free population in the Sudan. Ann Otorhinol Laryngol 71:727–743.

Rosen S, Preobrajensky N, Tbilisi SK, Glazunov I, Tbilisi Nk, Rosen HV (1970) Epidemiologic hearing studies in the USSR. Arch Otolaryngol 91:424–428.

Rosenhall U, Sixt E, Sundh V, Svanborg A (1993) Correlations between presbyacusis and extrinsic noxious factors. Audiology 32:234–243.

Schuknecht HF (1964) Further observations on the pathology of presbycusis. Arch Otolaryngol 80:269–382.

Schwartz J, Otto D (1987) Blood lead, hearing thresholds and neurobehavioral development in children and youths. Arch Environ Health 42:153–160.

Schwartz J, Otto D (1991) Lead and minor hearing impairment. Arch Environ Health 46: 300–305.

Sliwinska-Kowalska M, Zamyslowska-Szmytke E, Szymczak W, Kotylo P, Fiszer M, Wesolowski W, Pawlaczyk-Luszczynska M (2003) Ototoxic effects of occupational exposure to styrene and co-exposure to styrene and noise. J Occup Environ Med 45:15–24.

Susmano A, Rosenbush SW (1988) Hearing loss and ischemic heart disease. Am J Otol 9:403–408.

Torre P, Cruickshanks KJ, Klein BEK Klein R, Nondahl DM (2005) The association between cardiovascular disease and cochlear function in older adults. J Speech Lang Hear Res 48:473–481.

Turyk M, Anderson HA, Hanrahan LP, Falk C, Steenport DN, Needham LL, Patterson DG Jr, Freels S (2006) Relationship of serum levels of individual PCB, dioxin, and furan congeners and DDE with Great Lakes sport-caught fish consumption. Environ Res 100:173–183.

Unal M, Tamer L, Doğruer ZN, Yildirim H, Vayisoğlu Y, Camdeviren H (2005) N -acetyltransferase 2 gene polymorphism and presbycusis. Laryngoscope 115:2238–2241.

Van Eyken E, Van Camp G, Van Laer L (2007a) The complexity of age-related hearing impairment: contributing environmental and genetic factors Audiol Neurotol 12:345–358.

Van Eyken E, Van Camp G, Fransen E, Topsakal V, Hendrickx JJ, Demeester K, Van de Heyning P, Mäki-Torkko E, Hannula S, Sorri M, Jensen M, Parving A, Bille M, Baur M, Pfister M, Bonaconsa A, Mazzoli M, Orzan E, Espeso A, Stephens D, Verbruggen K, Huyghe J, Dhooge I, Huygen P, Kremer H, Cremers CW, Kunst S, Manninen M, Pyykkö I, Lacava A, Steffens M, Wienker TF, Van Laer L (2007b) Contribution of the N-acetyltransferase 2 polymorphism NAT2*6A to age-related hearing impairment. J Med Genet 44: 570–578.

Van Laer L, Van Eyken E, Fransen E, Huyghe JR, Topsakal V, Hendrickx JJ, Hannula S, Mäki-Torkko E, Jensen M, Demeester K, Baur M, Bonaconsa A, Mazzoli M, Espeso A, Verbruggen K, Huyghe J, Huygen P, Kunst S, Manninen M, Konings A, Diaz-Lacava AN, Steffens M,

Wienker TF, Pyykkö I, Cremers CW, Kremer H, Dhooge I, Stephens D, Orzan E, Pfister M, Bille M, Parving A, Sorri M, Van de Heyning PH, Van Camp G. (2008) The grainyhead like 2 gene (*GRHL2*), alias *TFCP2L3*, is associated with age-related hearing impairment. Hum Mol Genet 17:159–169.
Viljanen A, Era P, Kaprio J, Pyykkö I, Koskenvuo M, Rantanen T (2007) Genetic and environmental influences on hearing in older women. J Gerontol A Biol Sci Med Sci 62:447–452.
Willott JF (1991) Age-related sensorineural histopathology. In: Aging and the Auditory System: Anatomy, Physiology, and Psychophysics. San Diego, CA: Singular Publishing Group, pp. 39–42.

Chapter 10
Interventions and Future Therapies: Lessons from Animal Models

James F. Willott and Jochen Schacht

10.1 Introduction

The chapters in this book provide ample evidence that age-related hearing loss is caused by multiple factors combining genetic traits with a constant barrage of lifetime insults to the hearing organ. Such insults may include noise exposure in occupational settings or at leisure (from loud machinery to iPods or rock concerts), chemicals and solvents in the work place, life style (drinking, smoking), diseases (diabetes, infections), and even the adverse "ototoxic" effects of medications on the inner ear. It is not even necessary that the insults be severe enough to cause immediate damage. Kujawa and Liberman (2006) subjected adult mice to a noise level that did not induce any threshold shifts two weeks after exposure. However, as the animals aged, they showed a continuing primary neural degeneration and deterioration of neural responses. Age-related changes in the central auditory system add to the complexity of the problem. Determining the cause(s) of hearing difficulties in an aging patient is challenging to say the least, let alone the question of how to prevent or treat such hearing impairment.

Potential therapies and interventions that might ameliorate presbycusis can be viewed from a broad conceptual perspective: the auditory system is but one component of the nervous systemthat, in turn, interacts with other "nonneural" systems. These "nonneural" systems may undergo subtle or obvious alterations with age, indirectly affecting hearing. Familiar examples are the immune and cardiovascular systems that often decline with age and have concomitant detrimental effects on

J.F. Willott (✉)
University of South Florida and The Jackson Laboratory, Department of Psychology, University of South Florida, PCD4118G, Tampa, FL,33620
e-mail: jimw@niu.edu

J. Schacht
Kresge Hearing Research Institute, The University of Michigan, 5315 Medical Sciences Bldg I, Ann Arbor, MI, 48109-5616
e-mail: schacht@umich.edu

S. Gordon-Salant et al. (eds.), *The Aging Auditory System*, Springer Handbook of Auditory Research 34,
DOI 10.1007/978-1-4419-0993-0_10,

neural tissue in the auditory system and elsewhere (Willott 1999). In addition, cognitive processes typically slow or diminish, which may have negative effects on auditory perception (Schneider, Pichora-Fuller, and Daneman, Chapter 7). Many approaches to geriatric health and therapeutics have been aimed at variables such as general health, specific diseases, and genetic defects that modulate the rate of aging of physiological systems and their myriad cellular components (Gordon-Salant and Frisina, Chapter 1). These approaches may benefit the auditory system, perhaps even in unexpected ways.

This being said, the present chapter focuses on the auditory system as it pertains to presbycusis, and this requires a conceptual perspective as well. In thinking about interventions aimed at the aging auditory system, one must consider both the cochlea and the auditory regions of the brain as quasi-independent yet interacting components. On the one hand, the organ of Corti and other cochlear tissue are extremely vulnerable to degeneration with age. Virtually all older people exhibit some degree of degenerative changes in cochlear tissue including hair cells, supporting cells, stria vascularis, and spiral ganglion neurons or any combination thereof (Schmiedt, Chapter 2; Fitzgibbons and Gordon-Salant, Chapter 5; Cruickshanks, Zhan, and Zhong, Chapter 9).

This, of course, results in reduced auditory sensitivity and degraded neural information being relayed to the brain via the auditory nerve, many of whose axons may also be damaged or missing. Altered inputs from the cochlea can then cause secondary changes in the central auditory system, dubbed the "central effects of peripheral pathology" (Willott 1991). Denervation or diminished auditory-evoked synaptic input to central auditory nuclei includes both degenerative effects and plasticity, discussed in Section 5 and by Canlon. Illing, and Walton (Chapter 3).

Complicating the situation still further, the auditory system faces a variety of ills that may visit the aging central nervous system and are independent of cochlear events. These "central effects of biological aging" (Willott 1991) are manifested in numerous ways in the brain. For example, older neurons may be more susceptible to certain toxins and to excitoxicity by amino acid neurotransmitters (for example, damage from too much glutamatergic synaptic activity), programmed cell death, somatic mutations, oxidative damage, and more. Thus, even if the cochlea were not damaged, auditory perception can be degraded.

Given all of this, interventions aimed at the auditory system might focus on specific components (e.g., cochlear or brainstem prostheses, hearing aids), and/or variables that may impact both the peripheral and the central auditory system (e.g., pharmacological interventions designed to benefit hair cell or neuronal survival or neural circuit integrity; Willott et al. 2001a). For example, a deficient blood supply to auditory structures might benefit from repair of age-related damage specific to strial capillaries (if possible) or from treatment of cardiovascular disease threatening cerebral or cochlear blood vessels. Although directed at the same target, these interventions would be taking on the problem from rather different directions.

Most interventions aim at slowing, stopping, or reversing degenerative or pathological events. These goals are especially challenging when it comes to biological changes that intrinsically accompany aging. Such geriatric interventions are hampered by the dynamic, progressive, and diverse nature of age-related change and by the complex reactions and interactions of molecular, cellular, systemic, and behavioral processes. Thus slowing degenerative changes or attenuating their negative effects on auditory perception and the quality of life may currently be the best attainable outcome. And even then, it may be virtually impossible to predict the effects of generalized treatments on the aging auditory system.

10.2 Contributions of Animal Models

Because of the complexity of human presbycusis, research has long turned to animal models from which a large amount of our knowledge about age-related hearing loss has been derived (Ohlemiller 2006). The assumption is that confounding factors can be better controlled, and a clearer picture of true age-dependent changes should become apparent in the laboratory. No single animal model has yet emerged as the gold standard for presbycusis and a variety of species have been investigated to approximate the different forms of human hearing loss. The gerbil, e.g., is considered a model for strial degeneration. Rats and cats can show features of sensory or sensorineural hearing loss that may be the most common form of age-related hearing impairment. Mice are appealing because of the availability of molecular tools and genetic information. For example, molecular findings in aging CBA/J or CBA/CaJ mice suggest a close relationship between oxidative stress mechanisms of the general physiological aging process and aging in the auditory system (Jiang et al. 2007). An added advantage for research design is the different rate of hearing loss in different mouse strains, ranging from those such as CBA and CAST, which exhibit minimal hearing loss by 18 months of age, to those such as C57BL/6J (B6) and DBA/2J, which exhibit significant hearing loss by approximately eight and two months of age, respectively. Such diversity has helped elucidate genetic components of accelerated hearing loss, e.g., the *Ahl* locus on chromosome 10 that is associated with cadherin 23, a protein intimately involved in stereocilia structure (Noben-Trauth et al. 2003). Mice carrying the recessive allele develop disorganized hair bundles.

The present discussion focuses primarily on animal models in which some tantalizing albeit inconsistent progress has been made in delaying presbycusis. Although extrapolations from nonhuman species to humans must always be made with caution, animal models, particularly mammals, provide information about what can and does occur in species whose auditory systems are in many ways similar to our own. A case in point for the applicability of this notion to the auditory system is the recent prevention of gentamicin-induced hearing loss in a clinical trial that was based on the outcome of laboratory investigations (Sha et al. 2006). Thus interventions that ameliorate presbycusis in animal models might also be beneficial in humans. Moreover, ideas generated by our insights from animal research may inspire new clinical directions.

10.3 Oxidative Imbalance: Target for Interventions

10.3.1 Oxidative Stress

Investigations into the aging process in nematodes, fruit flies, and mammals have pointed to a relationship between longevity and resistance to oxidative stress. Oxidative stress can be caused by reactive oxygen species (free radicals) that are products of enzymatic reactions serving physiological functions (such as the signaling molecule nitric oxide) or by-products of physiological processes (such as superoxide "leaking" from respiring mitochondria). Although essential constituents of every cell, reactive oxygen species also have destructive properties. "Redox homeostasis" is the equilibrium a cell must achieve for its survival by containing reactive oxygen species within physiological levels by balancing their production with their removal. For this purpose, every cell is endowed with antioxidants, such as glutathione, that remove ("scavenge") reactive oxygen species and with enzymes, such as superoxide dismutase, that inactivate them or their precursors. If reactive oxygen species overwhelm the antioxidant capacity of a cell, oxidant stress ensues. This condition will activate signaling pathways and gene expression, first in an attempt to restore homeostasis and, if this fails, to initiate cell death.

This general concept of age-related oxidant stress may also apply to the auditory system, at least in models where hair cell loss is a primary cause of deafness. For example, in heterozygous mice lacking Cu/Zn (cytoplasmic) superoxide dismutase, age-related hearing loss developed earlier and with greater severity than in mice with this enzyme (McFadden et al. 1999). In the aging CBA mouse, reactive oxygen species are increased and antioxidant defense systems decreased in the cochlea (Jiang et al. 2007). Although oxidative stress may be produced by a multitude of physiological reactions, mitochondria are the major sources of reactive oxygen species during normal metabolism and even more so during enhanced metabolism, e.g. during noise exposure. Because the destructive actions of reactive oxygen species can, in turn, target the mitochondria themselves, a vicious cycle can commence in which reactive oxygen species damage the mitochondria and then those damaged mitochondria produce more reactive oxygen species. Deletions in mitochondrial DNA are an indicator of oxidative damage, and an accumulation of mitochondrial DNA mutations has been observed in essentially all aging tissues including the cochlea (Bai et al. 1997; Fischel-Ghodsian et al. 1997; Seidman et al. 1997; Pickles 2004). These signs of free radical-inflicted injury correlate well with a decreased hearing ability of rats (Seidman 2000), but the question whether there is indeed a causal (rather than a correlative) relationship remains unresolved. It is indicative of the complexity of the issue that Le and Keithley (2007) found no significant effect on age-related hearing loss in transgenic mice lacking the mitochondrial Mn-superoxide dismutase as compared to the B6 parent strain.

Another stress on mitochondrial respiration is exerted by hypoxia from vasoconstriction with subsequent reperfusion. Reperfusion, the restoration of blood flow, brings with it inflammatory factors and free radicals to the affected site, increasing cellular damage. This condition can arise in the cochlea for various reasons. Vascular abnormalities are a general pathological change, with aging potentially stressing oxygen supply and mitochondrial respiration (Picciotti et al. 2004), and vasoconstriction of the microvasculature in the lateral wall tissues has often been observed as a result of noise exposure (Hawkins 1971). In a more direct experimental approach, suppression of cochlear blood flow promoted mitochondrial DNA mutations and impacted hearing loss (Dai et al. 2004). Whether noise-induced or age-associated, vascular pathologies can be expected to add to oxidant stress in the inner ear, with potentially harmful and long-lasting consequences.

Another aspect of oxidative damage is "excitotoxicity," a phenomenon observed in the nervous system in response to a synaptic overload of glutamate. It may cause permeability changes in the postsynaptic membrane, leading to calcium influx, osmotic imbalance, dendritic edema, and eventual compromise of cell structure and function. Excess release of the neurotransmitter glutamate due to acute overstimulation of the inner hair cell or perhaps lower level chronic stimulation may trigger such excitotoxicity (Pujol et al. 1990). Pathological changes include vacuolizations, edema, and retraction of dendrites at the inner hair cell synapsethat, however, are reversible after an acute insult. It remains speculative as to what extent cumulative excitotoxic events could contribute to age-related hearing impairment, but they could perhaps promote neural degeneration without outright loss of hair cells.

10.3.2 Pharmacological Interventions

If oxidant stress has a causative relation to (sensory) presbycusis, a pharmacological intervention should be possible using supplementation by antioxidants to remove free radicals. Antioxidant treatment is a successful protective therapy in animals against some cochlear pathologies that are based on oxidant stress, namely, aminoglycoside and cisplatin ototoxicity (Rybak and Whitworth 2005) and noise trauma (Seidman et al. 1993; Ohinata et al. 2003). Although such positive intervention is encouraging, it should be considered that drug and noise trauma are rather acute and treatment can commence soon after or even before the ototoxic event. However, antioxidants can also be effective against chronic oxidant stress, increasing life span by protecting against oxidative damage in mitochondria (Miquel et al. 1995), and therefore may possibly be efficacious in presbycusis.

Rats kept on over supplementation with vitamin E or vitamin C during their life span indeed maintained better auditory sensitivity than their cohort on a placebo (Seidman 2000). For a translation to "real life," a lifetime on vitamin supplementation seems impractical, and a protective therapy would preferably begin at an age when hearing is still normal but just before the expected onset of presbycusic changes. This principle was also tested, albeit with less convincing results (Seidman 2000). A small contingent of rats was followed for six weeks during which a threshold shift of 4 dB was seen at 18 kHz and 7 dB at 3 kHz, a trend opposite to the expected greater high-frequency loss in presbycusis. Of the two antioxidants (mitochondrial metabolites) chosen, acetyl L-carnitine protected at 18 kHz, not at 3 kHz and lipoic acid protected at 3 kHz but not at 18 kHz. More encouraging was a study in aging dogs fed an antioxidant diet for the last three years of their lives (Le and Keithley 2007). Animals on the diet that included DL-α-tocopherol acetate (vitamin E), L-carnitine, DL-α-lipoic acid, and ascorbic acid (vitamin C) showed less neuronal degeneration in the cochlea than animals on a normal diet. The caveat here is that the dogs had a lifetime of exposure to excessive noise (from constant barking) in their kennels and that some of the damage and the antioxidant protection might relate more to acoustic trauma than to aging.

To what extent dietary manipulations can attenuate presbycusis in humans remains to be established, and some recent studies have shown divergent results. Folic acid and vitamin B_{12} are essential for general health, and reduced serum concentrations have been associated with increased cognitive decline as well as auditory dysfunction in aging individuals (Houston et al. 1999). Short-term supplementation with vitamin B_{12}, however, was unrelated to hearing status in a group of older adults (Park et al. 2006). In contrast, a clinical study in the Netherlands (Durga et al. 2007) provided some evidence for the efficacy of folic acid. Over a three-year period, the decline in low-frequency hearing was modestly (by 0.7 dB) but significantly attenuated in elderly participants with daily supplementation of folic acid as compared with a placebo group. High-frequency hearing was not affected, but a preservation of the lower frequencies could be important for speech discrimination. The result may not be generally applicable because basic folate levels in the Dutch population are low because folic acid fortification of food is prohibited in the Netherlands. Some other countries, including the United States, Canada, and the United Kingdom, have introduced folic acid fortification (primarily of flour products) to lower the incidence of congenital neural tube deformation. In any case, the Dutch study does signal the possibility that presbycusis may be amenable to pharmacological prevention.

In another pilot study, 23 patients with presbycusis received a combination of the radical scavenger rebamipide and vitamin C for eight weeks (Takumida and Anniko 2005). Hearing levels after treatment were improved ($\geq$10 dB) at 125, 250, 500, and 8,000 Hz in half the patients but remained unchanged at 1,000, 2,000, and 4,000 Hz. The large improvement in thresholds after such a short time is intriguing but unlikely to be related to an effect on sensory or sensorineural presbycusis where irreversible damage to hair cells and neuronal elements would have occurred.

Alcohol has emerged from several population studies as a boon and a bane. Moderate alcohol consumption was associated with better hearing at both low and high frequencies, whereas heavy drinkers had a tendency for greater high-frequency hearing loss (Popelka et al. 2000). Several other studies have corroborated such a fine line between the beneficial and the detrimental effects of alcohol consumption that might, at least in part, be related to effects on general health. We may suspect that other therapies may likewise have dose-dependent positive or negative effects when administered chronically, but such studies are still lacking. In any case, the sum of these results clearly suggests that presbycusis, despite some genetic predisposition, is not invariable but that its progression can be pharmacologically manipulated.

10.4 Dietary Restriction

A large body of gerontological literature on rodents and other species has shown that calorically restricted diets can extend longevity, slow certain age-related physiological declines, and decrease tumors and diseases (Heilbronn and Ravussin 2003; Bordone and Guarente 2005; Bengmark 2006). Although much of this research has focused on nonneural systems, there is ample evidence that dietary restriction modulates aging of the brain (Mobbs et al. 2001; Martin et al. 2006; Ingram et al. 2007). Interestingly, increased stress resistance by lowering reactive oxygen species is among the metabolic consequences of dietary restriction (Sohal and Weindruch 1996).

With respect to the effects of caloric restriction on the auditory system, the results are mixed. Several studies have used inbred strains of mice that exhibit genetic progressive hearing loss as models of presbycusis. Caloric restriction reduced age-related changes in some strains but not in others (Henry 1986; Sweet et al. 1988; Park et al. 1990; Willott et al. 1995). Willott et al. (1995) evaluated the effects of a well-controlled calorically restricted diet on age-related hearing loss in 15 mouse strains maintained on their diets from a young age. The strains included several with genetic progressive hearing loss, B6, DBA/2J, BALB/cByJ, and WB/ReJ plus a normal-hearing CBA/H-T6J strain and all 10 combinations of their F_1 hybrids. The following effects of dietary restriction were observed among the 15 strains: (1) no apparent effect on either longevity, the rate of progressive hearing loss from auditory brainstem response (ABR), or end-point cochlear pathology; (2) extended longevity, with cochlear pathology continuing at a rate predictable from the regression curve derived from mice with the typical, high-calorie diet. This was also observed in rats by Feldman (1984); (3) extended longevity, but with the rate of auditory degeneration accelerating compared with the linear regression prediction; (4) an ameliorative effect on cochlear damage for a portion of the extended longevity, followed by acceleration of cochlear damage; (5) slowing of the rate of hearing loss (indicated by ABR thresholds) but not the ultimate severity of cochlear pathology (i.e., severe hearing loss occurred later in life); and (6) in B6 mice, slowing of the rate of hearing loss, lessened cochlear pathology, and extended longevity. Henry (1986) also found less severe presbycusis in diet-restricted B6 mice.

Research on this topic using other species or different parameters of evaluation likewise shows varied effects. Seidman (2000) concluded that caloric restriction in rats had beneficial effects on presbycusis, including reduced damage to mitochondrial DNA that he linked to cochlear damage. Someya et al. (2007) reported that caloric restriction suppressed apoptotic cell death in the mouse cochlea, downregulating the expression of 24 apoptotic genes. In contrast, Torre et al. (2004) were unable to find evidence of an effect of caloric restriction using electrophysiological measures to evaluate hearing in rhesus monkeys.

These findings suggest that caloric restriction has the capacity to slow or mitigate the rate or severity of cochlear degeneration. However, this effect appears to be tied to genotype because dietary restriction often has no effect or even detrimental effects on the auditory system in different species or genetic strains. The ultimate application of some sort of long-term approach to humans may therefore have to deal with unknown variables and substantial individual differences.

10.5 Manipulation of Hearing-Loss Induced Central Plasticity

Neurons in adult brains retain the capacity for synaptic plasticity in response to injury or environmental manipulations, in some cases as effectively as young brains, and dendrites continue to be modifiable (e.g., Willott 1999; Syka 2002; Jones et al. 2006; Mahncke et al. 2006; Keuroghlian and Knudsen 2007; Mora et al. 2007). The implications for the auditory system are that degenerative tendencies may be counteracted by replacement of damaged synapses and repair of neural circuits.

This idea has been extensively tested in B6 mice (Willott 1986, 1996, 2006; Frisina and Walton 2001). B6 mice exhibit progressive cochlear pathology that begins during young adulthood (2-3 months of age) and is well developed by middle age (6-12 months). Sensorineural hearing loss begins with involvement of the basal cochlea (and, consequently, elevation of high-frequency thresholds) and progresses to include lower frequencies as well. As high-frequency hearing loss becomes more severe with age, changes in the representation of stimulus frequency by neurons in the B6 inferior colliculus (IC) occur. Neurons in the ventral IC that normally have low thresholds for high-frequency tones, of course, can no longer respond to them and the "tails" of the neuronal tuning curves normally have high thresholds (>80-dB SPL) for low-frequency tones. However, in middle-aged B6 mice with high-frequency impairment, ventral IC neurons come to respond to low-frequency sounds at intensities of <70- or even 60-dB SPL (i.e., activity from more apical regions of the cochlea now influences responses of these neurons). In the ventral IC, the "best frequencies" (the frequency for which a neuron has the lowest threshold) shift from high to midfrequencies (10 to 15 kHz) so that best frequencies tend to be similar throughout the IC. Importantly, the "shifted" best frequencies have thresholds that are lower than thresholds of the same frequencies in the tuning curve "tails" of neurons in the same IC location of young adult mice. The changes in frequency representation are even more pronounced in the auditory cortex (Willott 2006). Indeed, virtually the entire auditory cortex contains neurons with low-threshold best frequencies at the midfrequencies.

10.5.1 Functional Consequences

This type of hearing loss-induced plasticity appears to have functional consequences as indicated by behavioral research using the prepulse inhibition paradigm (PPI; summarized in Willott et al. 2001b). In PPI, a tone "prepulse" stimulus (S1) is presented 100 ms before a standard startle-evoking noise-burst stimulus (S2). If S1 is effective, the amplitude of the startle response evoked by S2 is reduced (inhibited), and the degree of amplitude reduction may be used as a measure of the behavioral salience of S1. PPI of B6 mice exhibited the expected weakening of PPI to a high-frequency (24 kHz) S1 between 1 and 12 months of age, consistent with the decline of high-frequency hearing. More importantly, the efficacy of lower-frequency S1s was significantly enhanced. In 6 month olds, enhancement was greatest for 12- to 16-kHz S1s, frequencies for which neural plasticity is pronounced. In 12 month olds, enhancement of startle modification was especially prominent for 4-kHz S1s. "Overrepresented" midfrequencies also became more potent stimuli in a fear-conditioning paradigm (fear-potentiated startle; Falls and Pistell, 2001). These studies indicate that hearing loss-induced plasticity leads to greater behavioral salience of sounds consisting of the overrepresented frequencies.

10.5.2 Diminished Inhibition

Some properties of hearing loss-induced plasticity suggest that neural-inhibitory processes are impaired in the IC and auditory cortex of middle-aged B6 mice, spawning the "diminished-inhibition" hypothesis (Willott 1996). For example, IC neurons of middle-aged B6 mice (but not normal-hearing CBA mice) exhibit increased spontaneous activity, which would be expected if inhibition of neurons was reduced (Willott et al. 1988). Fewer nonmonotonic rate-level functions are observed in IC neurons of middle-aged B6 mice. Nonmonotonicity is likely to result when inhibition is evoked by stimuli at higher sound pressure levels. Free-field studies provided evidence that the typical ipsilaterally evoked inhibition and binaural release from masking are weaker in middle-aged B6 mice (McFadden and Willott 1994). Also, in the cochlear nucleus, potency of the inhibitory neurotransmitter glycine is lessened (Willott et al.1997).

If a mechanism involving diminished inhibition were responsible for hearing loss-induced plasticity, it might be possible to alter it with pharmacological interventions or other means that affect the relative strength of excitatory and inhibitory synapses or circuits. But this begs the question as to what changes would be aimed for and whether hearing-loss-induced plasticity is generally beneficial or deleterious for auditory perception (Willott 1996). On one hand, more auditory neurons become devoted to the frequencies that can still be heard, and this might allow the auditory system to respond more vigorously to certain sounds. However, the altered responses are also inappropriate responses, in that an abnormally large number of neurons are excited by stimuli that did not activate them when the individual was young. If the auditory system cannot "recognize" that formerly high best-frequency neurons are now responding to midfrequency sounds and associate the neuronal responses with their new stimuli, perceptual processes could be affected. Moreover, if the physiologically amplified sounds are noise rather than signal, problems would ensue in noisy environments. The upshot is that we cannot predict the effects of hearing loss-induced plasticity on auditory perception and whether it might be altered beneficially. The therapeutic potential of this approach has yet to be determined.

10.6 Treatment with an Augmented Acoustic Environment

Treatment with an augmented acoustic environment (AAE) has the potential to alter the severity and time course of presbycusis. AAEs can be produced by introducing controlled ambient stimuli or by amplifying and/or "shaping" sounds using hearing aids or other devices. Of course, amplification is widely used clinically, albeit to treat the symptoms (loss of hearing), rather than as a potential therapeutic strategy to alter the course of hearing loss. Indeed, we know little about the positive or negative consequences of hearing aids or other forms of AAE on the peripheral and central contributors to presbycusis (but see Silman et al. 1984; Neuman 1996). Should AAEs be avoided in some circumstances because they will do more harm than good? Or might AAEs be used as an effective therapeutic strategy to mitigate presbycusis? Are some types of AAE better or worse than others? These questions must be answered before we can even consider using AAE treatment therapeutically.

Studies of AAE treatment have been performed on inbred mice including the B6 model of presbycusis (Willott et al. 2000, 2001b, 2006; Willott and Bross 2004). The AAE in these studies consisted of 12 hours of nightly exposure to 70 dB (re 20 μPa) and 200-ms noise bursts at a rate of 2/second using 1 of 3 noise bands: low (2-8 kHz), middle (8-24 kHz), and high (half-octave centered at 20 kHz). ABR thresholds for tones within the low- and midfrequency AAEs are minimally affected in B6 mice by 9 months of age, whereas thresholds for frequencies within the high-frequency AAE (around 20 kHz and above) exhibit more severe elevations. The frequency spectrum of the AAE as well as gender and the severity of hearing loss at the initiation of treatment can affect its outcome. Indeed, various combinations of these parameters can either ameliorate or exacerbate progressive sensorineural hearing loss and central degenerative changes in B6 mice.

10.6.1 Importance of Tonotopic Organization

As in the case of hearing lossinduced plasticity discussed in Section 5, tonotopic organization of the cochlea and central auditory system play key roles in determining the effects of AAE treatment. In the cochlea, frequency is mapped in an orderly fashion from high in the basal turn to low in the apex. Consequently, low-, mid-, and high-frequency AAEs maximally stimulate the apical, middle, and basal regions of the cochlea, respectively. In turn, spiral ganglion cells (SGCs) in the basal cochlea project via the eighth nerve to neurons in the dorsal region of anteroventral cochlear nucleus (AVCN), SGCs in the apical cochlea project to the ventral region and so on. Thus the AVCN has a ventral/low-frequency to dorsal/high-frequency tonotopic map (the opposite of the IC in which high frequencies are represented ventrally). Physiological dysfunction or degeneration in the basal cochlea, typical of presbycusis, results in attenuation or removal of auditory nerve input that is most pronounced in the dorsal AVCN. Of course, AAE treatment (low, mid, or high frequencies) also results in a higher level of auditory input to corresponding regions of the AVCN.

The tonotopic nature of AAE effects is especially important. On one hand, sensorineural hearing loss is almost always frequency dependent, typically affecting high frequencies most severely. Second, hearing aid amplification (a type of AAE) is usually shaped according to frequency. It seems important to gain a better understanding of the relationships among tonotopicity, acoustic stimulus frequencies, and central effects of hearing loss and amplification.

10.6.2 Ameliorative Effects of AAE Treatment

When the midfrequency band has been used on B6 and other inbred strains with genetic progressive hearing loss, ameliorative effects occur (Willott et al. 2001b; Willott and Bross 2004). These include slowing of progressive elevation of ABR thresholds, diminished loss of outer hair cells and SGCs, lessened age-related reduction in the volume and number of neurons in the AVCN, increased amplitude of the acoustic startle response, and stronger PPI for midfrequency prepulses. However, whereas AAE treatment can lessen and slow progressive sensorineural hearing loss, once damage has occurred, it cannot be reversed. Thus AAE treatment beginning before serious sensorineural damage is more effective in slowing the course of threshold elevations than delayed treatment. It seems likely that the effectiveness of early intervention is due to the healthier condition of the cochlea at the time treatment is begun, a rather general therapeutic principle, but one stressing the importance of early intervention.

By 9 months of age, B6 mice have lost ~15% of AVCN neurons and tissue volume has shrunk (Willott and Bross 1996). These degenerative changes are lessened by treatment with the low- and midfrequency AAEs. Consequently, these AAE effects are most prominent in the low- and midfrequency tonotopic regions of the AVCN, respectively, although sex differences are apparent (Willott et al. 2001b; Willott and Bross 2004)

10.6.3 Deleterious Effects of AAE Treatment

Despite the generally beneficial effects of treatment with the midfrequency AAE, neuronal loss was exacerbated in the ventral and dorsal extremes of the AVCN in male B6 mice aged 12-14 months (Willott and Bross 2004). Other negative effects of AAE treatment have also been observed. Treatment with the high-frequency AAE facilitated elevation of ABR thresholds for high-frequency tones as B6 mice aged. Hair cell loss was also exacerbated in the basal cochlea as was the loss of neurons in the dorsal, high-frequency region of the AVCN. Treatment with the low-frequency AAE also caused some loss of outer hair cells in cochlear apex.

The mechanism(s) underlying AAE-induced cell damage in the cochlea and AVCN is not known, but several hypotheses for AAE-induced cell death can be considered. In the cochlea, genetic vulnerability to noise-induced damage characteristic of B6 mice (Erway et al. 2001) might cause a normally benign (70-dB SPL) noise environment to damage the basal hair cells over time. Another possibility that might apply to the SGCs as well as the AVCN involves glutamate, the neurotransmitter at SGC and AVCN synapses. Excessive glutamatergic activity can be excitotoxic, so it could be that increased afferent activity from the AAE induces neurotoxicity. This hypothesis would predict the most severe damage in tonotopic regions that receive most of the AAE-activated input, and this was the case for the high-frequency AAE in the cochlea and AVCN. Yet another hypothesis stems from the observation that the midfrequency AAE causes cell loss in the nonstimulated dorsoventral extremes of the AVCN. The "limited-resources" hypothesis proposes that stimulation of the midfrequency AVCN regions usurps metabolic or other beneficial resources at the expense of understimulated, high- and low-frequency regions, causing cell death. Yet another hypothesis for AVCN cell loss with the high-frequency AAE is that cochlear damage functionally denervates AVCN cells to their detriment. If nothing else, the variety of hypotheses suggests several avenues by which an AAE might exacerbate presbycusis, underscoring the potential risks.

10.6.4 Summary of AAE Treatment

Certain AAE effects may share properties with other interesting phenomena such as "toughening" or "sound conditioning" (e.g., Canlon 1996; Henderson et al. 1996; Subramaniam et al. 1996); the beneficial effects of electrical stimulation on SGCs and central auditory neurons via cochlear implants (e.g., Leake et al. 1991; Miller et al. 1996); and/or the effects of "enriched" environments on neural structure and physiology (e.g., Cotman and Neeper 1996). In each case, periods of auditory system stimulation can be beneficial in various ways, and research on these topics and on AAE may inform one another. However, the AAE paradigm and its effects on slowly progressive hearing loss in adults are unique and especially relevant to presbycusis. Studies of AAE treatment help us learn about mechanisms of amelioration or damage that might be used in a clinical context beyond acoustic stimulation. For example, if it were found that an AAE produced benefits via a neurotrophin, perhaps that could be used rather than noise per se. This would alleviate problems with the potential for negative effects of noise.

10.7 Gonadal Hormones

A potentially important "nonauditory" variable that can be manipulated clinically is gonadal hormones. Sex differences in presbycusis in humans are well documented (Gordon-Salant and Frisina, Chapter 1; Cruickshanks, Zhan, and Zhong, Chapter 9) and occur in mice as well (Henry 2002; Willott and Bross 2004). If sex hormonal status plays a role in presbycusis, supplemental hormones or other manipulations might have a therapeutic potential.

10.7.1 Estrogen

Estrogen has neuroprotective effects in various neural systems (e.g., Garcia-Segura et al. 2001), suggesting the possibility that auditory tissue might be somewhat protected as well. α- And β-estrogen receptors (ERs) are present in the cochleae of mice and rats (Stenberg et al. 1999), suggesting a potential cochlear site of action. Making a case for neuroprotection, Mills et al. (1999) demonstrated that female rats were less susceptible to kanamycin ototoxicity than males. However, the opposite was observed in two strains of guinea pigs in which females were significantly more sensitive to gentamicin-induced hearing loss (Halsey et al. 2005). Cyclic fluctuations in auditory sensitivity during the menstrual cycle have been reported (Swanson and Dengerink 1988). In both humans and mouse models of Turner syndrome (ovarian dysgenesis, no estrogen production, infertility), sensorineural hearing loss and early presbycusis occur (Hultcrantz et al. 2000). Studies in humans also suggest acceleration of hearing loss in postmenopausal women (Pearson et al. 1995; Hederstierna et al. 2007; see Fitzgibbons and Gordon-Salant, Chapter 5). The mechanisms of a protective action are unresolved, but estrogen may affect ion transport in the stria vascularis (Lee and Marcus 2001) and there is some evidence for central auditory effects of estrogen. For example, blocking ERs can accelerate loss of the auditory efferent system with age in female CBA mice (Thompson et al. 2006). ABR latencies are shortened during high levels of estrogen in the menstrual cycle (Dehan and Jerger 1990; Elkind-Hirsch et al. 1992), in postmenopausal women who were treated with hormone replacement therapy (Caruso et al. 2000), and in rats receiving hormone replacement after ovariectomy (Coleman et al. 1994). Also, deleterious effects of progestin on peripheral and central auditory systems have been demonstrated in postmenopausal women (Guimaraes et al. 2006).

10.7.2 Androgen

There are potential routes by which androgen might modulate hearing loss, although the evidence is sparse at present. Androgens might be able to affect hearing by modulating blood pressure and/or cochlear blood flow (Laugel et al. 1987, 1988;

Flynn et al. 1990). Moreover, testosterone is enzymatically converted to estradiol in many brain regions, so protection could potentially be provided via actions on either ERs or androgen receptors (estrogen can be a male hormone). Thus it is of interest that androgens can have neuroprotective or neurotrophic effect on brain tissue (Dluzen et al. 1994; Pouliot et al. 1996; Kimonides et al. 1998, 1999; Pike 2001). The presence of androgen receptors in the AVCN of rats (Simerly et al. 1990) suggests the possibility of direct effects on AVCN tissue. On the other hand, testosterone increased glutamate neurotoxicity in ischemia and in vitro models (Yang et al. 2002), raising the possibility that should androgen affect the central auditory system, the effects might be positive or negative.

10.7.3 Sex Hormones and Noise-Induced Hearing Loss

Studies using gonadectomized B6 mice suggest that gonadal hormones may play a role in sex differences with respect to AAE treatment (Willott et al. 2006, 2008). When the low-frequency AAE was used, gonadally intact females had more severe AAE-induced hearing loss than ovariectomized mice and gonadectomized males had more severe loss than intact males. Thus it appeared that the combination of AAE treatment and male gonadal hormone(s) ameliorated the degenerative changes in the cochlea, whereas the combination of AAE treatment plus ovarian hormone(s) exacerbated these changes. The most likely mechanism is estrogen-facilitated noise-induced damage in the already noise-susceptible B6 mice. Central degenerative changes were similarly related to hormones in the ventral, low-frequency region of the AVCN. These findings raise the possibility that gonadal hormones might modulate contributions to presbycusis by long-term exposure to moderate ambient noise.

As in the case for other manipulations of the auditory system, the role of gonadal hormones appears to be complex. Although there may be ameliorative or palliative effects of hormone treatments, the potential for negative effects is also present.

10.8 Summary and Outlook

At this time, no consistently effective interventions are available for ameliorating presbycusis in humans. Indeed, despite some encouraging beginnings, reliable therapeutic approaches have not been developed, even for animal models, and various experimental treatments are fraught with potentially serious pitfalls. Nonetheless, research on animal models has provided important advances in our basic understanding of age-related hearing loss and pointed us in directions that hold promise for the future. Furthermore, research into other auditory pathologies has shown considerable progress in elucidating underlying mechanisms and establishing principles of intervention and even repair. Presbycusis will undoubtedly benefit from such developments, but its multifactorial nature poses specific challenges that have yet to be confronted.

Acknowledgments This work was supported by Grants R01 AG 07554 (to JFW) and P01 AG 025164 (to JS) from the National Institute on Aging, National Institutes of Health.

References

Bai U, Seidman MD, Hinojosa R, Quirk WS (1997) Mitochondrial DNA deletions associated with aging and possibly presbycusis: a human archival temporal bone study. Am J Otol 18:449–453.

Bengmark S (2006) Impact of nutrition on aging and disease. Curr Opin Clin Nutr Metab Care 9:2–7.

Bordone L, Guarente L (2005) Calorie restriction, SIRT1 and metabolism: understanding longevity. Nat Rev Mol Cell Biol 6:298–305.

Canlon B (1996) The effects of sound conditioning on the cochlea. In: Salvi RJ, Henderson DH, Colletti V, Fiorino F (eds) Auditory Plasticity and Regeneration. New York: Thieme Medical Publishers, pp. 118–127.

Caruso S, Cianci A, Grasso D, Agnello C, Galvani F, Maiolino L, Serra A (2000) Auditory brainstem response in postmenopausal women treated with hormone replacement therapy: a pilot study. Menopause 7:178–183.

Coleman JR, Campbell D, Cooper WA, Welsh MG, Moyer J (1994) Auditory brainstem responses after ovariectomy and estrogen replacement in rat. Hear Res 80:209–215.

Cotman CW, Neeper S (1996) Activity-dependent plasticity and the aging brain. In: Rowe JW, Schneider EL (eds) Handbook of the Biology of Aging. San Diego, CA: Academic Press, pp. 284–299.

Dai P, Yang W, Jiang S, Gu R, Yuan H, Han D, Guo W, Cao J (2004) Correlation of cochlear blood supply with mitochondrial DNA common deletion in presbyacusis. Acta Otolaryngol 124:130–136.

Dehan CP, Jerger J (1990) Analysis of gender differences in the auditory brainstem response. Laryngoscope 100:18–24.

Dluzen D, Jain R, Liu B (1994) Modulatory effects of testosterone on 1-methyl-4-phenyl-1,2,3,6-tetrahydropyridine-induced neurotoxicity. J Neurochem 62:94–101.

Durga J, Verhoef P, Anteunis LJ, Schouten E, Kok FJ (2007) Effects of folic acid supplementation on hearing in older adults: a randomized, controlled trial. Ann Intern Med 146:1–9.

Elkind-Hirsch KE, Stoner WR, Stach BA, Jerger JF (1992) Estrogen influences auditory brainstem responses during the normal menstrual cycle. Hear Res 60:143–148.

Erway LC, Zheng QY, Johnson KR (2001) Inbred strails of mice for genetics of hearing in mammals: searching for genes for hearing loss. In: Willott JF (ed) Handbook of Mouse Auditory Research: From Behavior to Molecular Biology. Boca Raton, FL: CRC Press, pp. 429–440.

Falls WA, Pistell PJ (2001) Learning and the auditory system: fear-potentiated startle studies. In: Willott JF (ed) Handbook of Mouse Auditory Research: From Behavior to Molecular Biology. Boca Raton, FL: CRC Press, pp. 91–96.

Feldman ML (1984) Morphological observations on the cochleas of very old rats. Assoc Res Otolaryngol Abstr 7:14.

Fischel-Ghodsian N, Bykhovskaya Y, Taylor K, Kahen T, Cantor R, Ehrenman K, Smith R, Keithley E (1997) Temporal bone analysis of patients with presbycusis reveals high frequency of mitochondrial mutations. Hear Res 110:147–154.

Flynn AJ, Dengerink HA, Wright JW (1990) Androgenic effects on angiotensin II-induced blood pressure and cochlear blood flow changes in rats. Hear Res 50:119–125.

Frisina RD, Walton JP (2001) Aging and the mouse central auditory system. In: Willott JF (ed) Handbook of Mouse Auditory Research: From Behavior to Molecular Biology. Boca Raton, FL: CRC Press, pp. 339–380.

Garcia-Segura LM, Azcoitia I, DonCarlos LL (2001) Neuroprotection by estradiol. Prog Neurobiol 63:29–60.

Guimaraes P, Frisina ST, Mapes F, Tadros SF, Frisina DR, Frisina RD (2006) Progestin negatively affects hearing in aged women. Proc Nat Acad Sci USA 103:14246–14249.

Halsey K, Skjönsberg A, Ulfendahl M, Dolan DF (2005) Efferent-mediated adaptation of the DPOAE as a predictor of aminoglycoside toxicity. Hear Res 201:99–108.

Hawkins JE Jr (1971) The role of vasoconstriction in noise-induced hearing loss. Ann Otol Rhinol Laryngol 80:903–913.

Hederstierna C, Hultcrantz M, Collins A, Rosenhall U (2007) Hearing in women at menopause. Prevalence of hearing loss, audiometric configuration and relation to hormone replacement therapy. Acta Otolaryngol 127:149–155.

Heilbronn LK, Ravussin E (2003) Calorie restriction and aging: review of the literature and implications for studies in humans. Am J Clin Nutr 78:361–369.

Henderson D, Subramaniam M, Spongr V, Attanasio G (1996) Biological mechanisms for the "toughening" phenomenon. In: Salvi RJ, Henderson DH, Colletti V, Fiorino F (eds) Auditory Plasticity and Regeneration. New York: Thieme Medical Publishers, pp. 143–154.

Henry KR (1986) Effects of dietary restriction on presbyacusis in the mouse. Audiology 25:329–337.

Henry KR (2002) Sex- and age-related elevation of cochlear nerve envelope response (CNER) and auditory brainstem response (ABR) thresholds in C57BL/6 mice. Hear Res 170:107–115.

Houston DK, Johnson MA, Nozza RJ, Gunter EW, Shea KJ, Cutler GM, Edmonds JT (1999) Age-related hearing loss, vitamin B-12, and folate in elderly women. Am J Clin Nutr 69:564–571.

Hultcrantz M, Stenberg AE, Fransson A, Canlon B (2000) Characterization of hearing in an X,0 'Turner mouse'. Hear Res 143:182–188.

Ingram DK, Young J, Mattison JA (2007) Calorie restriction in nonhuman primates: assessing effects on brain and behavioral aging. Neuroscience 145:1359–1364.

Jiang H, Talaska AE, Schacht J, Sha SH (2007) Oxidative imbalance in the aging inner ear. Neurobiol Aging 28:1605–1612.

Jones S, Nyberg L, Sandblom J, Stigsdotter Neely A, Ingvar M, Magnus Petersson K, Backman L (2006) Cognitive and neural plasticity in aging: general and task-specific limitations. Neurosci Biobehav Rev 30:864–871.

Keuroghlian AS, Knudsen EI (2007) Adaptive auditory plasticity in developing and adult animals. Prog Neurobiol 82:109–121.

Kimonides VG, Khatibi NH, Svendsen CN, Sofroniew MV, Herbert J (1998) Dehydroepiandrosterone (DHEA) and DHEA-sulfate (DHEAS) protect hippocampal neurons against excitatory amino acid-induced neurotoxicity. Proc Natl Acad Sci USA 95:1852–1857.

Kimonides VG, Spillantini MG, Sofroniew MV, Fawcett JW, Herbert J (1999) Dehydroepiandrosterone antagonizes the neurotoxic effects of corticosterone and translocation of stress-activated protein kinase 3 in hippocampal primary cultures. Neuroscience 89:429–436.

Kujawa SG, Liberman MC (2006) Acceleration of age-related hearing loss by early noise exposure: evidence of a misspent youth. J Neurosci 26:2115–2123.

Laugel GR, Dengerink HA, Wright JW (1987) Ovarian steroid and vasoconstrictor effects on cochlear blood flow. Hear Res 31:245–251.

Laugel GR, Wright JW, Dengerink HA (1988) Angiotensin II and progesterone effects on laser Doppler measures of cochlear blood flow. Acta Otolaryngol 106:34–39.

Le T, Keithley EM (2007) Effects of antioxidants on the aging inner ear. Hear Res 226:194–202.

Leake PA, Hradek GT, Rebscher SJ, Snyder RL (1991) Chronic intracochlear electrical stimulation induces selective survival of spiral ganglion neurons in neonatally deafened cats. Hear Res 54:251–271.

Lee JH, Marcus DC (2001) Estrogen acutely inhibits ion transport by isolated stria vascularis. Hear Res 158:123–130.

Mahncke HW, Bronstone A, Merzenich MM (2006) Brain plasticity and functional losses in the aged: scientific bases for a novel intervention. Prog Brain Res 157:81–109.

Martin B, Mattson MP, Maudsley S (2006) Caloric restriction and intermittent fasting: two potential diets for successful brain aging. Ageing Res Rev 5:332–353.

McFadden SL, Willott JF (1994) Responses of inferior colliculus neurons in C57BL/6J mice with and without sensorineural hearing loss: effects of changing the azimuthal location of an unmasked pure-tone stimulus. Hear Res 78:115–131.

McFadden SL, Ding D, Reaume AG, Flood DG, Salvi RJ (1999) Age-related cochlear hair cell loss is enhanced in mice lacking copper/zinc superoxide dismutase. Neurobiol Aging 20:1–8.
Miller JM, Altschuler RA, Dupont J, Tucci D (1996) Consequences of deafness and electrical stimulation on the auditory systerm. In: Salvi RJ, Henderson DH, Colletti V, Fiorino F (eds) Auditory Plasticity and Regeneration. New York: Thieme Medical Publishers, pp. 378–391.
Mills CD, Loos BM, Henley CM (1999) Increased susceptibility of male rats to kanamycin-induced cochleotoxicity. Hear Res 128:75–79.
Miquel J, Ferrandiz ML, De Juan E, Sevila I, Martinez M (1995) *N*-acetylcysteine protects against age-related decline of oxidative phosphorylation in liver mitochondria. Eur J Pharmacol 292:333–335.
Mobbs CV, Bray GA, Atkinson RL, Bartke A, Finch CE, Maratos-Flier E, Crawley JN, Nelson JF (2001) Neuroendocrine and pharmacological manipulations to assess how caloric restriction increases life span. J Gerontol A Biol Sci Med Sci 56:34–44.
Mora F, Segovia G, del Arco A (2007) Aging, plasticity and environmental enrichment: structural changes and neurotransmitter dynamics in several areas of the brain. Brain Res Rev 55:78–88.
Neuman, AC (1996) Late-onset auditory deprivation: A review of past research and an assessment of future research needs. Ear Hear 17:3S-13S.
Noben-Trauth K, Zheng QY, Johnson KR (2003) Association of cadherin 23 with polygenic inheritance and genetic modification of sensorineural hearing loss. Nat Genet 35:21–23.
Ohinata Y, Miller JM, Schacht J (2003) Protection from noise-induced lipid peroxidation and hair cell loss in the cochlea. Brain Res 966:265–273.
Ohlemiller KK (2006) Contributions of mouse models to understanding of age- and noise-related hearing loss. Brain Res 1091:89–102.
Park JC, Cook KC, Verde EA (1990) Dietary restriction slows the abnormally rapid loss of spiral ganglion neurons in C57BL/6 mice. Hear Res 48:275–279.
Park S, Johnson MA, Shea-Miller K, De Chicchis AR, Allen RH, Stabler SP (2006) Age-related hearing loss, methylmalonic acid, and vitamin B12 status in older adults. J Nutr Elderly 25:105–120.
Pearson JD, Morrell CH, Gordon-Salant S, Brant LJ, Metter EJ, Klein LL, Fozard JL (1995) Gender differences in a longitudinal study of age-associated hearing loss. J Acoust Soc Am 97:1196–1205.
Picciotti P, Torsello A, Wolf FI, Paludetti G, Gaetani E, Pola R (2004) Age-dependent modifications of expression level of VEGF and its receptors in the inner ear. Exp Gerontol 39:1253–1258.
Pickles JO (2004) Mutation in mitochondrial DNA as a cause of presbyacusis. Audiol Neurootol 9:23–33.
Pike CJ (2001) Testosterone attenuates beta-amyloid toxicity in cultured hippocampal neurons. Brain Res 919:160–165.
Popelka MM, Cruickshanks KJ, Wiley TL, Tweed TS, Klein BE, Klein R, Nondahl DM (2000) Moderate alcohol consumption and hearing loss: a protective effect. J Am Geriatr Soc 48:1273–1278.
Pouliot WA, Handa RJ, Beck SG (1996) Androgen modulates *N*-methyl-D-aspartate-mediated depolarization in CA1 hippocampal pyramidal cells. Synapse 23:10–19.
Pujol R, Rebillard G, Puel JL, Lenoir M, Eybalin M, Recasens M (1990) Glutamate neurotoxicity in the cochlea: a possible consequence of ischaemic or anoxic conditions occurring in aging. Acta Otolaryngol Suppl 476:32–36.
Rybak LP, Whitworth CA (2005) Ototoxicity: therapeutic opportunities. Drug Discov Today 10:1313–1321.
Seidman MD (2000) Effects of dietary restriction and antioxidants on presbyacusis. Laryngoscope 110:727–738.
Seidman MD, Shivapuja BG, Quirk WS (1993) The protective effects of allopurinol and superoxide dismutase on noise-induced cochlear damage. Otolaryngol Head Neck Surg 109: 1052–1056.
Seidman MD, Bai U, Khan MJ, Quirk WS (1997) Mitochondrial DNA deletions associated with aging and presbyacusis. Arch Otolaryngol Head Neck Surg 123:1039–1045.
Sha SH, Qiu JH, Schacht J (2006) Aspirin to prevent gentamicin-induced hearing loss. N Engl J Med 354:1856–1857.

Silman S, Gelfand SA, Silverman CA (1984) Late-onset auditory deprivation: effects of monaural versus binaural hearing aids. J Acoust Soc Am 76:1357–1362.
Simerly RB, Chang C, Muramatsu M, Swanson LW (1990) Distribution of androgen and estrogen receptor mRNA-containing cells in the rat brain: an in situ hybridization study. J Comp Neurol 294:76–95.
Sohal RS, Weindruch R (1996) Oxidative stress, caloric restriction, and aging. Science 273:59–63.
Someya S, Yamasoba T, Weindruch R, Prolla TA, Tanokura M (2007) Caloric restriction suppresses apoptotic cell death in the mammalian cochlea and leads to prevention of presbycusis. Neurobiol Aging 28:1613–1622.
Stenberg AE, Wang H, Sahlin L, Hultcrantz M (1999) Mapping of estrogen receptors alpha and beta in the inner ear of mouse and rat. Hear Res 136:29–34.
Subramaniam M, Henderson D, Henselman L (1996) Toughening of the mammalian auditory system: spectral, temporal, and intensity effects. In: Salvi RJ, Henderson DH, Colletti V, Fiorino F (eds) Auditory Plasticity and Regeneration. New York: Thieme Medical Publishers, pp. 128–142.
Swanson SJ, Dengerink HA (1988) Changes in pure-tone thresholds and temporary threshold shifts as a function of menstrual cycle and oral contraceptives. J Speech Hear Res 31:569–574.
Sweet RJ, Price JM, Henry KR (1988) Dietary restriction and presbyacusis: periods of restriction and auditory threshold losses in the CBA/J mouse. Audiology 27:305–312.
Syka J (2002) Plastic changes in the central auditory system after hearing loss, restoration of function, and during learning. Physiol Rev 82:601–636.
Takumida M, Anniko M (2005) Radical scavengers: a remedy for presbyacusis. A pilot study. Acta Otolaryngol 125:1290–1295.
Thompson SK, Zhu X Frisina RD (2006) Estrogen blockade reduces auditory feedback in CBA mice. Otolaryngol Head Neck Surg 135:100–105.
Torre P III, Mattison JA, Fowler CG, Lane MA, Roth GS, Ingram DK (2004) Assessment of auditory function in rhesus monkeys (*Macaca mulatta*): effects of age and calorie restriction. Neurobiol Aging 25:945–954.
Willott JF (1986) Effects of aging, hearing loss, and anatomical location on thresholds of inferior colliculus neurons in C57BL/6 and CBA mice. J Neurophysiol 56:391–408.
Willott JF (1991) Aging and the Auditory System: Anatomy, Physiology, and Psychophysics. San Diego, CA: Singular Publishing Group.
Willott JF (1996) Auditory system plasticity in the adult C57BL/6J mouse. In Salvi RJ, Henderson DH, Colletti V, Fiorino F (eds) Auditory Plasticity and Regeneration, New York: Thieme, pp. 297–316.
Willott JF (1999) Neurogerontology: Aging and the Nervous System. New York: Springer.
Willott JF (2006) Neural reorganization following age-related and slowly-developing hearing loss. In: Lomber SG, Eggermont JJ (eds) Reprogramming the Cerebral Cortex: Plasticity Following Central and Peripheral Lesions. New York: Oxford University Press, New York, pp. 182–193.
Willott JF, Bross LS (1996) Morphological changes in the anteroventral cochlear nucleus that accompany sensorineural hearing loss in DBA/2J and C57BL/6J mice. Brain Res Dev Brain Res 91:218–226.
Willott JF, Bross L (2004) Effects of prolonged exposure to an augmented acoustic environment on the auditory system of middle-aged C57BL/6J mice: cochlear and central histology and sex differences. J Comp Neurol 472:358–370.
Willott JF, Parham K, Hunter KP (1988) Response properties of inferior colliculus neurons in middle-aged C57BL/6J mice with presbycusis. Hear Res 37:15–27.
Willott JF, Erway LC, Archer JR, Harrison D (1995) Genetics of age-related hearing loss in mice: II. Strain differences and effects of caloric restriction on cochlear pathology and evoked response thresholds. Hear Res 88:143–155.
Willott JF, Milbrandt JC, Bross LS, Caspary DM (1997) Glycine immunoreactivity and receptor binding in the cochlear nucleus of C57BL/6J and CBA/CaJ mice: effects of cochlear impairment and aging. J Comp Neurol 385:405–414.
Willott JF, Turner JG, Sundin VS (2000) Effects of exposure to an augmented acoustic environment on auditory function in mice: roles of hearing loss and age during treatment. Hear Res 142:79–88.

Willott JF, Hnath Chisolm T, Lister JJ (2001a) Modulation of presbycusis: current status and future directions. Audiol Neurootol 6:231–249.

Willott JF, Sundin V, Jeskey J (2001b) Effects of an augmented acoustic environment on the mouse auditory system. In: Willott JF (ed) Handbook of Mouse Auditory Research: From Behavior to Molecular Biology. Boca Raton, FL: CRC Press, pp. 205–214.

Willott JF, VandenBosche J, Shimizu T, Ding DL, Salvi R (2006) Effects of exposing gonadectomized and intact C57BL/6J mice to a high-frequency augmented acoustic environment: auditory brainstem response thresholds and cytocochleograms. Hear Res 221:73–81.

Willott JF, VandenBosche J, Shimizu T, Ding D, Salvi R (2008) Effects of exposing C57BL/6J mice to high- and low-frequency augmented acoustic environments; auditory brainstem response thresholds, cytocochleograms, anterior cochlear nucleus morphology and the role of gonadal hormones. Hear Res 235:60–71.

Yang SH, Perez E, Cutright J, Liu R, He Z, Day AL, Simpkins JW (2002) Testosterone increases neurotoxicity of glutamate in vitro and ischemia-reperfusion injury in an animal model. J Appl Physiol 92:195–201.

Index

MIX
Papier aus verantwortungsvollen Quellen
Paper from responsible sources
FSC® C105338

If you have any concerns about our products,
you can contact us on
ProductSafety@springernature.com

In case Publisher is established outside the EU,
the EU authorized representative is:
Springer Nature Customer Service Center GmbH
Europaplatz 3, 69115 Heidelberg, Germany

Printed by Libri Plureos GmbH
in Hamburg, Germany